METAL MEDIATED TEMPLATE SYNTHESIS OF LIGANDS

M E T A L
MEDIATED TEMPLATE
S Y N T H E S I S   O F
L I G A N D S

Otilia Costisor
Institute of Chemistry Timisoara, Academy of Science, Romania

•

Wolfgang Linert
Institute of Applied Synthetic Chemistry, TU- Vienna, Austria

NEW JERSEY • LONDON • SINGAPORE • BEIJING • SHANGHAI • HONG KONG • TAIPEI • CHENNAI

Published by

World Scientific Publishing Co. Pte. Ltd.
5 Toh Tuck Link, Singapore 596224
USA office: Suite 202, 1060 Main Street, River Edge, NJ 07661
UK office: 57 Shelton Street, Covent Garden, London WC2H 9HE

Chemistry Library

British Library Cataloguing-in-Publication Data
A catalogue record for this book is available from the British Library.

METAL MEDIATED TEMPLATE SYNTHESIS OF LIGANDS

ISBN 981-238-813-3

Editor: Tjan Kwang Wei

Printed in Singapore by World Scientific Printers (S) Pte Ltd

Preface

This book, "METAL MEDIATED TEMPLATE SYNTHESIS OF LIGANDS", has been written in order to fulfil the need for an up-to-date text on metal-directed ligand formation. The subject has already been covered in several existing treatises and simple duplication would be unwarranted. The present volume has, therefore, several distinct purposes: The first of these is the collecting of information concerning the basic problems of metal-template synthesis. The metal-template route is now an important area of organic synthesis that has widened since the mid 1960's by the development of many new types of reaction (for example, metal-template Mannich or alkylation reactions) as well as the development of new molecules of different shapes (macrocycles, clusters, knots, rotaxanes, etc.). At the same time, the theoretical interpretation of metal-template reactions has been extended. Consequently, an enormous expansion of the published literature has taken place.

The second goal of the book is to provide an organisational framework in which the vast amount of accumulated literature could be organised in a logical manner. The organisational criteria employed are based on the type of organic chemical reactions that are controlled by transition-metal ions – namely Schiff and Mannich condensations and alkylation reactions. In this presentation, the stereochemistry and special properties of the systems discussed are stressed, and the shapes of the molecules obtained for each type of ligand is presented. The so-called covalent template reactions in which metalloid elements such as tin, silicon, boron, and antimony form covalent bonds with oxygen, nitrogen, or sulphur during the course of the reaction are not, however, discussed. Many excellent books and reviews have already been published on the use of alkali metals as templates applied to the synthesis of crown ethers and so this subject is not referred to here.

This book is addressed to students and young researchers working for their Ph.D. as well as to those working in the field of coordination chemistry. This results in the third goal of the book, namely to encourage further work in this field (with the subsequent aim of making the book itself obsolete!). From the point of view of increasing interdisciplinary of scientific research it is recommended as further reading on all topics.

In order to facilitate the development of these goals, the book is organised in chapters that refer to the basic features of template reactions and to the specific types of reactions that have been developed in each area to date.

The Authors' understanding of this subject owes much to having had the good fortune to work with and discuss this subject with Professors Maria Brezeanu and Marius Andruh in Romania as well as Markus Holzweber, Matthias Bartel and Annegret Kleiner in Austria, for whose great efforts we are very grateful. We also wish to express our thanks to the Romanian Academy and to the Austrian Science Foundation for financial support when preparing the manuscript.

O. Costisor and W. Linert
Vienna, September 2003

Contents

Preface	**v**
The Template Effect	**1**
1.1. Types of Template Effects	1
1.2. The Template Effect as a Molecular Organizer Effect	4
1.3. Factors Affecting the Product of a Template Reaction	5
1.3.1. Coordination of ligands	5
1.3.2. The chelate effect	9
1.3.3. Macrocyclic effect	12
1.4. The Negative Template Effect	14
1.5. Advantages of Metal Template Reaction	16
Alkylation Reactions	**18**
2.1. Alkylation of the Nitrogen Atom	18
2.2. Alkylation of the Sulfur Atom	19
2.2.1. Open chain systems	19
2.2.2. Macrocyclic ligands	22
Schiff Condensation	**24**
3.1. Mechanistic Aspects	24
3.2. Open-chain Ligands	30
3.3. Macrocyclic Ligands	40
3.3.1 Diimine macrocycles	40
3.3.1.1. Aliphatic monocarbonylic precursors	40
3.3.1.2. Oxamido carbonyls as precursors	42
3.3.1.3. Dicarbonylic precursors	44
3.3.1.3.1. Aliphatic precursors	48
3.3.1.3.2. Unsaturated and aromatic precursors	51
3.3.2. Tetraimine macrocycles	65
3.3.2.1. Aliphatic carbonylic precursors	65
3.3.2.2. Unsaturated and aromatic precursors	73
3.3.2.3 Pendant-arm macrocycles	89
3.4. Cage Ligands	93
3.5 Compartmental Ligands	96
3.5.1. Closed-chain ligands	98
3.5.1.1. N_2O donor precursors	98
3.5.1.2. N_2S donor precursors	112
3.5.1.3. Multicompartmental ligands	113
3.5.2. Open chain ligands	118
3.5.2.1. End – off ligands	118
3.5.2.2. Side – off ligands	124

Mannich Condensation   **126**
4.1. Mechanistic Aspects   127
4.2 Acyclic Ligands   132
    4.2.1. Polyamine ligands   132
    4.2.2. NO donor ligands   134
    4.2.3. SN donor ligands   136
4.3. Monocyclic Ligands   138
    4.3.1. Tetraaza macrocycles   138
    4.3.2. Pentaaza macrocycles   146
    4.3.3. Hexaaza macrocycles   150
    4.3.4. Octaaza macrocycles   150
    4.3.5. Azaether macrocycle   152
    4.3.6. Azathioether macrocycle   152
    4.3.7. Reinforced macrocycles   153
    4.3.8. Macromonocyclic dicompartmental ligands   159
4.4. Isolated Dimacrocycles   160
4.5.Condensed Polymacrocyclic Ligands   167
    4.5.1. Carbon and nitrogen caped amine ligands   167
    4.5.2. P-, As amine cage ligands   182
    4.5.3. Thioamine cage ligands   183
    4.5.4. Cage complexes of other metal ions   184
Self Condensation of Nitriles   **187**
5.1. Phthalocyanines   187
    5.1.1. Mechanism   192
5.2. Porphyrins   195
5.3. Corrole   201
Self-Assembled Systems   **208**
6.1. Catenanes   209
    6.1.1. Two interlocked rings   209
    6.1.2. Three interlocked rings   217
6.2. Rotaxanes   219
6.3. Helicates   228
    6.3.1. Acyclic helicates   228
    6.3.2. Circular helicates   240
6.4. Knots   243
6.5. Macrocycles and Cages   248
6.6. Racks, Ladder and Grids   252
References   **257**
Index   **295**

Chapter 1

The Template Effect

In 1840 Ettling[1] isolated a dark green crystalline product from the reaction of cupric acetate, salicylaldehyde and aqueous ammonia. Few years later, Schiff[2] established that the salicylaldimine complex resulted by reaction of preformed metal-salicylaldehyde complex with ammonia. Pfeiffer and co-workers have realised the systematic study of Schiff's base complexes in the period 1930-1940 and noticed the role of the metal ions. Einchorn and Latif[3] studied the self-condensation of *o*-aminobenzaldehyde in the presence of divalent transition metal ions but they were unable to separate and identify the macrocyclic complexes in the reaction mixture. However, the importance of the metal ion became certain and the accumulated qualitative information allowed to notice that a new synthetic way was discovered, that which involves the metal ions in organic synthesis and the first example of a deliberated template synthesis was described by Busch[4] in 1964.

1.1. Types of Template Effects

The ability of metal ions to affect the course of some organic reactions has been systematically studied at the beginning of the sixties[5]. Since than, various and amazing molecular architectures have been created as a result of employing this effect. This effect has been termed the *metal-template* effect. Metal-template reactions are ligand reactions, which are dependent on, or can be significantly enhanced by a particular geometrical orientation imposed by metal coordination.

Two classes of chemical templates have been recognised: kinetic templates and thermodynamic templates.

When the template effect arises from the stereochemistry imposed by metal ion coordination of some reactants, promoting a series of controlled steps, a coordination or kinetic template effect occurs. This effect provides routes to products which are not formed in the absence of the metal ion. Kinetic templates influence the mechanistic pathway. Experimental data lead to the following types of kinetic metal template reactions[6]:

- the molecules are coordinated and assembled around a metal cation in a single step (Scheme 1.1) :

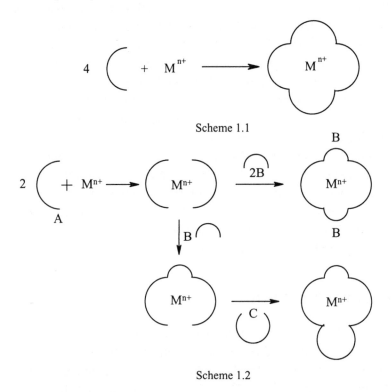

Scheme 1.1

Scheme 1.2

- the coordinated molecules react with an external molecule which bridges the ends of the ligated ones (Scheme 1.2).

It can be noticed that the most frequent products are cyclic ligands. However, acyclic, linear ligands, have also been obtained, mostly when the reaction is stopped along half a building stage.

Busch[7] showed first the kinetic template effect on studying the planar diamagnetic complex 2,3-butane-bis(mercaptoethylimino)-nickel(II) which contains two terminal cis-mercaptide groups. It reacts with monofunctional or bifunctional alkylating agents resulting in products, which contain a coordinated[a] thioether.

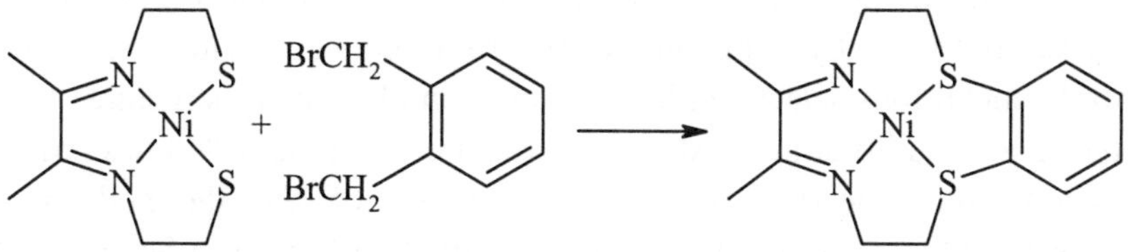

Scheme 1.3

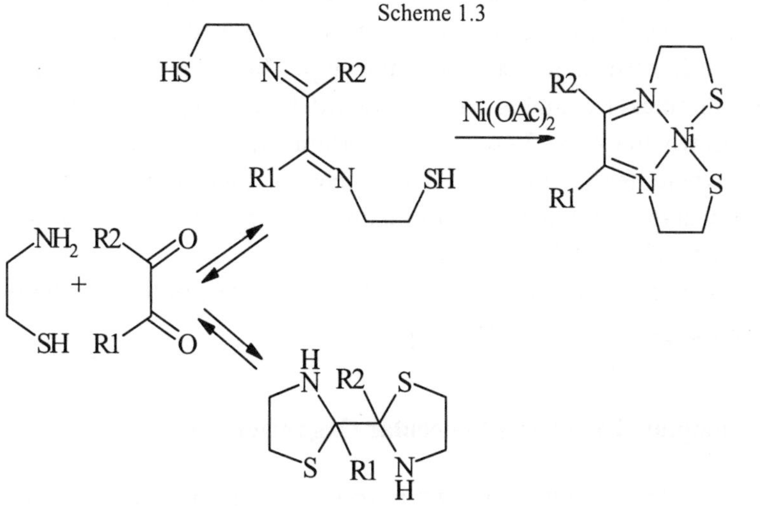

Scheme 1.4

When the bifunctional reagent α,α'-dibromo-*o*-xylene is employed, a new chelate ring is formed and the product of reaction is a macrocyclic

---

[a] Coordination of ligands involves coordinative bonding which implies the problem of correct and consistent graphical representation. At the one hand arrows are often used, at the other hand coordinative and covalent bonds are physical identical, so straight lines can be drawn. This, however leads to uncommon numbers of valences which can be bypassed by dashed lines. Each of these representations has advantages and diadvantages and is used accordingly.

complex (Scheme 1.3). The probability of ring closure is assured by the presence of the metal atom, which holds the mercapto groups in cis position, and the reaction occurs in a single rate-determining step. It is demonstrated also that this mechanism arises because the reagent α,α'-dibromo-*o*-xylene creates a new chelate ring *in situ*.

Thermodynamic templates refer to reactions that do proceed in the absence of the metal ion. In this case, metal ion promotes formation of desired products by removing them from the equilibrium[8]. In other words, metal ions select and bind certain complementary structures from among an equilibrating mixture of structures. This is a "sequestration" phenomenon. The first recognised example of the thermodynamic template effect refers to the formation of the Schiff base in Scheme 1.4. In the absence of nickel(II) the reaction between β-mercaptoethylamine and a α-diketone, thiazolines and mercaptals are formed in competition with the Schiff base of the corresponding type. In the presence of the metal ion, the nickel complex with the Schiff base as ligand is formed with a yield, which excess 70%. An equilibrium template effect has been identified[9] where aspects of both effects mentioned above combine, the distinctive characteristic being the formation of different products in the metal assisted and metal free reactions.

This classification says nothing about the importance of the structural factors affecting the template reactions.

1.2. The Template Effect as a Molecular Organizer Effect

Molecular organisation is the rule in *in-vivo* chemistry. The study of the simple chemical systems must be able to manifest the same organizational principles. The knowledge of these principles helps us to understand and exploit those in vivo.

The template effect involves the organisation of an assembly of atoms, with respect to one or more geometric positions, in order to achieve a particular linking of atoms. This effect recognises thus, a molecular organisation. Further, the chemical template is recognised as an evoluate and a complex molecular organisers, this in contradiction with a primitive system where no organisation does exist.[10] The template effect involves the presence of other effects and phenomena, which are also, manifestation of molecular organisation processes: coordination of ligands, chelate effect, macrocyclic effect, cryptate tethering effect. Elementary structural factors result in and control these phenomena and, consequently control template effects too. These are topologic, metric, geometric shape, rigidity and complementarily between the interacting species.

1.3. Factors Affecting the Product of a Template Reaction

1.3.1. Coordination of ligands

The coordination of ligands to metal ions involves electronic factors as well as geometric relationship between the two parts. Ahrland *et al.*[11] divided the acceptor metal ions into two classes depending on whether they form more stable complexes with the smaller donor atoms as nitrogen, oxygen or fluorine or the large ones, which include sulfur, chloride or phosphorus. Regarding the formation of the complex compounds as Lewis acids and bases, Pearson[12] elaborated the theory and principles of hard and soft acids and bases (HSAB) as it is shown in Table 1.1. In accord with HSAB, the acids of class *a* react preferentially with the *a* class bases and the acids of class *b* prefer the *b* class bases. Although the classification is very useful, there are many metal ions, which lie in a borderline region.

According to HSAB theory, the preference of alkaline and alkaline earth metal ions for the oxygen donors and their template activity to form crown ethers can be understood and also the preference of the transition metal ions toward nitrogen donors. The transitional metals widely used as templates are shown grey in Table 1.2. where the most capable metals are marked in dark squares.

Metal-template involves the reaction of the coordinated ligands. An effective template metal ion binds strongly to the donor atom of the precursor as well as of the resulted ligand and in this frame the above theories work also. The polarisation induced by the coordination of the ligands to the metal ions favours the nucleophilic attack of reactants. Therefore, the terminal groups must be able to undergo a characteristic addition with the new strap forming groups and for this, special steric requirements for the pre-existing ligands are also needed. They have to occupy a conveyable *cis* position.

The metal ion exerts a preference for a particular kind of environment and some ligands are better able to conform to that environment than others. In addition, each ligand has its own preference for a particular geometrical arrangement. The organisation of metal ions with specific electronic properties within a potentially template ligand system can lead to the appropriate orientation of substrate molecules, (the pre-existing ligands in the precursor molecules) required for particular reactions.

Table 1.1 Classification of Acids and Bases according to HSAB

Acids	
Hard	Soft
H^+, Li^+, Na^+, K^+	Cu^+, Ag^+, Au^+, Tl^+, Hg^+
Be^{2+}, Mg^{2+}, Ca^{2+}, $Sr^{2+}Ba^{2+}$	Pd^{2+}, Cd^{2+}, Pt^{2+}, Hg^{2+}
Al^{3+}, Sc^{3+}, Ga^{3+}, In^{3+}, La^{3+}	$CH_3{}^+Hg$, $Co(CN)_5{}^{2-}$, Pt^{4+}
Gd^{3+}, Lu^{3+}, Cr^{3+}, Co^{3+}, Fe^{3+}, As^{3+}	Te^{4+}, Br^+, I^+
Si^{4+}, Ti^{4+}, Zr^{4+}, Hf^{4+}, Th^{4+}, U^{4+}	
Pu^{4+}, Ce^{4+}, Wo^{4+}, Sn^{4+}	
UO^{2+}, VO^{2+}, MoO^{3+}	

Borderline
Fe^{2+}, Co^{2+}, Ni^{2+}, Cu^{2+}, Zn^{2+}, Pb^{2+}, Sn^{2+}, Sb^{3+}, Bi^{3+}, Rh^{3+b}, Ir^{3+b}, $B(CH_3)_3$

Bases	
Hard	Soft
H_2O, OH^-, F^-, $CH_3CO_2{}^-$, $PO_4{}^{3-}$	R_2S, RSH, RS^-, I^-, SCN^-
$SO_4{}^{2-}$, Cl^-, $CO_3{}^{2-}$, $ClO_4{}^-$, $NO_3{}^-$	$S_2O_3{}^{2-}$, R_3P, R_3As, $(RO)_3P$
ROH, RO^-, R_2O, NH_3, RNH_2	CN^-, RNC, CO, C_2H_4, H^-, R^-
NH_2NH_2	

Borderline
$C_6H_5NH_2$, C_5H_5N, $N_3{}^-$, Br^-, $NO_2{}^-$, N_2, $SO_3{}^{2-}$

In the design of the ligand, the coordination sphere of the metal ion is considered to serve as a flexible template within which the coordination positions are in juxtaposition to the functional groups of the protracted ligand molecule. In there terms, the type of the obtained product depends on the metal cation, namely on the size, favoured coordination sites and coordination polyhedron.

The template effect may lead to bi- or tridimensional systems according to the number and positions of the reactive sites, which are organised by the template metal ion. In the first case, the metal ion organises two reactive sites at an edge of its idealised coordination polyhedron. In this case, a new chelate ring is formed and the usual product is a macrocyclic ligand Fig.1.1, *a*.

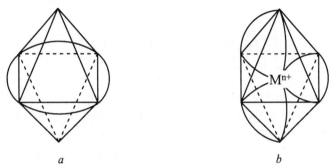

a   *b*

Fig. 1.1. The corners and faces of idealized coordination polyhedron

For example, the formation of macrocycles containing four nitrogen atoms as donors is favoured by the preference for square planar geometry of the nickel(II) or copper(II). The metal-template can also create tridimensional systems, namely cage complexes when metal ion organises three reactive sites on the face of its idealised coordination polyhedron Fig. 1.1. *b*.

The ligands obtained via metal-template reactions are multidentate ones and donor atoms can be present in a number of different functional groups. Nitrogen atoms are by far the most frequent donors and they appear as secondary and tertiary amino or imino groups. Other atoms, like oxygen, sulfur, phosphorus or arsenic atoms also can act as donor atoms. Oxygen atoms act as donor atoms in phenolate, alkoxide or ether groups; sulfur atoms acts mostly as thiolate or thioether groups and phosphorus or arsenic can act as phosphino or arsino groups.

Table 1.2. Periodic system of elements with the highlight of elements with high (dark) and low (light) ability to act as template

H																	H
Li	Be											B	C	N	O	F	N
Na	M											Al	Si	P	S	Cl	Ar
K	Ca	Sc	Ti	V	Cr	Mn	Fe	Co	Ni	Cu	Zn	Ga	Ge	As	Se	Br	Kr
R	Sr	Y	Zr	N	M	Tc	Ru	Rh	Pd	A	Cd	In	Sn	Sb	Te	I	X
Cs	Ba	La	Hf	Ta	W	Re	Os	Ir	Pt	A	H	Tl	Pb	Bi	Po	At	R
Fr	Ra	A															

C	Pr	Nd	Pm	S	Eu	Gd	Tb	D	H	Er	T	Y	L
T	Pa	U	Np	Pu	A	C	Bk	Cf	Es	F	M	N	Lr

Bond distribution about non-terminal atoms influences the geometry of the ligand in the metal complex and consequently the configuration of the metal complex itself. Among the driving forces for metal template reactions is the achievement of stable metal complexes, either by the forming of new chelate rings or macrocycles or the modification of the existing ones. Both chelate effect and macrocyclic effect are largely reflections of topologic and metric factors and in its simpler examples, the kinetic coordination template effect derives from the same metric and topological factors as chelate effect. For this reason the chelate and the macrocyclic effect will be presented in detail in the following parts. Specific requirements for a metal ion template will be presented for each type of the chemical reaction.

The chemical nature of the precursors determines the type of the chemical reactions. The most frequent metal template reactions are the Schiff condensation and Mannich condensation, which lead to multidentate ligands. Nowadays, the self-assembled processes around metal ions are used to build complicated structures with interesting properties. It has been recognised for linear, macrocyclyc, polymacrocyclic, knots, catenanes, rotaxanes ligands.[13] The resulting ligands are saturated as well as unsaturated ones.

1.3.2. The chelate effect

Empirical observations as well as quantitative studies have established that metal chelate complexes are more stable than those of related unidentate ligands. This is called the *chelate effect*. In terms of chemical thermodynamic the chelate effect is described by the following equation[14]:

$$\log K_1 \text{ (polydentate)} = \log \beta_n \text{ (unidentate)} + (n-1) \log 55.5 \qquad (1)$$

where $\log K_1$ (polydentate) refers to the stability of the complex of an n-dentate polydentate ligand, $\log \beta_n$ (unidentate) refers to the stability of the complex containing n unidentate analogues of the polydentate ligand, and 55.5 is the molarity of water. Actually, this last value represents the entropy of translation of 1 mol of solute generated at $1m$ concentration. Schwarzenbach has previously explained the chelate effect as an entropic one. It was considered the result of the restricted volume in which the second donor atom of the chelate ring could move once the first donor atom had been coordinated. The Adamson[15] approach suggests that the chelate effect arises because of the way the standard reference state is defined and that disappears once the formation constant is expressed in terms of mole fractions. Fundamentally, there is no difference between the explanations of the nature of the chelate effect as it represents the consequence of an entropic result. Quantitative results as well as theoretical approach has however emphasized the large contribution of the entropic nature of the chelate effect. For example, for polyamine ligands, the formation constant can be expressed by the equation (2):

$$\log K_1 \text{ (polyamine)} = 1.152 \log \beta_n \text{ (NH}_3) + (n-1) \log 55.5 \qquad (2)$$

The first term should refer to an aliphatic amine as unidentate analogues. When ammonia is considered the reference, the inductive factor of 1.152 should be considered. It reflects the greater basicity of an aliphatic amine ($pK_a = 10.6$) compared to that of ammonia (pK_a for NH_3 9.22): $10.6/9.22 = 1.152$. It is a god agreement between the values obtained using the equation (2) and experimental ones[16] for a series of complexes.

The factors which affect the stability of chelate rings are essentially those which act in heterocyclic rings and additionally, special constrains appear due to the metal ion which takes part in the heterocycle. They are metric factors and among them, the size of the chelate ring is of special importance. Five- and six-membered chelate rings are by far the most common. For most metal ions and in the case of saturated structures, five-membered chelate rings are the most stable. For example, the stability constants for ethylenediamine are greater than those of the corresponding trimethylene derivatives. The differences in stability were associated to less favourable enthalpy contributions rather than to entropy one, as it result from the data in Table 1.3.

It has been demonstrated in the case of the nickel(II) complexes with ligands in the Scheme 1.5 (where a large quantity of data on the formation constants and enthalpies of complex formation of these complexes exist) that the enthalpy effect is largely due to steric strain. Thus, the change in strain energy, ΔU, on going from the free components (ligand and metal ion) to complex formation established by means of molecular mechanics (MM) calculation and the changes in enthalpy of complex formation are very close.

Table 1.3. Thermodynamic parameters for some Cu(II), Ni(II) and Cd(II) complexes

metal ion	complex	L= EN			L= TN		
		ΔG	ΔH	ΔS	ΔG	ΔH	ΔS
Cu (II)	ML	- 14.3	-12.6	6	-13.2	-11.4	6
	ML$_2$	- 26.7	-25.2	5	-22.9	-22.4	2
Ni(II)	ML	- 10.0	-9.0	3	- 8.6	-7.8	3
	ML$_2$	- 18.3	-18.3	0	-14.3	-15.0	- 2
Cd (II)	ML	- 7.4	- 6	5	- 6.1	- 5	4
	ML$_2$	-13.3	-13.3	-1	- 9.8	-10	- 1

The special constrains which appear when the metal ion take part in a heterocycle relate complex stability to metal ion size and refers to the specific bond lengths and angles. Empirical observations lead to the conclusion that an increase of the chelate ring size leads to large drops in complex stability for the complexes of large metal ions and may even lead to increases in complex stability for small metal ions. The simplest example is offered when the change in complex stability $\Delta logK$, on going from 2,2,2-TET complex to the 2,3,2-TET complex for a wide variety of metal ions is compared. Use of MM to explore the potential energy surface for the EN chelate ring shows that the lowest strain energy will occur for a metal ion with a M-N bond length of 2.50 Å, and a N-M-N bond angle of 60°. The same calculations for the isolated TN ring indicate that the minimum strain energy will occur for a metal ion with a M-N bond length of 1.6 Å, and a N-M-N bond angle of 109.5°.

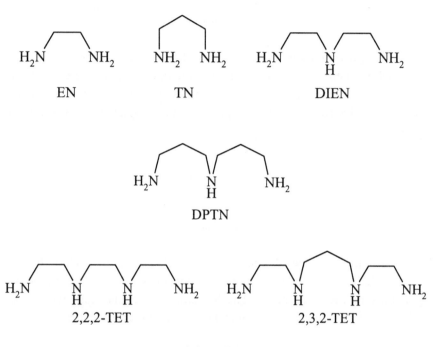

Scheme 1.5

It has been noticed that Ni(2,2,2-TET) is less stable than Ni(2,3,2-TET) which forms a six membered chelate ring. The greater stability of the latter complex was assigned to a steric effect associated with the release of steric strain in the 2,2,2-TET complex, which is to short to span the nickel(II) ion effectively, on adding another methylene group to give 2,3,2-TET.

The figure in Scheme 1.6*a* shows the shape of the chelate rings in the tris(1,2-diaminoethane)cobalt(III) ion. Similarly, the six-membered chelate ring in a 1,3-diaminopropane complex would be expected to adopt a chair conformation, as it is shown in Scheme 1.6*b*.

a   b

Scheme 1.6

Although well known, chelate rings of five, four and six members are less stable whereas the eight- and nine- membered are unexpected. In the case of unsaturated five- and six-membered rings in which resonance occurs, al the ring atoms are approximately coplanar as is shown in Scheme 1.7.

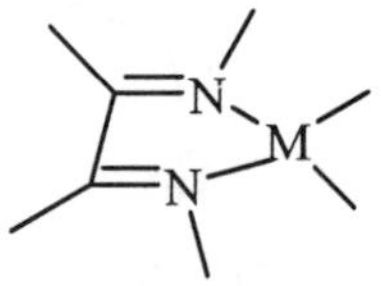

Scheme 1.7

The number of chelate rings is important. For similar metal chelates, those which contain the greater number of stable chelate rings will generally be the more stable. The formation of chelate rings reduces the lability of metal complexes with respect to simple substitution.

1.3.3. Macrocyclic effect

A metal template synthesis has been often used to obtain macrocyclic ligands[17,18] in a simple way avoiding the organic routes which are - as a rule - a multistep task. The macrocycles thus obtained are mostly 13- to 16-membered ring ligands. The most common are the polyamine ligands with four nitrogen donors situated in a plane although they can contain five or six nitrogen atoms in the ring closing a macrocycle around a labile metal centre with the formation of new five- or six-membered chelate rings. Other donor atoms, like oxygen, sulfur, phosphorus or arsen can participate as donors in a macrocycle.

Macrocyclic ligands render high thermodynamic stability and exceptionally kinetic inertness of their metal complexes against ligand substitution or dissociation compared to their open-chain analogues. This is the largest known "macrocyclic effect".[19] Both enthalpy and entropy term contribute to the stability of the macrocyclic complex. For example, the enthalpy of formation of the Cu(II) complexes[20] follows sequence [15]aneN$_5$ ≈ [16]aneN$_5$ > [17]aneN$_5$ but for the Ni[15]aneN$_5$ complex the entropy term has the major contribution.[21]

Stability of the complexes with macrocyclic ligands is strongly affected by the dimension of the macrocycle. For triaza, tetraaza and pentaaza it has been established that increase in the macrocyclic ring size results in decrease in the formation constants. However, several exceptions from this rule have been noticed. This is the case of the [X]aneN3 complexes with Cu(II) or that of the [X]aneN5 with Ni(II). These exceptions should be considered by taking in mind the special steric demands of the metal ions.

Among the general factors influencing a metal-ion-controlled synthesis[22] of the macrocyclic ligands, the relationship between the size of the metal ion and the opening in the middle of the ring clearly is important. Natural complexes demonstrate this relationship. So, iron porphyrins involve a 16-membered ring while the cobalt in vitamin B$_{12}$ occupies a 15-membered ring. Also, cyclic polyfunctional ethers show sharp selectivity towards alkali metal ions as a function of ring size. Especially the relationship between the size of the metal ion and the size of the cavity and the stability of the precursors should be considered. The subject has been deeply investigated[23] and it was found that there is an ideal ring size for any metal ion having a given metal-donor atom distance. When mismatch in these sizes exists, distortions from ideal geometry appear. The ideal metal-nitrogen bond distances and the average deviation from planarity of the macrocyclic ligands have been estimated by molecular mechanics and it was found that the most metal-nitrogen linkages fall within the range of 1.8 - 2.4 Å. A regular increase of ideal Me-N distance as the number of ring member increases was observed. Busch et al.[24] has found a minimum in strain energy for the 14-membered ring. The greater strain energies of the 12- and 16-membered rings parallel the observed resistance of the rings toward coordination in a planar fashion. According to Hancock the relative stability of the possible conformers that have different metal ion size preference should be considered. Based on molecular mechanics calculations and on the formation constant studies, Hancock and co-workers show that the selectivity patterns of tetraaza macrocycles are controlled more by chelate ring size than by the macrocyclic ring size.[25] They have reported the best-fitting size and geometry for coordination in triaza and tetraaza macrocycles as can be seen in Table 1.4.

Table 1.4. Correlation between geometry and bond length.

Macrocycle	Geometry	M-N bond length, Å
16-aneN$_4$	flated tetrahedron	1.81
14-aneN$_4$	square-planar	2.06
12-aneN$_4$	square pyramidal	2.15

High ligand field has been observed for macrocyclic complexes compared with the open chain analogous. This was attributed to the compression of the metal ion in the too-small cavity of the macrocyclic ligand. However, this assertion is not always true. The larger donor strength of the nitrogen atoms along the series - primary, secondary and tertiary amines have been considered to be responsible for the high ligand field strength, although in the last case, steric effects mask it.[26]

1.4. The Negative Template Effect

Two complementary roles for a template have been described. A positive template brings together two reactive parts of a single molecule. The above presented template reactions belong to this type. A negative template has also been noticed. In this case the template holds the reactive groups apart, thereby suppressing the desired reaction and encouraging the intermolecular one.[27] Scheme 1.8 shows the two template effects when a molecular cyclization is hindered and promoted, respectively.

The formation of cyclidene in Scheme 1.9 is an example. Here, the nickel(II) - which already acted as template, confers to the precursor for the further cyclization a rigid structure favourable to the second ring closure and, also, masks the donor atoms so that they can not act as nucleophiles during the course of. Metal template has inhibited the competing reactions: polymerization and formation of other undesired products. When macrocycles are the desired product, metal template offer the oportunity for a selective cyclization. The two template roles are types of kinetic template effect.

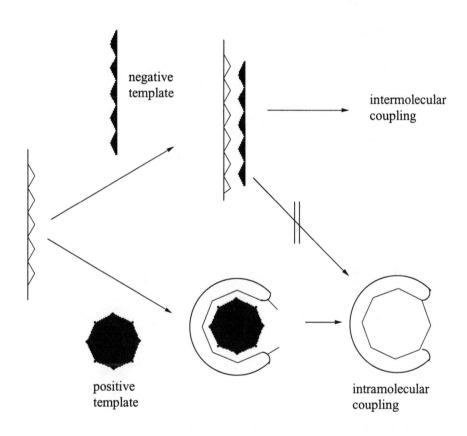

negative template

intermolecular coupling

positive template

intramolecular coupling

Scheme 1.8

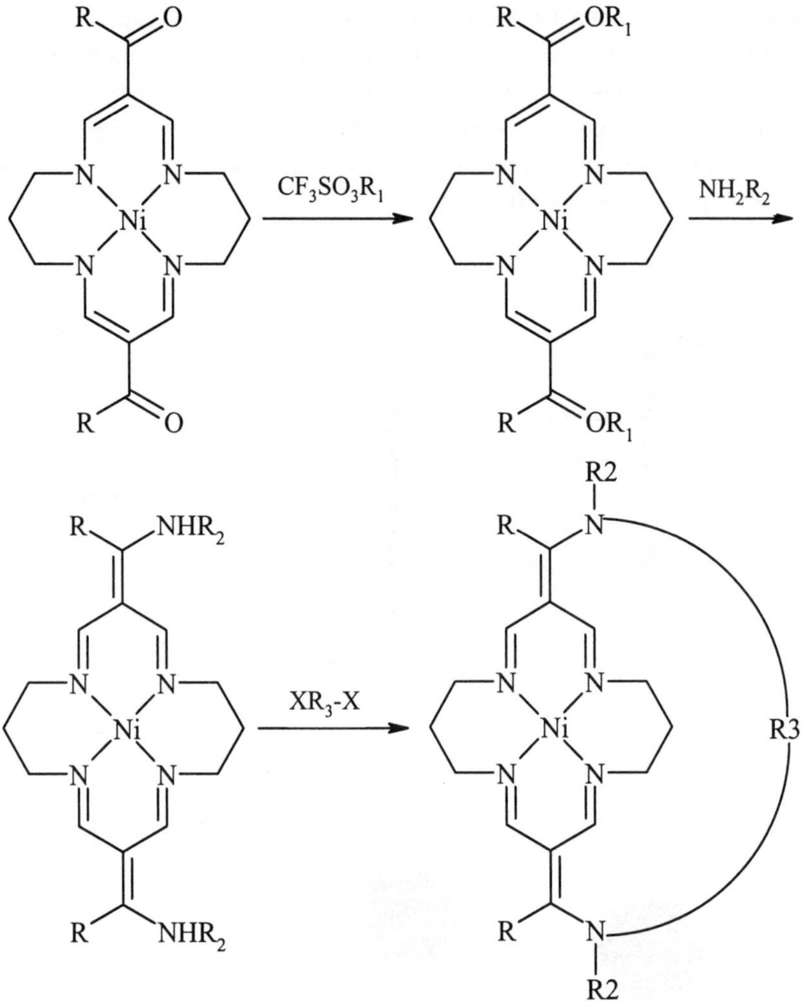

Scheme 1.9

1.5. Advantages of Metal Template Reaction

Metal template reactions offer simple ways to obtain organic molecules which otherwise involve complicated organic routes, high amounts of solvents, small yields and high costs. The organic molecules act as ligands and the high stability of the complexes allows reactions at the coordinated ligands without complex destruction. For example, the coordinated imino groups can be reduced to give coordinated amine groups using very effective agents like hydrogen, using platinum catalyst sodium borohydride, cathodic reduction. Other reactions including derivatization, functionalising and isotopic exchange reactions are also possible. Even specific isomers can be obtained.

The ligands obtained by a metal template route can be released by demetallation in a reaction with cyanide. The free ligand can then be used further to obtain complexes of other metal ions by a direct synthesis.

Sometimes, these molecules are of low stability toward hydrolysis or isomerization. Also, not all ligands can be isolated as such and some only exist in their metal complexes.

Macrocycle have aroused great interest for several reasons. Particularly, nitrogen-donor macrocycles display usual properties in that they complex most strongly with transition metal-ions. Macrocycles have a cavity, which might enable the ligand to display selectivity towards metal ions, which fit best into it.

Chapter 2

Alkylation Reactions

The study of the alkylation reactions directed by metal ions were pioneered and developed by Busch, as it has been shown in Chapter 1. In these reactions the alkylating agent are the alkyl halides and the mechanism is that of an electrophylic attack of the alkyl group to a metal-bonded donor atom which acts as a nucleophyle. It is expected that the coordinated ligands have thereby lost some of their nucleophylicity. However, the geometrical factors can overcome this disadvantage. Additionally, the appropriate chose of the halide ions and that of the substrates can result in the desired product of high purity and convenient yields, avoiding the undesired products.

Metal-directed alkylation reactions are mainly applicable to amines and thiols, and provide amines of superior a degree and thioethers.

2.1. Alkylation of the Nitrogen Atom

Alkylation of the nitrogen atoms occurs when it belong to a primary or secondary amine group as in these cases they have the non-bonding pairs of electrons and thereby the nucleophilic character enough strong. Additionally, the closing of a new chelate ring facilitates the formation of a macrocyclic ligand around a metal ion. Uhlemann and Plath[28] used the alkylated Schiff bases as high metal selective reagent. Thus, nickel(II) complexes of the Schiff bases **2.1** (R = H, CH_3 or C_6H_5) were alkylated with ethylenedibromide in $CHCl_3$ to yield macrocyclic complexes **2.2**, according to Scheme 2.1. Further, palladium(II) complexes were obtained by transmetallation.

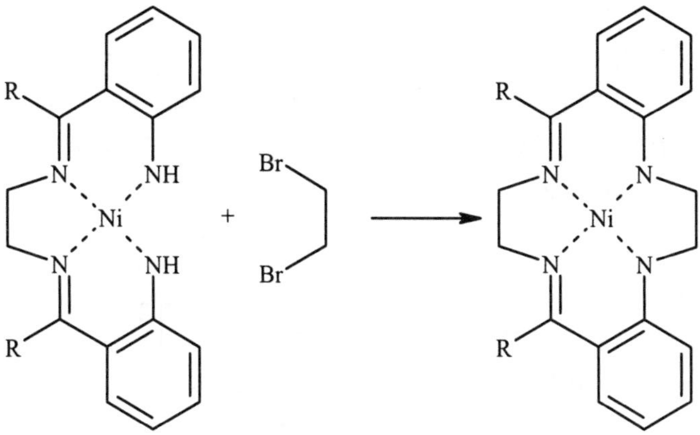

2.1   2.2

Scheme 2.1

Alkylation of the coordinated amine has been developed using formaldehyde and an appropriate acid component. This is a Mannich type reaction and will be developed in the Chapter 4.

2.2. Alkylation of the Sulfur Atom

2.2.1. Open chain systems

The alkylation of the coordinated sulfur has been studied by Busch's group.[29] The sparingly soluble square-planar nickel(II) and palladium(II) complexes with 2-aminoethanethiol were treated with usual alkylating agents in dimethylformamide (dmf). It has been shown - on the basis of elemental analyses, molecular weights, spectral properties and magnetic measurements – that the reaction occurred at the coordinated mercaptide groups producing thioether groups, according to Scheme 2.2. A detailed study of the reaction with methyl iodide shows that the product $[Ni(NH_2CH_2CH_2SCH_3)_2I_2]$ is formed in the first stage, which further leads to the trimer as a result of displacement of the thioether group, thus suggesting that bridging sulphur atoms are stronger ligands than thioether groups.

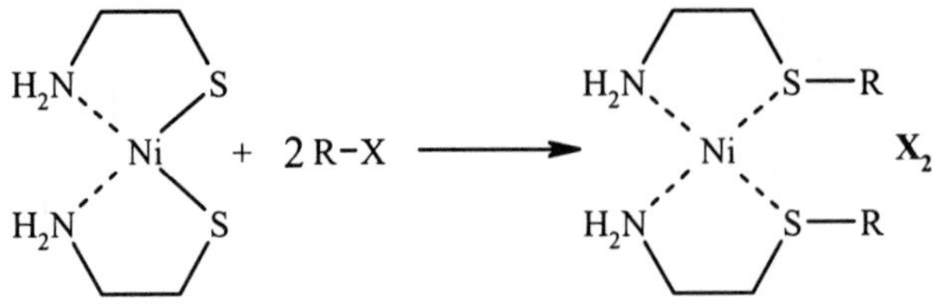

2.3

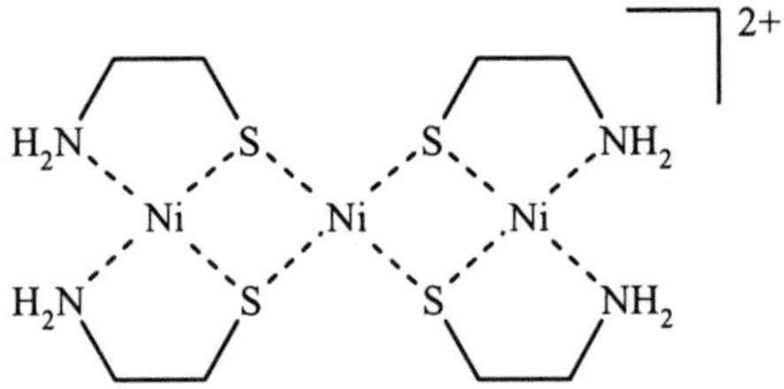

2.4

Scheme 2.2

The trimer **2.4** was isolated and it was found that its reaction with methyl iodide resulted in a product which is identical with that obtained by methylation of $[Ni(NH_2CH_2CH_2S)_2]$.

Following the same way, the dibenzilated complexes $[Ni(NH_2CH_2CH_2SCH_2C_6H_5)_2X_2]$ were obtained and it was observed the increase of reaction rate in the order Cl < Br < I. Magnetic moment values of 3.18-3.09 BM are indicative for the presence of two unpaired electrons, characteristic for octahedral nickel ion.

Alkylation of $[Pd(NH_2CH_2CH_2S)_2]$ with methyl iodide resulted in $[Pd(NH_2CH_2CH_2SCH_3)I_2]$ whereas with benzyl bromide leads to a mixture of $[Pd(NH_2CH_2CH_2SCH_2C_6H_5)_2Br_2]$ and $[Pd(NH_2CH_2CH_2SCH_3)Br_2]$. This behaviour was explained in terms of lesser coordinating ability toward palladium(II) of bromide ion as compared to iodide ion.

Differences have been noticed between the terminal and the bridge sulphur atoms. For example, the alkylation of the dimer **2.5** either with benzil bromide or methyl iodide resulted in the compound **2.6** which shows that only the terminal sulphur has reacted with alkyl halide.

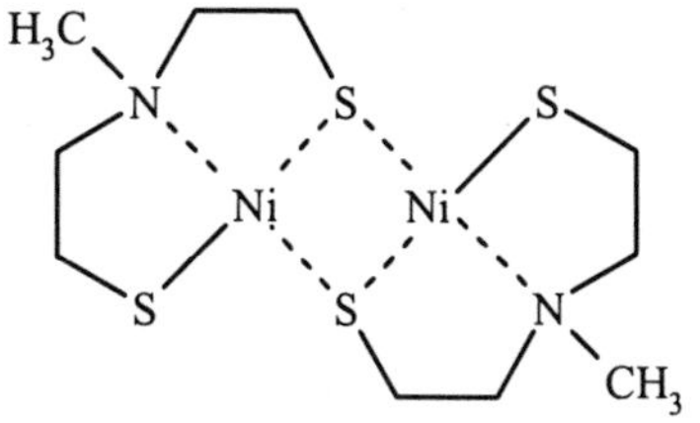

2.5 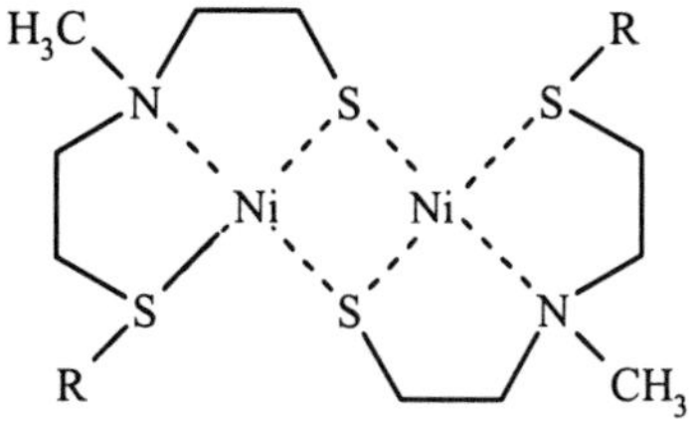   2.6

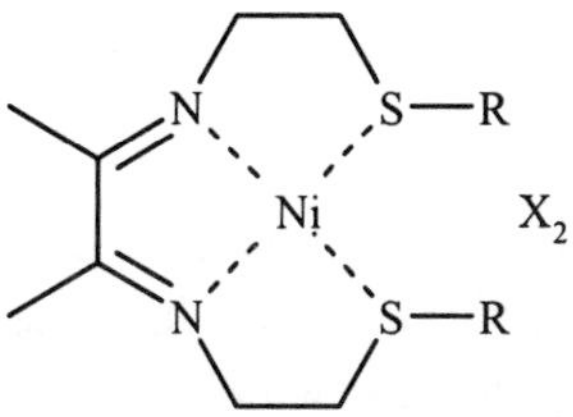

2.7

Compared with the above presented precursors, the α-diketobismercaptoethylimine complexes show no tendency to bind by sulphur bridges and their alkylation proceeds at a fairly rapid rate. For example, the alkylation of the planar diamagnetic 2,3-butanebis-(mercaptoethylimino)nickel(II) with methyl iodide, benzyl bromide and ethyl bromide lead to the corresponding thioether derivatives 2.7. The alkylated complex shows that the anions are coordinated to the metal ion in 1,2-dichloroethane solution and in the solid state as well.

Coordinated acyclic Schiff bases which contain two terminal *cis*-mercaptide groups, already obtained by metal-directed condensation, undergo alkylation reactions at the coordinated sulphur atoms. When monofunctional alkylating agents are used, open-chain ligands containing coordinated thioether can be obtained.

Reaction of the zinc(II) and cadmium(II) complexes of the type 2.8 with methyl iodide in acetone resulted in the yellow S-methylated compound 2.9 according to Scheme 2.3.

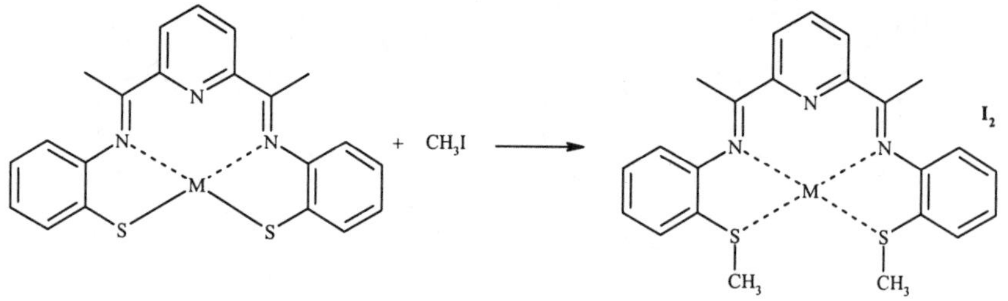

2.8   + CH₃I   **2.9**

Scheme 2.3

2.2.2. Macrocyclic ligands

The formation of macrocyclic ligands which involves the completely enclose of a planar central metal ion has been realised following the alkylation reaction of a precursor which has to response to the following requirements: *i*, the original ligand must be tetradentate and chelate in a planar or *cis*-octahedral geometry; *ii*, the terminal groups must undergo a characteristic reaction with ring formation. With these requirements fulfil, the using of suitable difunctional alkyl halides creates the possibility to bridge the sulphur donors.

Macrocyclic complexes have been obtained from biacetyl-bis(mercaptoethylimino)nickel(II), 2,3-octanedione-bis(mercaptoethyl-imino) -nickel(II), and 2,3-pentanedionebis(mercaptoethylimino)-nickel(II) by reaction with α,α'-dibromo-*o*-xylene. For example, Scheme 1.3 presents the reaction of 2,3-butanebis-(mercaptoethyl-imino)nickel(II). The probability of ring closure is assured by the presence of the metal atom which holds the mercapto groups in *cis* position and the reaction occurs in a single rate-determining step. It is demonstrated also that this mechanism arises because the reagent α,α'-dibromo-*o*-xylene creates a new chelate ring *in situ*. The authors have called this directive kinetic coordination effect.

It has been demonstrated that the terminal thiolo groups in metal complexes retain sufficient of their nucleophilic character to react with alkyl halides and produce the corresponding coordinated thioether complexes. Kinetic studies indicate that the sulphur atom remains coordinated during such S-alkylation reaction.

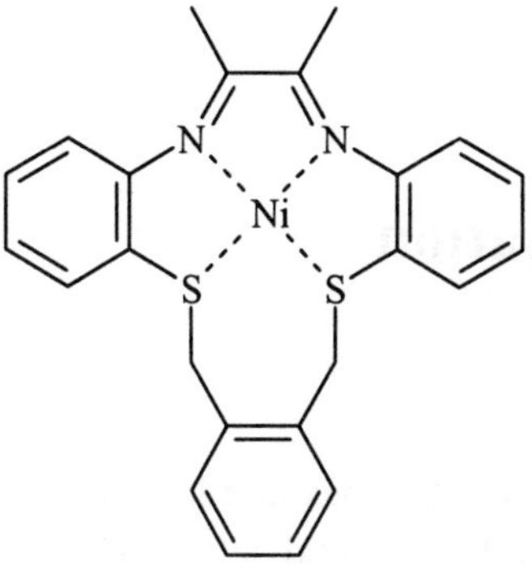

2.10 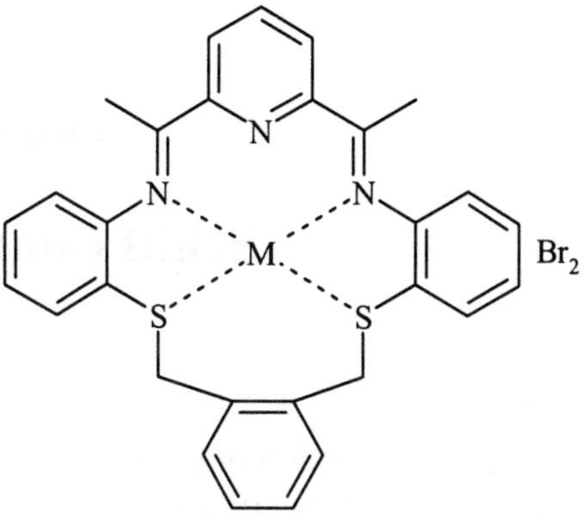   2.11

Similar alkylation reaction take place around nickel(II),[30] zinc(II) and cadmium(II)[31] resulting in the ring closure to **2.10** and **2.11**. The alkylation reaction is accompanied by changes in configuration of the precursors which are motivated by the requirement for a strain-free bridge between the sulphur atoms and the preference for square-planar coordination geometry. An additional factor is the removal of the formal negative charges on sulphur atoms. For example, the dithio precursor of **2.11** has a helical configuration which partially collapses towards a planar configuration during cyclization reaction, although it has the required distance between sulphur bridge (3.83 Å).

Chapter 3

Schiff Condensation

The condensation of primary amines with aldehydes and ketones was first reported by Schiff which also observed that a metal salicylaldehyde complex react with a primary amine resulting in a metal-imine complex. Hundred years later, at the beginning of 1960's, the chemistry of metal template cyclization to form Schiff-base macrocyclic ligands was developed by Curtis, Busch and Jäger. The method was quickly developed mostly with the aim to obtain unsaturated macrocycles but also open systems which would not readily obtained in the absence of metal ions have been formed as unsaturated multidentate ligands. The high quantity of experimental data allows to consider metal template Schiff condensation a general method for macrocycle formation in which the cyclization process involves the formation of a carbon-nitrogen double bond. The nature of the obtained ligands has been controlled by choosing appropriate organic precursors and by the metal ion which acts as template. Mononuclear and polynuclear complexes, containing different donor groups along with the imines ones and with various shapes can be obtained. We refer to transitional metal ions with template properties. However, a special attention is paid to the rare earth and actinide elements. The mechanism of the Schiff condensation in the presence of metal ions is also discussed.

3.1. Mechanistic Aspects

Metal template synthesis involving the condensation of primary amines with carbonyl-containing functional groups lead to the molecules in which metal ions are coordinated to the imine. Accumulated experimental data show that the yield of the template reaction is quite sensitive to a number of factors including order and timing of reagents additions, hydrogen ion concentration, anion of metal salts, solvent and temperature. On this basis, it has been postulated that either (*i*) organic intermediate, (*ii*) even the product, namely, ligand itself, must be formed prior the point where the metal salts is added or (*iii*) the coordination of the organic precursors occur. These are arguments for a kinetic or a thermodynamic template effect. In all the cases, reactions are presumed to proceed via carbinolamine intermediates.[32] Several examples will support these ideas.

In the course of the cobalt(II) directed condensation of 1,3-diaminopropane with dibenzyl, the precursor 1,2,8,9-tetraphenyl-3,7-diazaduohepta-2,7-diene-1,9-dione, **3.1**, has been isolated prior to the addition of $Co(OAc)_2 \cdot 4H_2O$. Mixing the precursor and $Co(OAc)_2 \cdot 4H_2O$ followed by addition of diprotonated diamine 1,3-diamino-2-hydroxypropane results in the corresponding macrocyclic ligand complex $[Co(3.104)Br_2]Br$.

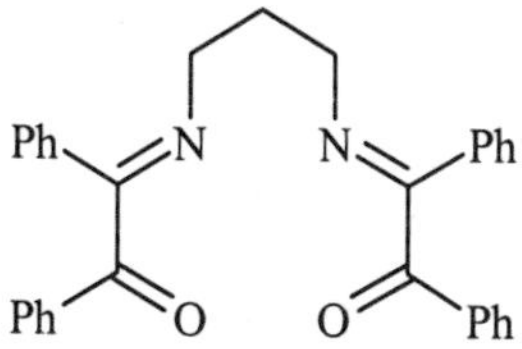

3.1

The mixed ligands copper complex of glycine and salicylaldehyde has been prepared and it has been shown that it readily converted into a Schiff base complex.[33]

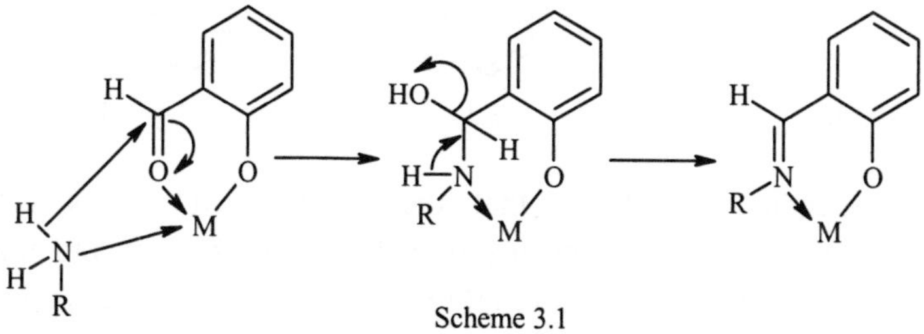

Scheme 3.1

The systematic study of the mechanism has shown that the Schiff base formation depends strongly on the order in which the reactants are mixed. It is concluded that the equilibrium is achieved most rapidly when the metal ion is added last.[34]

The interaction of either bis(salicylaldehydato)cobalt(II) or nickel(II) with 2,3-diaminoanthraquinone resulted in 1:1 adducts which converted into Schiff base complexes by heating. Thus, it has been suggested that the initial step in the formation of metal-Schiff base complexes is rapid coordination of the amine to the metal followed by the rate-determining Schiff base formation step according to Scheme 3.1. Coordination would have two effects; first, it would facilitate removal of the NH protons and secondly it would enhance nucleophilic attack at the carbon of the carbonyl group.

The formation of the binuclear copper(II) complexes with 1,3,5-triketones[35, 36] or like those depicted in formula **3.2 – 3.4**[37, 38] are well known and for some of them, the structure has been resolved. These complexes react with primary amines resulting in coordinated Schiff bases.

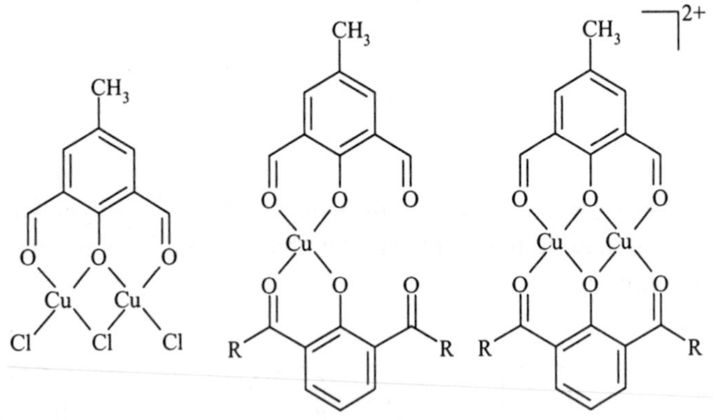

3.2   **3.3**   **3.4**

Experimental data shows that the presence of excess diamine determines whether or not the macrocyclic complex form and that the yield of the product depends strongly on the presence of excess diamine and the total reaction time. On this basis, a general-base-catalysed addition of an amine to a carbonyl compound has been proposed (Scheme

3.2) and it has been brought proves that the condensation take place around the metal ion. Thus, in the course of the preparation of a series of diamagnetic square- planar nickel(II) complexes [Ni(**3.5**)] according to Scheme 3.3, it has been isolated the intermediates which contains the coordinated dicarbinolamine moiety in the ligand molecule.[39] The second step involves the elimination of a water molecule, a process which is affected by the pH. Thus, a neutral and slight alkaline medium which is assured by an excess of diamine enhances the stability of the carbinolamine.

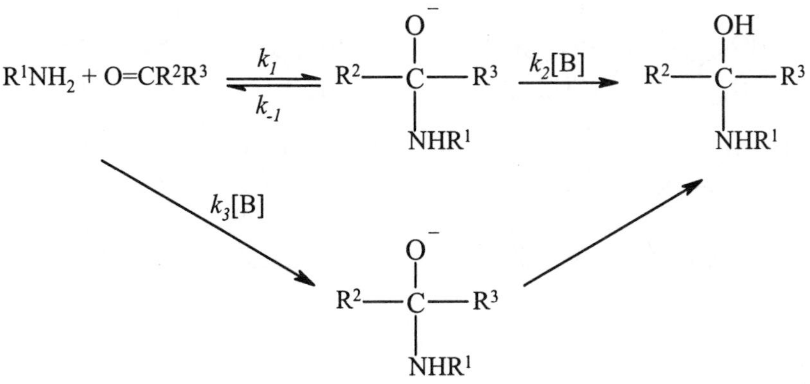

$$R^1NH_2 + O{=}CR^2R^3 \underset{k_{-1}}{\overset{k_1}{\rightleftharpoons}} \underset{NHR^1}{\overset{\overset{\displaystyle O^-}{|}}{R^2{-}C{-}R^3}} \xrightarrow{k_2[B]} \underset{NHR^1}{\overset{\overset{\displaystyle OH}{|}}{R^2{-}C{-}R^3}}$$

$$\xrightarrow{k_3[B]} \underset{NHR^1}{\overset{\overset{\displaystyle O^-}{|}}{R^2{-}C{-}R^3}}$$

Scheme 3.2

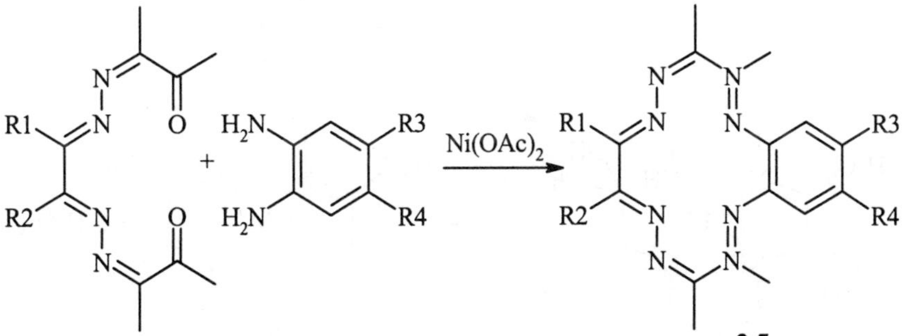

3.5

Scheme 3.3 ($R_1 = R_2 = H$ and $R_3 = R_4 = H$; $R_1 = H$, $R_2 = CH_3$ and $R_3 = H$, $R_4 = CH_3$)

Tasker *et al.*[40] have reported the isolation and X-ray crystal structure of the macrocyclic dicarbinolamine complex **3.6** which is the intermediate in the Zn(II) template condensation of 2,6-diformyl pyridine and dihydrazinobipyridine, **3.7**.

The size of the cation used as a template has proved to be of importance in controlling the synthetic pathway in the Schiff base system.[41] Thus, transition metals are generally used for the smaller rings and rare earth metals for the larger ones.

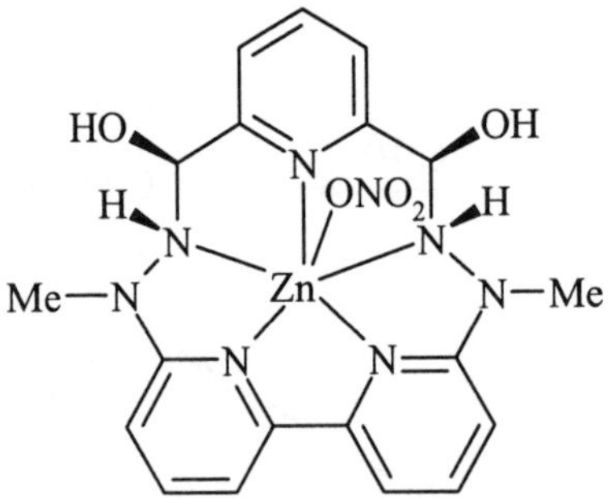

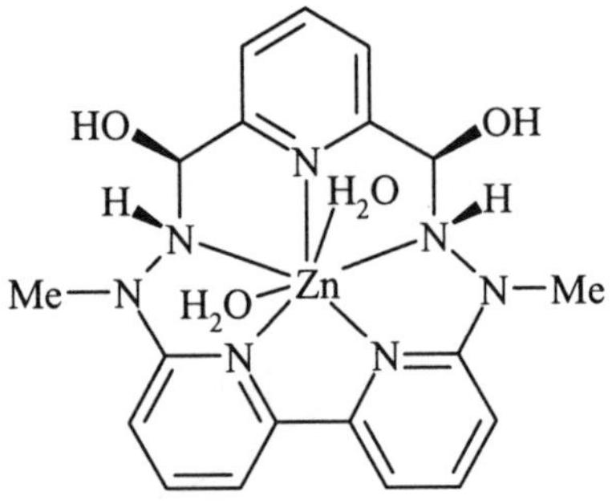

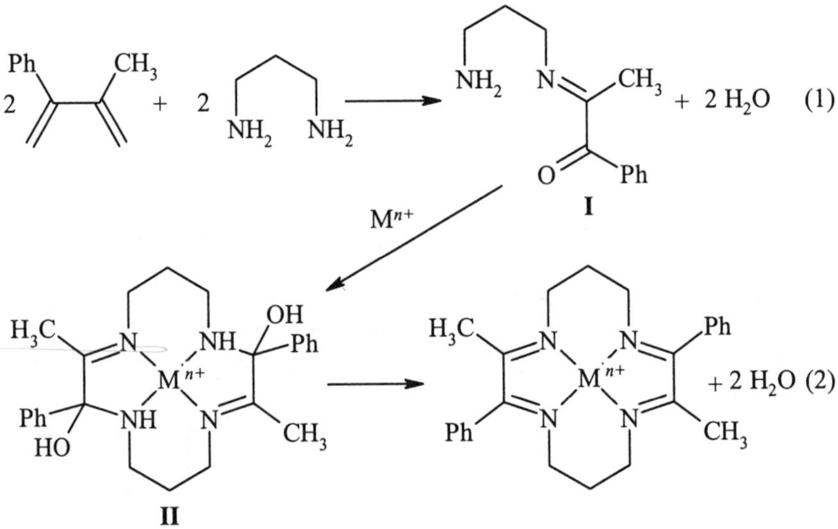

Scheme 3.4

The metal ions are involved in the mechanism of Schiff base condensation in a way which leads to a specific isomer from a mixture of isomeric products. Two examples are further presented. The template condensation of 1-phenyl-1,2-propanedione (AB) with 1,3-diaminopropane with metal ions iron(II), cobalt(II) or nickel(II) yields only one of the two possiblemacrocyclic products, the *trans*-MePhTIM, **3.99**, although the *cis*- isomer is also a possible product.[42] The proposed mechanism (Scheme 3.4) shows that the metal ion acts at the point where only *trans* isomer can be formed.

Two fact support this mechanism: (*i*) the Schiff base **I** is formed in good yield in the absence of metal ion and (*ii*) the coordinated carbinolamine **II** complex of cobalt(II) was isolated and characterized.

The second example refers to the Curtis reaction where a mixture of *cis*- and *trans*-[Ni(14-diene)]$^{2+}$ species are formed. The template mechanism in the Curtis type reaction, was studied, using N-monosubstituted ethylenediamine with a bulky group, namely Ph one, instead of a hydrogen atom in en. Although several products could be potentially obtained, in the presence of Ni(II) only the product **3.8**, has been isolated.[43]

3.8

It is suggested that the mechanism involves a *cis*-condensation between two adjacent acetoimino groups. For this aim, the tris-N-benzylethylenediamine nickel(II) cation must be the starting material. It is considered that the key step involves the deprotonation of a methyl group in acetone. Thus, an equilibrium is present between the tris-complex and the bis-complex and free base molecule. This free base molecule realizes the deprotonation of a methyl group in acetone thereby making the quaternary carbon of the neighboring acetoimine more accessible to a nucleophylic attack.

As a rule, metal directed Schiff condensation occurs in solution and the nature of the solvent plays an important role. However, a solid-state template synthesis has also been reported. For example, the thermal reaction of $Cu(3-NO_2-acac)_2 \cdot en_2$ produced $Cu(NO_2-N_4 \cdot [14])$ intramolecularly according to Scheme 3.5.[44] It should be noticed that the strong electron-withdrawing nitro groups enhance the reactivity of the coordinated carbonyl groups. This fact explains the failure of the intramolecular condensation in the case of $Cu(acac)_2 \cdot en_2$.

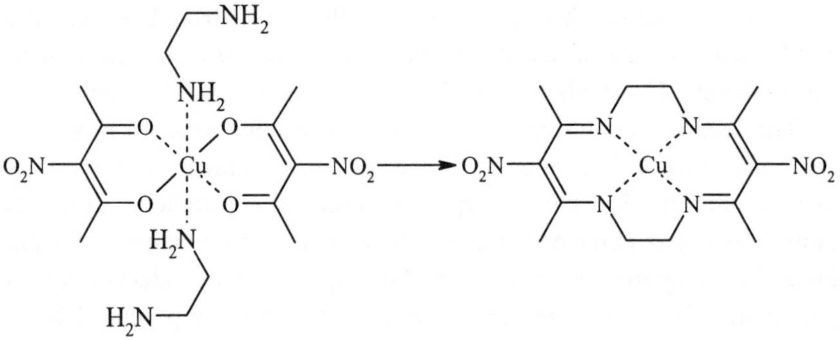

Scheme 3.5

3.2. Open-chain Ligands

Although the metal directed Schiff condensation resulted mainly in closed systems, open-chain ligands have been obtained too. Acyclic structures **3.9** have been obtained when acetone reacted with 1,2-ethanediamine (en) complexes of nickel(II) or copper(II).[45]

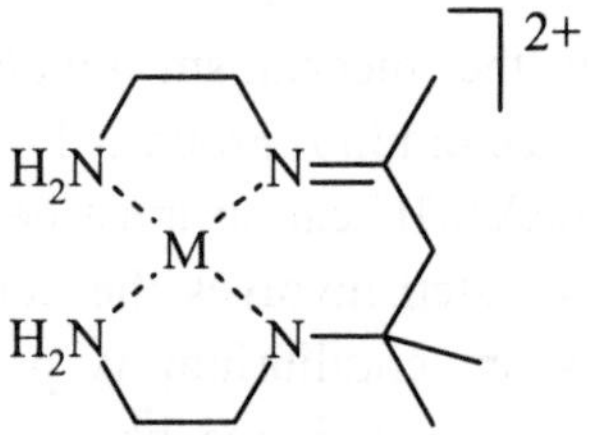

3.9

The presence of an additional potential donor group enhances the ability of a ketone to act as precursor. For example, Schiff bases have been obtained by template condensation of florinated ketoalcohol 5,5,5-trifluoro-4-(trifluoromethyl)-4-hydroxy-2-pentanone   (HFDA)   with primary amines and polyamines. In this case, the precursor contains an additional potential donor group, namely, alkoxy, and resulted ligand is an iminoakoxy one. The nature of the formed complexes and the coordination of the ligands depend on the identity of the metal ions and that of amine component and, on the size of potential chelate rings in the complex. The metal which act as template in these reactions are Cu(II), Ni(II) and, sometimes cobalt(II). When alkyl monoamines are used, it results uninegative ligands in mononuclear square-planar complexes of the type **3.10**. When diamines $H_2N\text{-}(H_2C)_n\text{-}H_2N$ are used, mononuclear or dinuclear complexes were formed, depending on the chain length.[46] Thus, mononuclear species **3.11** were obtained for $n = 2$ or 3, whereas complexes **3.12** were obtained when $n = 6$ for M = Ni(II) or $n = 5$ for M = Cu. The chemistry of these Schiff bases has been enriched when additional potential groups were introduced in the diamine precursors. Thus, the polyamines $H_2N\text{-}(H_2C)_2\text{-}HN\text{-}(H_2C)_2\text{-}H_2N$ and $H_2N\text{-}(H_2C)_3\text{-}HN\text{-}(H_2C)_2\text{-}HN\text{-}(H_2C)_3\text{-}H_2N$ leads to the mononuclear complex of five-coordinate nickel(II) **3.13** and of six-coordinate one **3.15**. Nickel(II) and copper(II) acted as template in the condensation reaction of HFDA with bis(2-aminoethyl)ether, $H_2N\text{-}(H_2C)_2\text{-}O\text{-}(H_2C)_2\text{-}NH_2$ when complexes of **3.17** result.

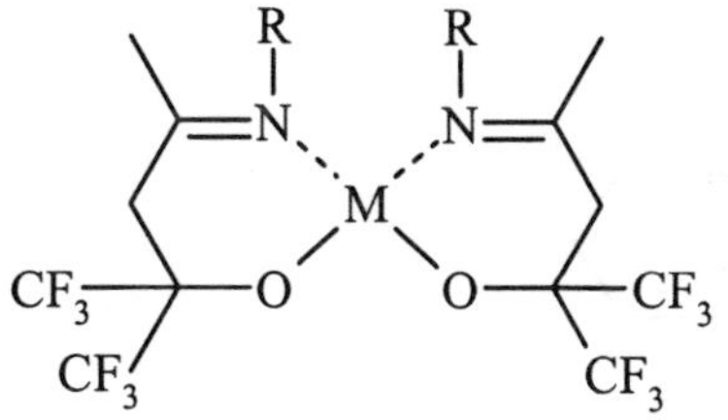

3.10

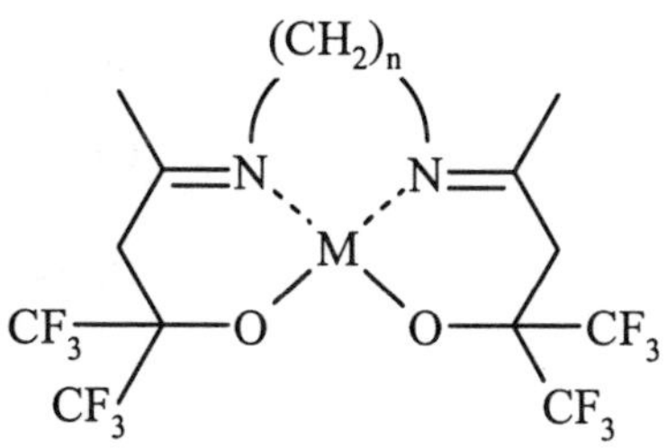

3.11

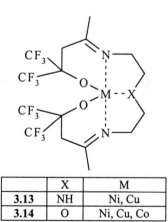

3.12

The mass spectra of both nickel(II) and copper(II) complexes show peaks corresponding to dinuclear species. On the basis of spectral and magnetic properties, it has been proposed similar structures to **3.12** in which two coordinated metal ions of square-planar geometry, are held together by two five-membered $-(H_2C)_2-O-(H_2C)_2-$ chains. The weaker coordinating ability of oxygen atom than that of nitrogen and the reduced flexibility of the iminoalkoxy chelate ring can explain this structure. However, in the presence of cobalt(II) as template, for which steric preferences are less significant, the mononuclear **3.18** complex is formed in which oxygen is coordinated. Hexadentate ligands have been obtained when the chain connecting the two amine groups contain an additionally ether donor sites.

	X	M
3.13	NH	Ni, Cu
3.14	O	Ni, Cu, Co

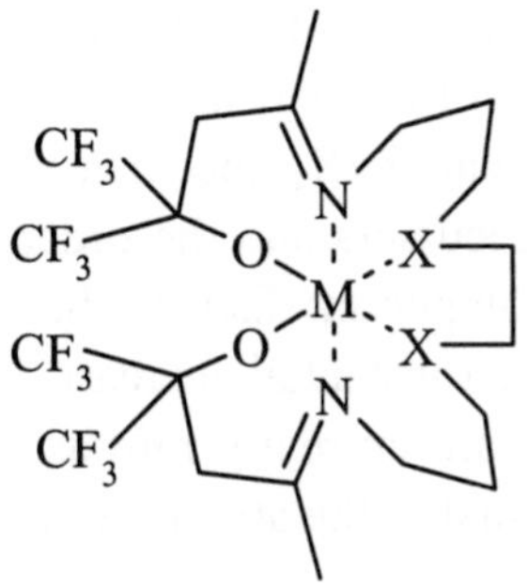

	X	M
3.15	NH	Ni,
3.16	O	Ni, Cu, Co

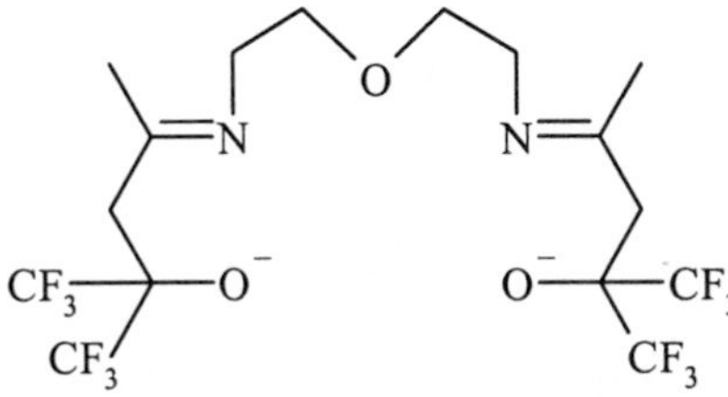

3.17 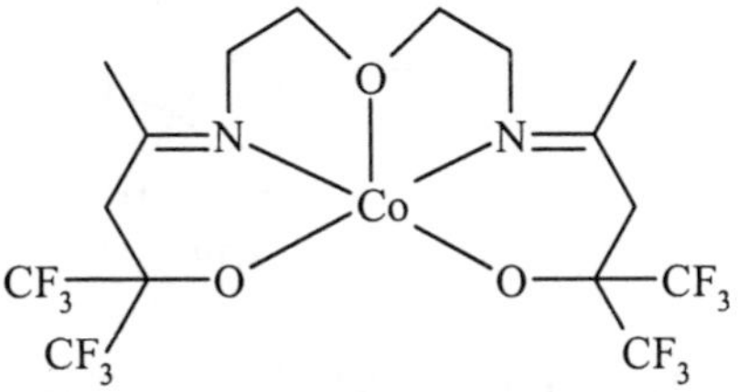   **3.18**

The simplest Schiff base has been obtained from salicylaldehyde and $(CH_3)_2CH-NH_2$, as complexes of the general formula $Me(L)_2$, **3.19**, when Co, Ni, Cu, Zn and Fe(II) acted as template. The X-ray analyses show the four coordinate metal ion with distorted tetrahedral geometries. The bond length and angles as well as the distortions are caused by the nature of the metal-ligand bonding, the electronic structure of the metal atoms and the structural requirement of the ligand.[47] Orange microcrystalline iron(II) complexes of N-isopropylsalicylaldimine, have been obtained by treatment of bis(salicylaldehydato)iron(II) with isopropylaldehyde in tetrahydrofuran and its structure along with the Zn analogue.

Sometimes, open-chain ligands appear as by-products in the synthesis of macrocyclic Schiff bases. These ligands are referred to as "half-units" and are potential precursors for symmetrical or non-symmetrical tetradentate Schiff bases. Thus, tridentate ligands can be formed during the template syntheses of hexdentate Shiff bases.[48, 49, 50]

For example, in the presence of Zn(II) ion, the complexes [ZnX$_2$L], **3.20**, where X = Cl or Br, of the half unit L has been obtained when the acetylpyridine is treated with ZnX$_2$ and then with ethane-1,2-diamine.[51] The molar ratio of the precursors and the order in which the precursors are added are the most important factors. In this case, respecting the above mentioned order, a substantial increase in yields and purity was noticed. Such complexes are currently rare as they contain an exposed primary amine group. Actually, the attempt to obtain the free Schiff base L was unsuccessful.

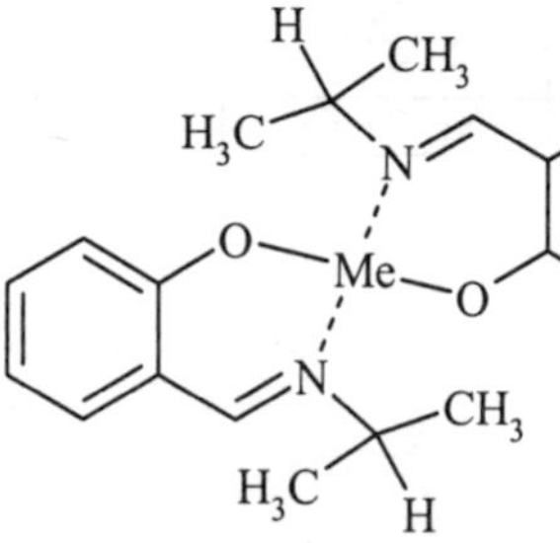

3.19 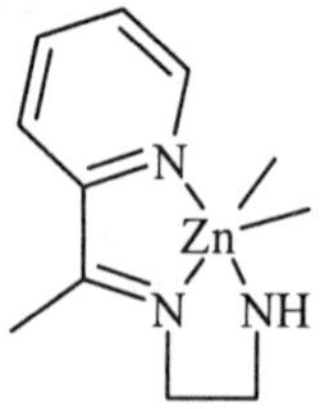   **3.20**

N$_4$ donors. Metal β-diketonates were used in template condensation reactions for the synthesis of open and closed systems. For example, iron(II) acts as template in the synthesis of the open chain ligand **3.21** by successive additions of the tridentate precursor 4-oxo-2-pentylidene-2-pyridylmethanamine, MeCOCH$_2$C(Me)=NH-CH$_2$C$_5$H$_4$N, as well as 2-aminomethyl-pyridine to an acetonitrile solution of [Fe(MeCN)$_6$]$^{2+}$.[52] The resulted red complex [Fe(**3.21**)(MeCN)$_2$][ClO$_4$]$_2$ shows an equilibrium between $^1A_{1g}$ and $^5T_{2g}$ states characterized by the thermodynamic parameters of ΔH = 26 kJ mol^{-1}, ΔS = 7326 kJ mol^{-1} and T_c = 351 **K**.

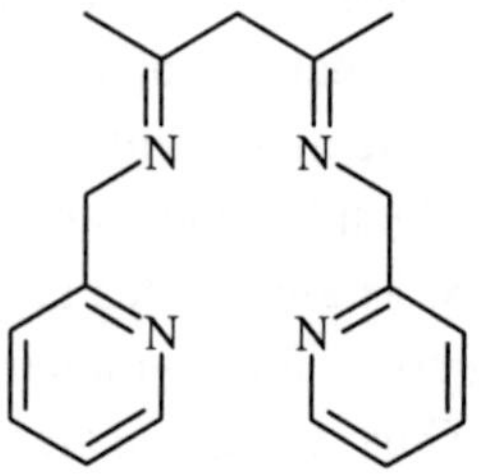

3.21

One of the first deeply studied reaction of the coordinated ligands is the condensation of an α-dicarbonylic compound with an amine in the presence of a metal ion. A new α-diimine linkage is formed as a part of a five-membered chelate ring whose "aromatic character" lead to a high stability of the system. Iron(II) complex containing α-diimine ligand by condensation of glyoxal or biacetyl with metylamine in the presence of ferrous salts have been studied. Two red complexes were isolated as crystallized iodides and they correspond closely to the composition of tris-(biacetyl-bis-N-metylimine)ferrous-   and   tris-(glyoxal-bis-N-metylimine)ferrous iodide, respectively. The high stability of the complexes was attributed to the formation of the five-membered chelate rings which include d-electrons of iron(II).

N$_5$ donors. The use of Cd(ClO$_4$)$_2$.6H$_2$O as template affords the complexes [Cd(**3.22**)(en)](ClO$_4$)$_2$ and [Cd(**3.22**)(opd)](ClO$_4$)$_2$ where **3.22** is the open-chain ligands derived from the condensation of one molecule of diacetylpyridine (dap) with two molecules of 1,2-ethanediamine (en) and **3.23** is the corresponding ligand derived from diformylpyridine (dfp) and o-phenylenediamine (opd).[53] The infrared spectra of the complexes show the presence of the ν(C=N) and the X-ray structure confirms its formation.

The ligand **3.23** acts by its five nitrogen atoms forming a pentagonal girdle.

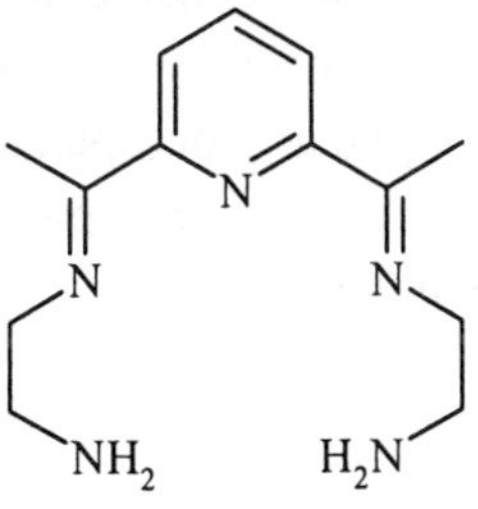

3.22 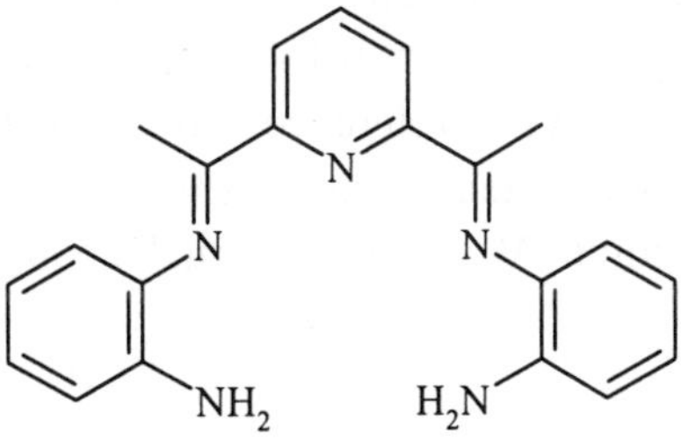   3.23

The coordination number 6 is fulfilling with a nitrogen atom of the opd, thereby rare pentagonal pyramid geometry is forming. [Cd(**3.22**)(NCS)](ClO$_4$)   is   obtained   by   methathesis   of [Cd(**3.22**)(en)](ClO$_4$)$_2$.

Constable *et al.*[54] reported the template condensation of 2,6-diacetylpyridine with 2-chloro-6-hydrazinopyridine in the presence of cobalt(II), copper(II), and zinc(II) resulting in complexes of the ligand **3.24**.

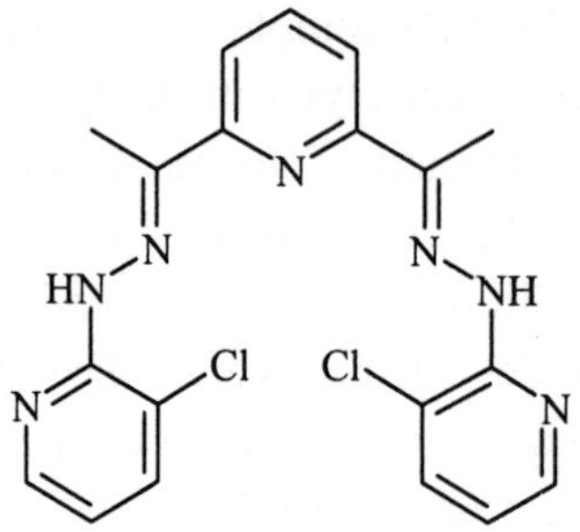

3.24

The dark brown paramagnetic (4.67 BM) cobalt complex, [Co(**3.24**)(H$_2$O)$_2$][PF$_6$], contains the capped octahedral geometry of the cation which is imposed by the bulky chlorosubstituents on the terminal pyridine rings. For the copper complex [Cu(**3.24**)Cl(H$_2$O)][PF$_6$], the pentagonal bipyramid arrangement was proposed with the pentadentate ligand occupying the equatorial sites and the coligands in the axial positions. The ESR spectrum exhibit the *g* values $g_\perp > g_{II} \approx 2$ with precise values of $g_\perp$ =2.202, g_{II} =2.046, A$_{II}$ = 123.3G and support the proposed structure. The magnetic moment of 1.51 BM suggest a considerable spin-exchange interaction between adjacent centres. The new pentadentate ligand shows preference for pentagonal planar or penatagonal bipyramidal geometries. This explains why no complexes of oxovanadium(V), or chromium(III) could be isolated.

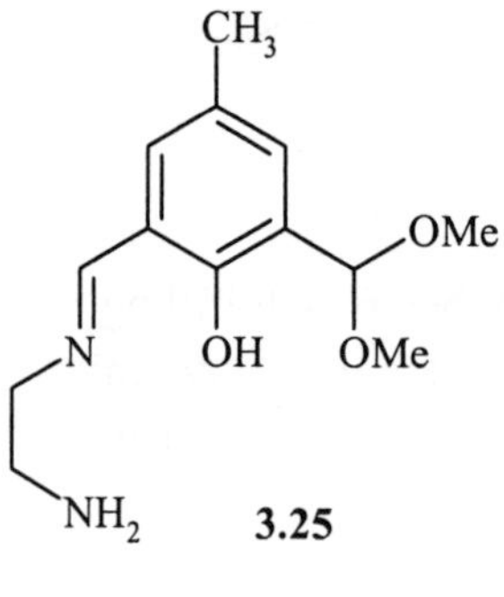

3.25

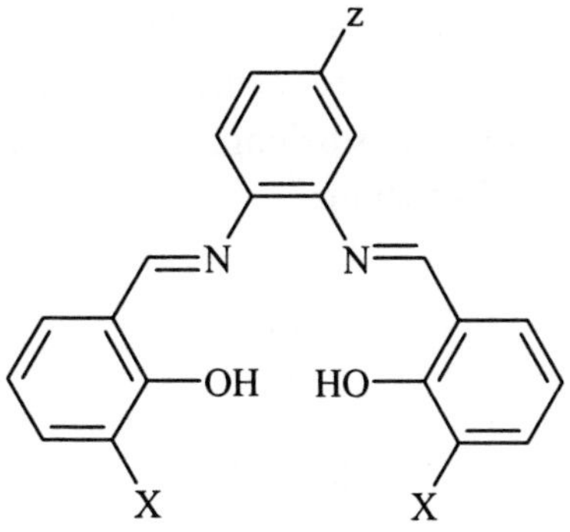

	Z	X
3.26	Me	EtO
3.27	Cl	MeO
3.28	H	EtO

N$_2$O donors. Ligands containing N$_2$O, **3.25**, and N$_2$O$_2$, (**3.26** - **3.28**), donor atoms have been obtained when oxovanadium(IV)[55] and iron(II),[56] respectively, were used as template. The interaction of oxovanadium(IV) salts, 2,6-diformyl-4-methylphenol and en in methanol yields the unexpected acetal **3.25** as the complex [VO$_2$(**3.25**)]$_2$, instead of the corresponding macrocycle. The oxidation of VO^{2+} to VO$_2^+$ takes place as the reaction occurs in an oxygen atmosphere and the X-ray structure shows the oxo-bridged dimeric structure with each Schiff base ligand being tridentate.

An open – chain ligand was prepared by the reaction of 2,6-diacetyl-4-methylphenol with 1,3-diamino-2-hydroxypropane in the presence of UO$_2$(NO$_3$)$_2$.6H$_2$O.

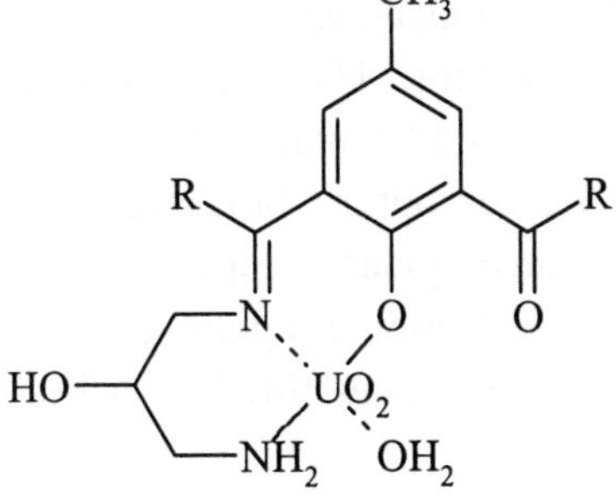

3.29

The resulted bright orange complex **3.29** contains seven coordinated uranium atoms which could be achieved through interaction with carbonyl or aminoalcohol groups belonging to the neighbouring molecule resulting in an aggregated structure.[57]

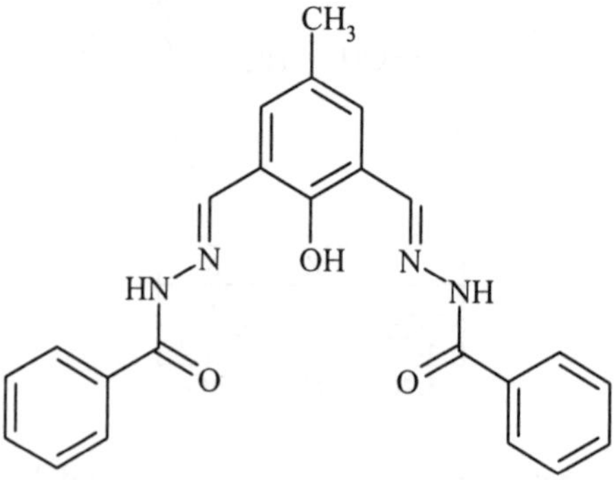

3.30

The complexes $[Cu_2(3.30)(\mu-N_3)](ClO_4)_2 \cdot 1/2C_2H_5OH$, $[Cu_2(3.30)(\mu-N_3)(H_2O)(C_2H_5OH)](ClO_4)]$ and $[Cu_2(3.30)(\mu-N_3)] \cdot DMF$ were obtained from the reaction of 2,6-diformyl-4-methyl-phenol with two equivalents of benzoylhidrazine and copper(II) perchlorate and acetate, respectively, and sodium azide in ethanol.[58] In these complexes, the pentadentate binucleating ligand can lose one, two or three hydrogen atoms to form the keto form (like **3.30**) mixed keto-enol form and enol form. The crystal structure of the complex $[Cu_2(3.29)(\mu-N_3)(H_2O)(C_2H_5OH)(ClO_4)]$ has been resolved.

NOS$_2$ Open chain ligands H$_2$pit, **3.31**, and H$_2$bit, **3.32**, have been obtained[59] using Ni(II) as template when pentanedione or salicylaldehyde reacted with aminoethanethiol. The X-ray structures[60] show that the binuclear [Ni(pit)]$_2$.CHCl$_3$ and trinuclear [Ni(bit)]$_3$ complexes have been formed directly from the appropriate building blocks. In both the cases, each nickel atom has a square planar geometry and has an identical NOS$_2$ ligand environment. The dianionic ligand occupies three of the four coordination sites, the fourth being filled by a bridging sulfur from the neighboring ligand.

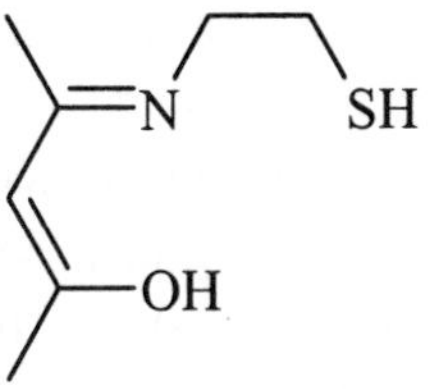

3.31

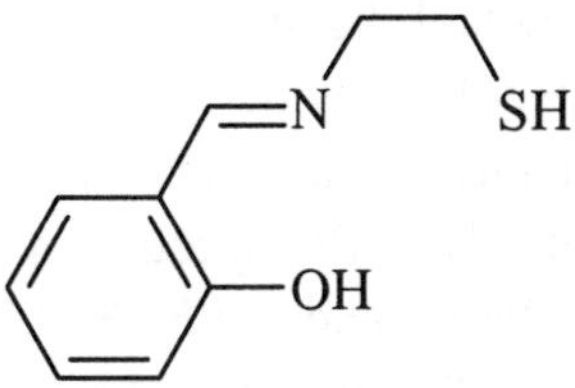

3.32

N₃S₂ donors. The quinquedentate ligand 2,6-diacetylpyrydine-bis-(thiosemicarbazide) **3.33** has been obtained by condensation of 2,6-diacetylpyridine and thiosemicarbazide in the presence of one of the following metal salt as template: manganese(II), cobalt(II) or nickel(II) chloride, iron sulfate, zinc(II) acetate or potassium platinumtetrachloride.[61] A detailed analysis of the infrared spectra of the complexes offered information regarding the bonding mode of the ligand. Infrared spectra of the complexes show that the $v_{as}(NH)$ remains unchanged on coordination, relative to that of the free ligand and $v_s(NH)$ is raised by about 60 cm^{-1}. Also, scissors deformations of the $v_{as}(CN)$ and $v_s(CN)$ of the -N-C(=S)-N- groups ($\Delta v_{as}(CN) \cong 40$ cm^{-1}, $\Delta v_s(CN) \cong 25$ cm^{-1}) were observed. This was interpreted as a lack of the direct bonds between the nitrogen atoms of the amino groups and the metal atom. The coordination of the azomethine nitrogen atom and of the sulfur atom is proved by the displacement of the $v(N-N)$ and $v(CS)$ bands by approximately 40 and 100 cm^{-1} respectively.

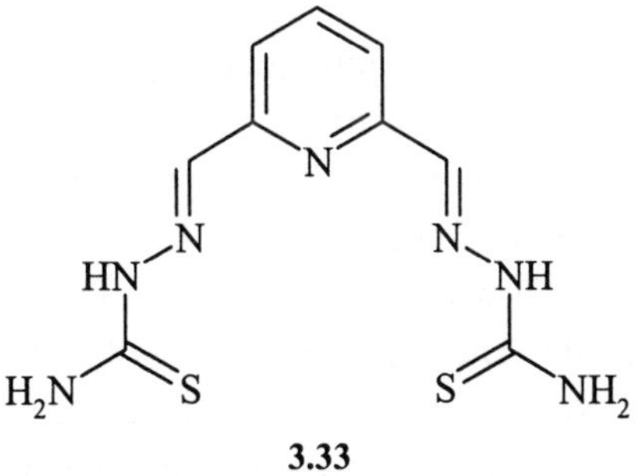

3.33

Complexes comparable with those obtained by α- and β-aminoacids have been obtained by using α- and β-aminosulphonic acids. Non-compartmental ligands resulted from condensation of 2,6-diformyl-4-chlorophenol with aminomethanesulphonic acid or 2-aminoethanesulfonic acid in the presence of nickel(II) acetate, in methanol. The IR spectra of the resulted bright green complexes Ni(**3.34**)·2H$_2$O and Ni(**3.35**)·2.5H$_2$O show the band at $\approx$ 1665 cm^{-1} attributed to ?(C=O) and a band at 1636 cm^{-1} to ?(C=N).[62]

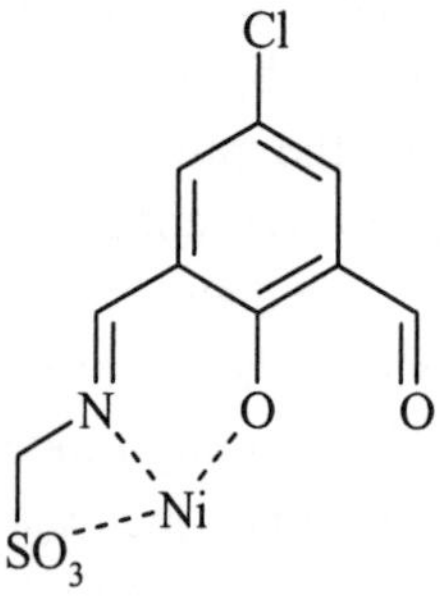

3.34 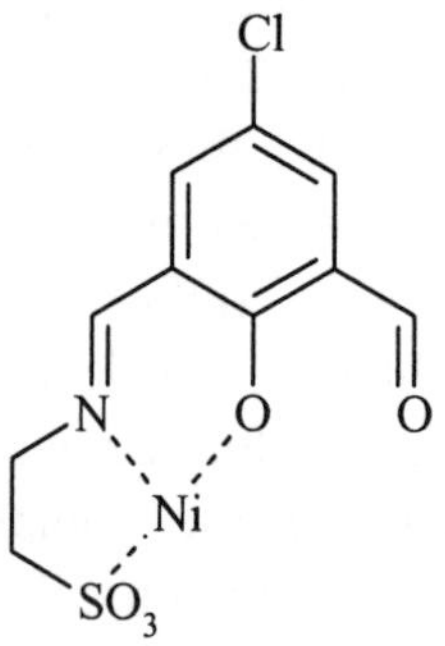   3.35

3.3. Macrocyclic Ligands

3.3.1 Diimine macrocycles

3.3.1.1. Aliphatic monocarbonylic precursors

Curtis[63] and Bush prepared complexes by condensation of ketones with transition-metal complexes of bi, tri or tetraamines when mostly tetraaza macrocyclic ligands were formed. Cyclic structures have been obtained through reaction of acetone with 1,2-ethanediamine complexes of nickel(II) or copper(II) when the formation of either the *cis*- or *trans*-isomers **3.36** is possible. The *trans* structure is favoured by the presence of the monodentate extracyclic ligands whereas the macrocycle is forced into folded form to produce *cis* structure by the bidentate extracyclic ligands (like C$_2$O$_4$). The unsaturated structure induces inflexibility of the ligand and further the preference for a planar structure.

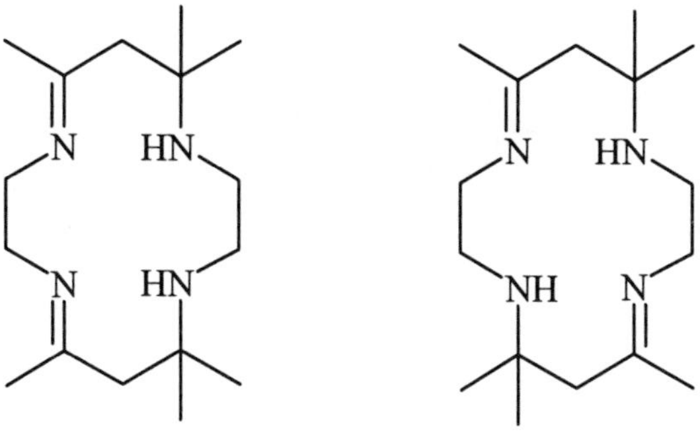

cis   trans

3.36

The complexes containing 14-membered macrocyclic ligands [Cu(*trans*-[14]dieneN₄)](ClO₄)₂ and [Cu(*cis*-[14]dieneN₄)](ClO₄)₂ have been deeply studied by physical and chemical methods[64] whereas the nickel analogues have been characterized by X-ray crystal structure.[65] The 15- and 16-membered tetraazamacrocyclic complexes have also been obtained using appropriate diamine complexes. It has been established that their properties are not remarkably different from those of the 14-membered systems except that they are more susceptible to acid hydrolysis. The large 18-membered macrocycles, *trans*-[18]dieneN₄, **3.37** and *cis*-[18]dieneN₄ **3.38** have been obtained as paramagnetic copper(II) complexes by condensation of acetone with 1,4-butanediamine in the presence of metal perchlorate (Scheme 3.6).[66] When nickel(II) is the template, only the yellow *trans*-isomer is formed. The X-ray structures of the copper(II) complexes show that the four nitrogen atoms coordinate copper(II) in a pseudotetrahedral manner in the purple *cis*-isomer and in a close to square-planar geometry in the orange-red *trans*-isomer.[67] The distortion observed for the *cis*-isomer is larger than that found for the [14]diene analogues and this can be attributed to the large, about 40°, dihedral angle between the six- and seven-membered chelate rings. The distortion and the strain in the macrocyclic ligand were estimated by deviation of bond angles and torsional angles from ideal values. It results that they depend on both the chelate ring size and the placement of the imine bond.

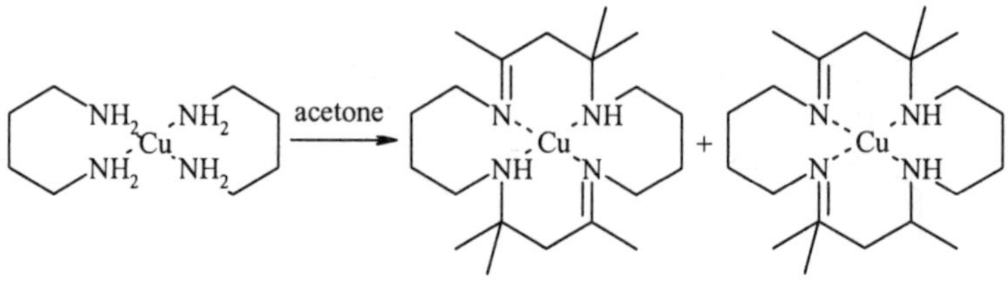

3.37   **3.38**

Scheme 3.6

3.3.1.2. Oxamido carbonyls as precursors

Oxamido aldehydes and ketones have been used as precursors for macrocyclic metal complexes.[68] The use of nickeldimethylmalonamido bisphenylglyoxylamide, **3.39**, and 1,2-diaminobenzene as precursors, in the presence of nickel(II) as template, resulted in the 15-membered macrocyclic nickel(II) complex **3.40** (Scheme 3.7). Considering this strategy, a special attention has been paid to obtain stable precursor for macrocyclic complexes. A range of substituted pyrrolidinyl amide complexes **3.41** have been obtained and characterized by means of IR spectra. These bear various substituents in the 5C position of the aryl ring. Precursors **3.42** containing glyoxal fragments have been obtaining following either the isatin[69] or by reaction of the appropriate aminoketone with oxalyl chloride.[70] Further, the 14-membered macrocyclic complexes **3.43** and **3.44** and **3.45** containing glyoxylic moiety have been obtained using nickel acetate as template (Scheme 3.8).

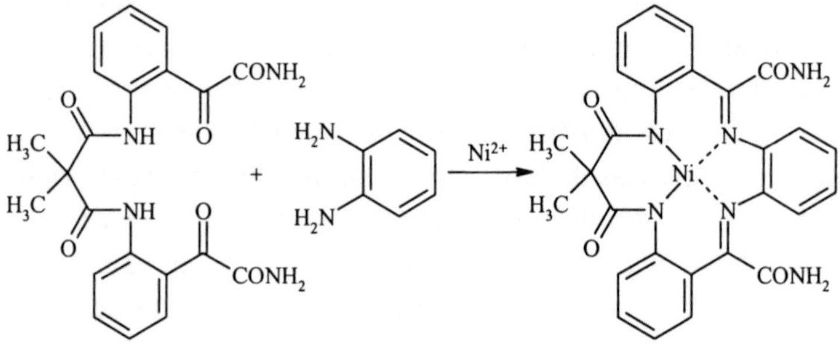

3.39   Scheme 3.7   **3.40**

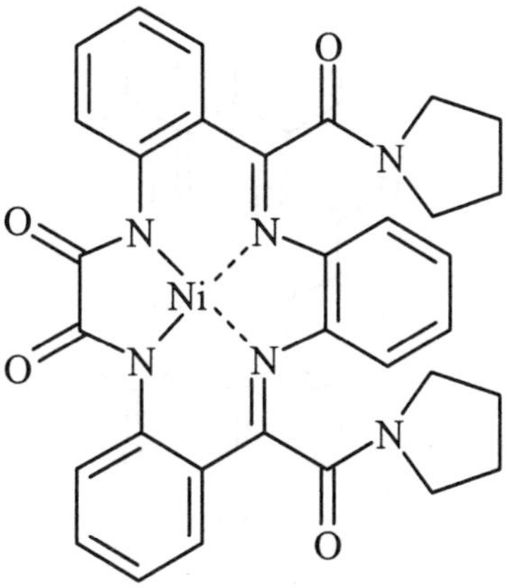

3.41

The similar copper(II) complexes **3.46**, **3.47** and **3.48** were prepared by the template reactions of 2,2'-(oxalyldiimino)-bis(phenyl glyoxalate) with appropriate diamines and copper(II) acetate in the presence of triethylamine. Differences in the ring size induce significant differences in molecular structure, ESR spectra and redox properties. Thus, **3.46**, a $[15]N_4$ macrocycle has a larger cavity than the $[14]N_4$ macrocycle and it contains a six-membered diimine chelate ring. Therefore, the copper(II) is located right in the cavity in [Cu(**3.46**)] and it is 0.048 Å out of the plane in [Cu(**3.47**)]. However a tetrahedral distortion of the coordination sphere is observed for [Cu(**3.46**)]. The g_{II} (2.169) of [Cu(**3.46**)] is large and the A_{iso} and A_{II} values are significantly smal. The complexes which incorporate both oxamido and imine groups undergo quasireversible reduction to Cu(I) and oxidation to Cu(III) proving that these macrocycle can stabilize both Cu(I) and Cu(III).

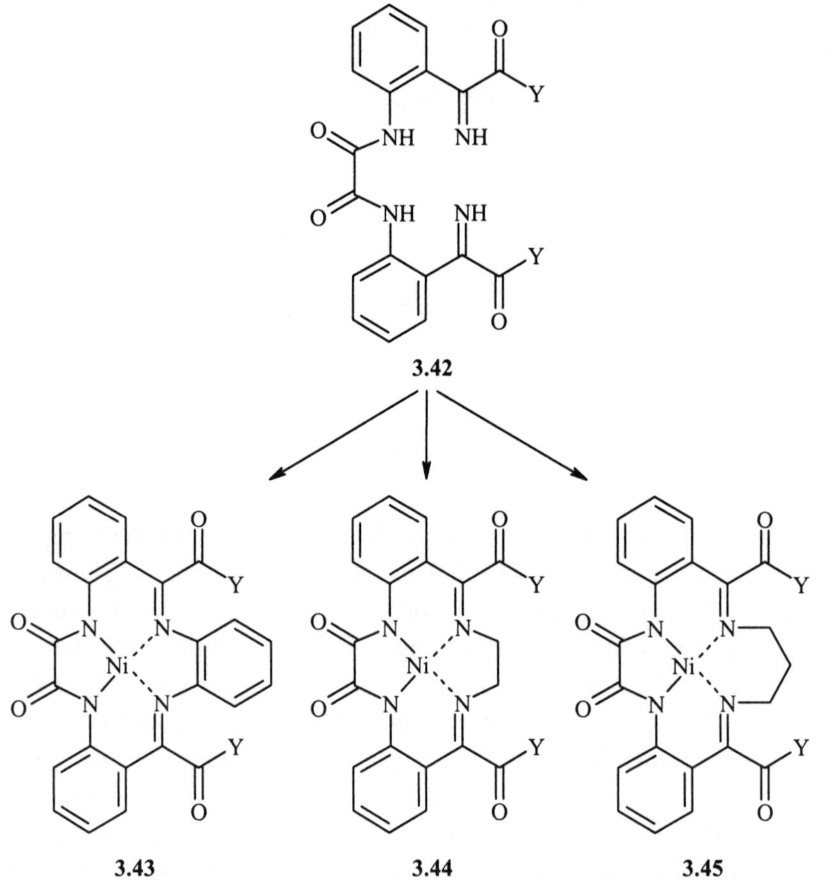

3.42

3.43 **3.44** **3.45**

Scheme 3.8

Y = *a)* OCH$_3$; *b)* OC$_2$H$_5$; *c)* OCH(CH$_3$)$_2$; *d)*. OC(CH$_3$)$_3$; *e)* NH$_2$; *f)* (CH$_2$)$_3$CH$_3$;
g) NH(CH$_2$)$_{11}$CH$_3$; *h)* NHCH$_2$Ph; *i)* NHCH$_2$CO$_2$C$_2$ H$_5$; *j)* NHOCH$_3$; *k)* NHPh;
l) NHC$_6$H$_4$N(CH$_3$)$_2$-*p*; *m)* NHC$_6$H$_4$NO$_2$-*p*; *n)* NHC$_6$H$_4$OCH$_3$-*p*.

3.3.1.3. Dicarbonylic precursors

Metal ion template assistance in the condensation of compounds containing dicarbonyl functional groups with primary diamines to give macrocyclic imines in which the template ion is found coordinated to the imine is well documented.

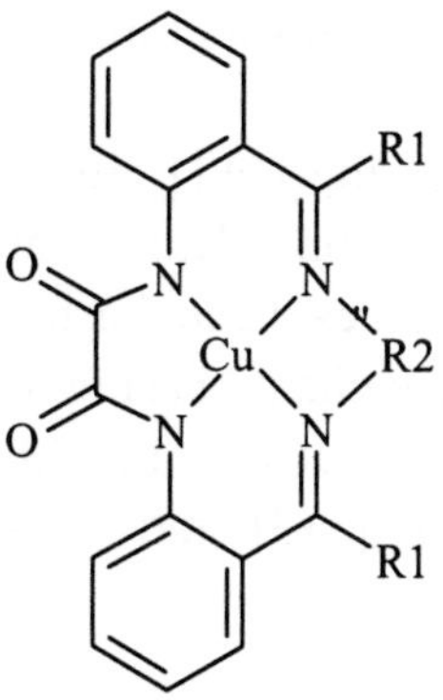

	R1	R2
3.46	-COOEt	-(CH2)3-
3.47	-COOEt	-(CH2)2-
3.48	-COOEt	-CH (CH3)CH2 -

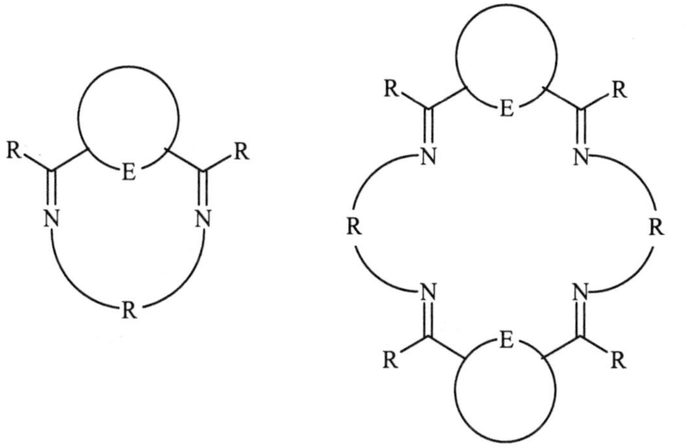

A   B

Scheme 3.9.

The dicarbonyl compounds form the "head" of the ligands and diamines are the straps which bond the carbonyl groups belonging either to the same head or to a different one. The cyclocondensation products are of the type presented in Scheme 3.9. The type **A** results in 1:1 condensation (termed [1+1] macrocycles) of precursors whereas the type

B is formed when the precursors are in 2:2 molar ratio ([2+2] macrocycles).

The metal complexes of the [1+1] are mononuclear whereas [2+2] macrocycles may be mono- or binuclear in nature. Dicarbonyl precursors and a wide range of diamines have been used for the synthesis of macrocycles. As the diamine chain became longer a [1+1] condensation is preferred. With certain precursors (i.e. 2,6-diacetylpyridine and 1,3-diamino-2-hydroxypropane), [3+3] and [4+4] macrocyclic complexes have been synthesized too.

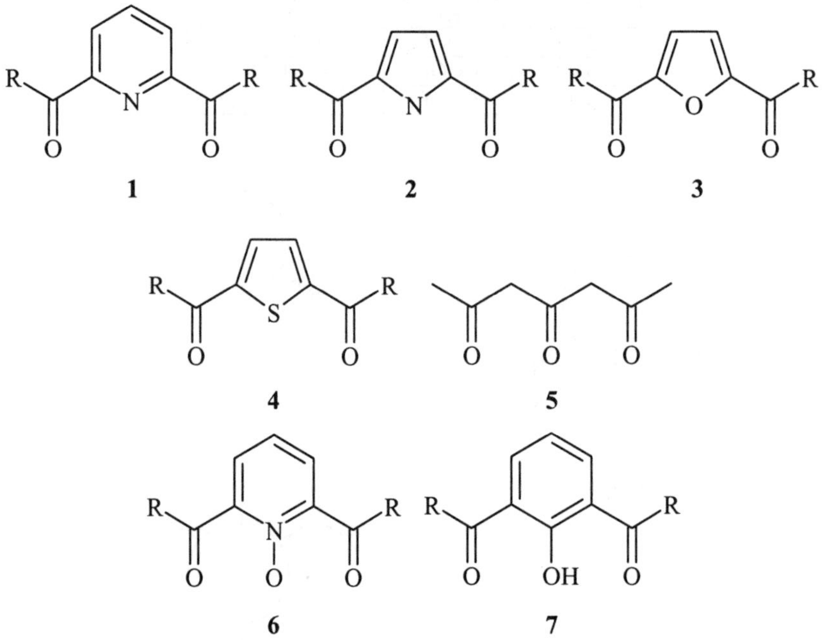

Scheme 3.10

The head units consist of molecules in which carbonyl groups are included in an aliphatic chain or are bonded to an unsaturated heterocycle. The most used head units are presented in Scheme 3.10. As it can be noticed, the head units contain an additional potential donor atom/group. When this donor group bridges two metal centres, compartmental ligands are formed.

Schiff bases derived from aliphatic diamines as well as aromatic ones have been obtained. The dimension of the bridge between their nitrogen atoms determine if the [1+1] or [2+2] condensation occurs and thereby if a diimine or a tetraimine macrocycle is obtained. As the number of imine groups increases the degree of unsaturation increases and the macrocycle become more rigid. Additional atom and groups donors can be part of these bridges and their nature is also important in Schiff base condensation reactions directed by metal ions. Several amines are presented in Scheme 1.6.

Metal complexes derived from these types of Schiff base will have five- or six-membered central chelate rings, respectively which have a major contribution to stabilize the systems. The ligands derived from aromatic diamines present special characters compared with that derived from diamines.[71] In these cases, thermodynamic data are reduced because of the insolubility of the compounds in water which is the most common solvent for potentiometric determination of stability constants and also to the possible hydrolysis of some compounds to give the starting organic fragments. The Schiff bases derived from *o*-phenylenediamines, the proximity of the nitrogen atoms allows simultaneous coordination of both to the same metal cation, leading mainly to monomer species whereas, *m*- or *p*- isomers can only coordinate a nitrogen atom to any one metal ion. The ligands based on aromatic amines are more rigid.

Of course, the nature of the head units as well as that of the bridges determines the dimension of cavity of the macrocycle and the extend of its unsaturation but - in template synthesis – the size and electronic structure of the metal ions controls the synthetic pathway. Transition metals are generally used for the smaller rings and rare earth metals for the larger ones. The compatibility between the radius of the templating cation and the "hole" of the macrocycle contributes to the effectiveness of the synthetic pathway and to the geometry of the product complex.

Taking in mind these factors, several types of macrocyclic ligands obtained by Schiff base condensation will be presented.

3.3.1.3.1. Aliphatic precursors

The 14-membered diimino rings **3.49-3.59** have been obtained in a cyclisation reaction of glyoxal or biacetyl, respectively, with the appropriates tetradentate amines in the presence of nickel(II), copper(II) or cobalt(II) acetates.[72] The resulted nickel(II) complexes are enough stable to be reduced with sodium borohydride to the corresponding cyclam whereas both cobalt and copper ones affect the oxidation state of the metal ions. The 15- membered macrocyclic complex containing **3.60** and **3.61** using cobalt acetate has been obtained in moderate yields but the attempt to obtain 13-membered macrocycles in this way failed.

The presence of additionally donor sites in the precursor molecule facilitates the formation of the diimine systems and can assure the geometrical requirement of the metal ions. For example, nickel(II) ion requires essentially planar quadridentate structure and, as a result, quadridentate or sexadentate diimine complexes have been obtained by template reactions.

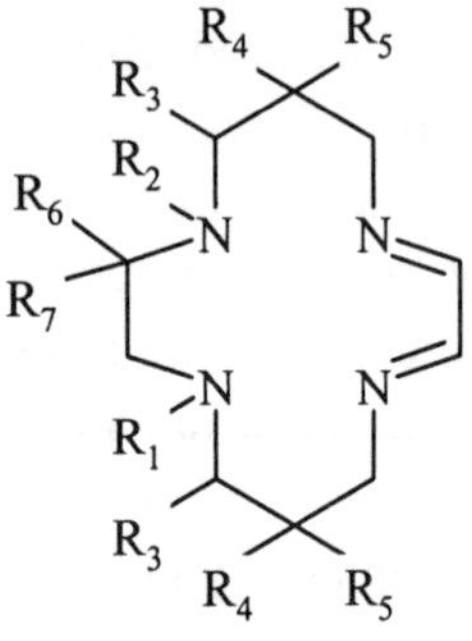

	R_1	R_2	R_3	R_4	R_5	R_6	R_7
3.49	H	H	H	H	H	H	H
3.50	H	CH_3	H	H	H	H	H
3.51	CH_3	CH_3	H	H	H	H	H
3.52	C_2H_5	C_2H_5	H	H	H	H	H
3.53	H	$C_6H_5 CH_2$	H	H	H	H	H
3.54	$C_6H_5CH_2$	$C_6H_5 CH_2$	H	H	H	H	H
3.55	H	H	H	H	H	CH_3	H
3.56	H	H	H	H	H	CH_3	CH_3
3.57	H	H	CH_3	H	H	H	H
3.58	H	H	H	CH_3	CH_3	H	H
3.59	H	H	H	H	OH	H	H

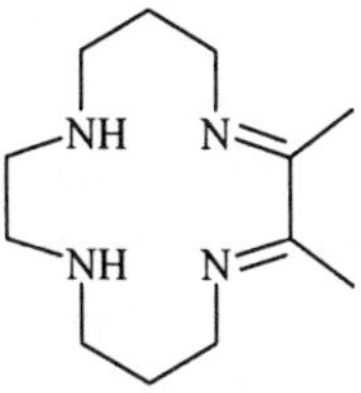

3.60 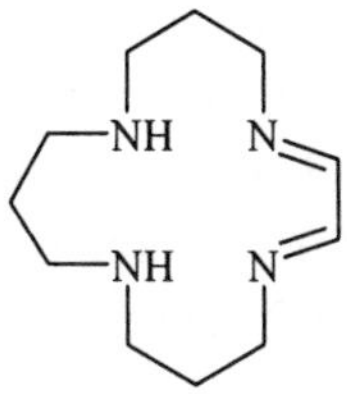   **3.61**

2,4-pentanedione (acetylacetone) is one of the most utilized precursor for the aliphatic macrocyclic ligands. Condensation of acetylacetone with triethylenetetramine in the presence of nickel(II)[73] or copper(II)[74] lead to the 13-membered macrocyclic metal complexes **3.63**, where $x = y = 2$ (Scheme 3.11). Single crystal X-ray structure of Ni(**3.63**)(ClO$_4$) has proved the formation of the ligand, and the founded bond lengths support the presence of a partially delocalized planar six-membered chelate ring, as in formula **3.62**. The nature of the ligand is affected by the acidity of the system. Thus in acidic solutions, protonation of the coordinated ligand occurs at the γ-methyne carbon atom, resulting in the complex containing the neutral macrocyclic β-diimine ligand 11,13-dimethyl-1,4-7,10-tetraazacyclotrideca-10,13-diene, **3.63**. The complexes containing the two forms of the macrocyclic ligand can be interconverted in solution by reversible protonation of the coordinated ligand. It has been observed that the relative acidity of the γ-methyne carbon of the coordinated ligand **3.63** decreases on going from the copper(II) complex ($K_a = 10^{-9}$) to the nickel(II) one ($K_a = 10^{-6}$). The difference has been attributed to steric strain of the six-membered chelate ring which comprises the slightly larger copper(II) ion. In this case, the steric strain is somewhat relieved by protonation of the methyne carbon because a change in hybridization from sp^2 to sp^3 occurs.

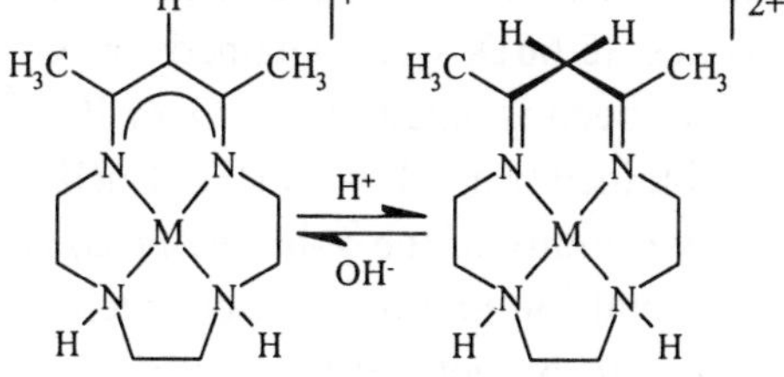

3.62   Scheme 3.11   **3.63**

The cyclization reaction has been extended to 14-membered ligands following the reaction of acetylacetone (acac) with 2,3,2-tet(= N,N'-bis(2-aminoethyl)-1,3-propanediamine in the presence of the appropriate metal acetate resulting in compexes **3.66**. Attempt to use 3,2,3-tet-(N,N'-bis(3-aminopropyl)-1,2-ethane-diamine to obtain a 15-membered macrocycle succeeded only in the case of Ni(II) when the complex **3.67** was isolated. The attempt to obtain a 16-membered ligand failed and the failure was mainly attributed to steric crowding effects.

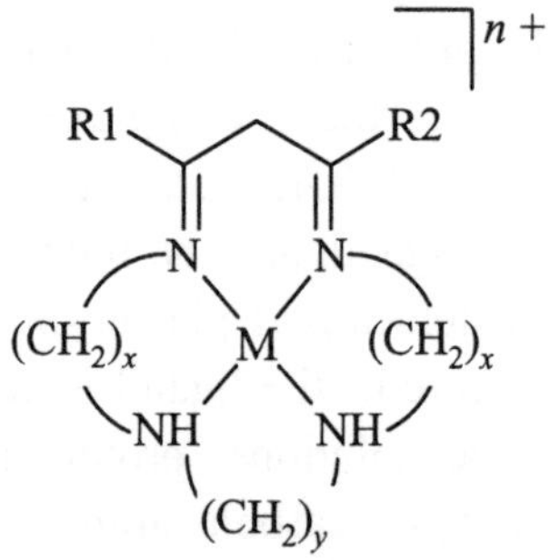

	R1	R2	x	y
3.64	H	H	2	2
3.65	CH$_3$	CF$_3$	2	2
3.66	CH$_3$	CH$_3$	2	3
3.67	CH$_3$	CH$_3$	3	2

It has been found that the nature of the complexes containing 14-membered macrocyclic ligands depends on the pH of the reaction solution at the time of compound isolation. Thus, when pH is 10, the deprotonated ligand is formed whereas at pH 2, the complexes containing the neutral ligand have been isolated. The nature of the counterion also seems to be important. For example, 14-membered macrocyclic nickel(II) or copper(II) complexes, **(3.66)**X were obtained for X = NO$_3^-$, Cl$^-$, Br$^-$ or PF$_6^-$, and **(3.66)**X$_2$ have been isolated for X = I$^-$ or PF$_6^-$.[75] The complexes [**(3.66)**](PF$_6$) were obtained for M = Ni(II), Cu(II) or Co(II) and it resulted that the relative acidity of the γ-bridge proton depends markedly on the nature of the metal ion and to a lesser extend on the size of the macrocyclic ring. It has to be noticed that the rigid, unsaturated ligand [14]dieneN$_4$ forces cobalt(II) to lie in the four-coordinate square-planar environment created by the macrocyclic ligand. The most convincing proof for the formation of the macrocyclic ligands is obtained from the infrared spectra. Evidence for reaction is demonstrated by the absence of the absorption bands that can be attributed to the free or coordinated C=O or NH_2 modes and the presence of the bands assignable to C=N or C-N, C-C stretching vibrations. Magnetic moments of the cobalt complexes fall in the range μ_{eff} = 1.88 - 2.33, close to that expected for low-spin cobalt(II).[76] Single crystal X-ray structure analysis of [(**3.66**)]$(PF_6)_2 \cdot H_2O$, M=Co, shows that Co(II) ion is six coordinated by the macrocyclic nitrogens, the oxygen atom of the water and a fluorine atom of the PF_6 ion. The four nitrogen atoms of the macrocyclic ligand are virtually coplanar and the cobalt(II) ion is displaced from this plane by 0.015 Å toward the coordinated water molecule which is bonded at a distance of 2.283(9) Å. The structure of the orange, diamagnetic nickel(II) complex shows also that the metal ion lies in the plane created by the nitrogen atoms although a small tetrahedral distortion was observed. The same distortion observed in 14-membered and in corrin (15-membered) complexes of nickel(II), copper(II) and even in the above cobalt one, allows to conclude that this type of distortion is not a function of macrocyclic size but of internal strain in the ligands.

The metal chelates of acetylacetone derivatives with electron-withdrawing substituents have been known to possess high reactivity towards amines, producing Schiff base compounds like **3.65**.[77] Also, the 14-membered tetraaza macrocycle resulted by reaction of Cu(3-NO_2acac)$_2$ and ethylenediamine in xylene as well as in solid state was reported.

3.3.1.3.2. Unsaturated and aromatic precursors

The most employed head units belonging to this class have been presented in Scheme 3.8. Among them, the pyridine one is the most used for producing Schiff base macrocyclic ligands with different nitrogen atom donors and different ring size.[78]

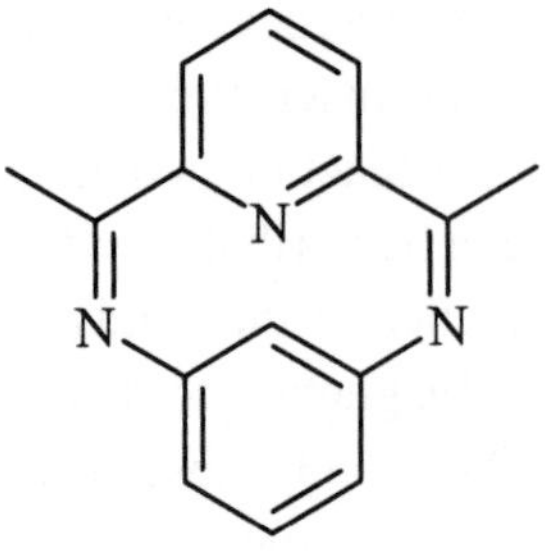

3.68

Relatively few reports of complexes containing tridentate macrocyclic ligands exist when compared with the large number of papers involving tetradentate macrocycles. Condensation of 2,6-diacetylpyridine with *m*-phenylenediamine and the use of $Y(NO_3)_3 \cdot 6H_2O$ as template agent gave the new 10-membered triaza macrocyclic complex of stoichiometry $[Y(\textbf{3.68})(NO_3)_3] \cdot 6H_2O$.[79] Attempts to obtain a free macrocycle in the reaction of 2,6-diacetylpyridine with *m*-phenylenediamine in the absence of the yttrium salt resulted in an acyclic imine.

The symmetrical, tetraaza macrocyclic ligands **3.69** – **3.71** resulted from condensation of diformyl- or diacetylpyridine and corresponding triamines with Ni(II), Cu(II) or Zn(II) perchlorates have been obtained according to reaction presented in Scheme 3.12.[80,81] Experiments show that the condensation process involve the formation of triamine-metal ion complex in a first stage and that the required minimum ring size is given for $m = n = 3$.

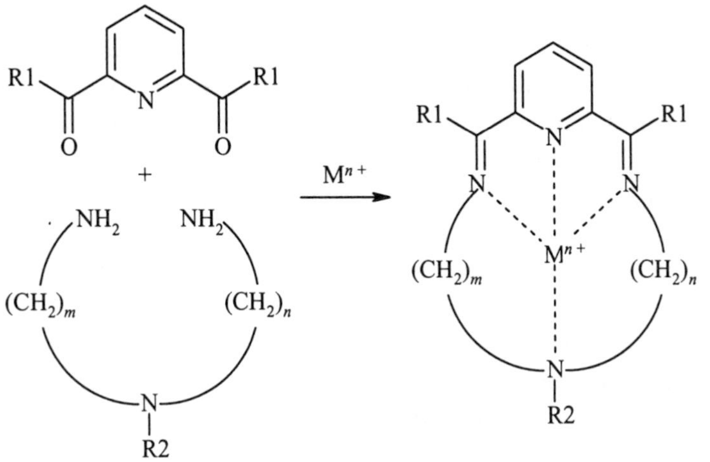

	m	n	R_1	R_2
3.69	3	3	CH_3	H
3.70	3	3	CH_3	CH_3
3.71	3	3	H	H

Scheme 3.12

The nickel(II) complexes of **3.69** are diamagnetic proving the square–planar geometry. The complexes are very stable so that they undergo reduction with sodium tetrahydroborate and further support the attachment of some bulky groups, like benzyl, to obtain pendant groups. The attempt to obtain a 2,3 macrocycle failed and this was interpreted in terms of the smaller aperture than in the case of $m = n = 3$ derivatives for coordination of the metal ions.[82] The complexes containing $m = 3$ $n = 4$ macrocycle have also been obtained but, in lower yield than that of the 3,3 macrocycle. This fact along with its lower stability is attributed to the presence of a seven-membered chelate ring.

The 14-membered macrocycle containing N4 donor set, **3.69**, has also been prepared with samarium(III) or praseodymium(III) as template.

Pentaaza ligands. Considerable effort has been devoted to the synthesis and study of pentaaza macrocycles formed by Schiff base condensation reactions in the presence of a metal ion which acts as a template. 2,6-diacetyl-pyridine and 2,6-diformyl-pyridine are the most widely studied precursor molecules which react with a wide variety of tetra-amines to give 15-, 16- or 17-membered macrocycles which in absence of the metal ions would not be isolated. The planarity of pyridine ring and that of the adjacent nitrogen atoms confer rigidity to the resulted ligands which force the donor atoms to keep in the same plane. Metal ions having appropriate size and electronic structures for the seven-coordination, are effective as template for these ligands. Thus, the spherical electron distribution of the high-spin iron(III) and manganese(II), respectively, favours this geometry, and therefore, these metal ions are effective templates.

Curry and Busch[83] reported the condensation of 2,6-diacetylpyridine with triethylenetetramine and tetraethylene pentamine in the presence of iron(II) salts when penta- and hexadentate macrocycles **3.72** and **3.73** have been obtained as ligands. The complexes were formulated as [Fe(**3.72**–H)](ClO$_4$)$_2$ and [Fe(**3.73**)(OH)]I$_2$, respectively. The experimental data show that the coordination of the precursors through the groups which are not involved into condensation occurs before the condensation. In that account, it has been considered that the nitrogen atom of the pyridine ring of the 2,6-diacetylpyridine and the secondary amines of the polyamine serve to position the components on the metal ion template during the macrocycle closure. It is significant that the iron(II) acts as template and, after condensation, the air oxidation proceeds to iron(III). Magnetic moment values for the iron atoms show that the spin paired states are present in both complexes and hence, that the ligands are powerful ones. In the proposed structures, Fig. 3.1 and Fig. 3.2 respectively, the N atoms at position (1) are the pyridine nitrogen atoms and the donors adjacent to the pyridine ring are coplanar with that ring. Although the proposed structures are highly strained, they are also supported by the increased acidity on the part of the proton originally on the nitrogen atom (6). The complexes [Fe(**3.72**)X$_2$]ClO$_4$ (X = halide or pseudohalide) have been prepared by template route and the anions were changed by metathetical reactions from initially formed products. The pentagonal bipyramidal coordination geometry of the complexes was established based upon various physical properties and was confirmed - in the case of X = NCS, by single-crystal X-ray study.[84,85] It has been established that the five macrocycle nitrogen atoms are approximately planar and describe the equatorial plane which includes iron atom too.

The thiocyanate nitrogen atoms occupy the axial positions. The maximum deviation of a contributing atom from the FeN$_5$ least-square plane is 0.11 Å and the Fe-N bond length of 2.198(13) - 2.257(13) Å are long enough to keep the iron ion in the plane.

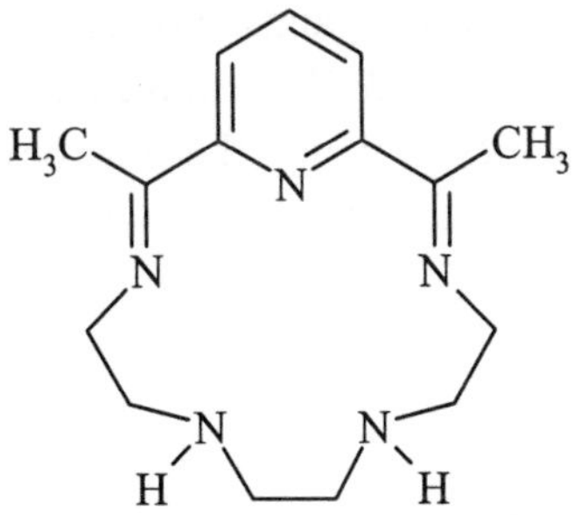

3.72

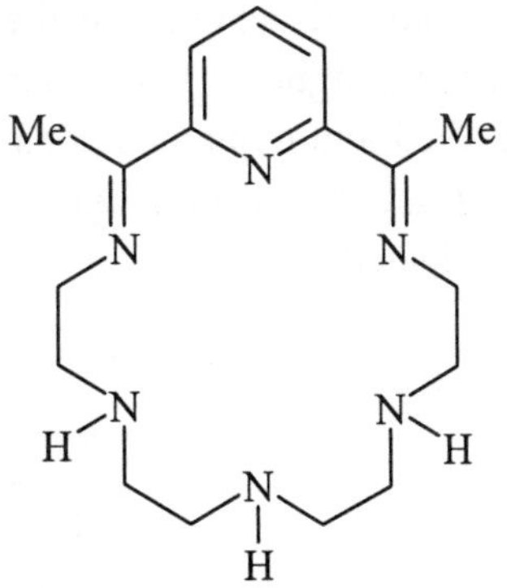

3.73

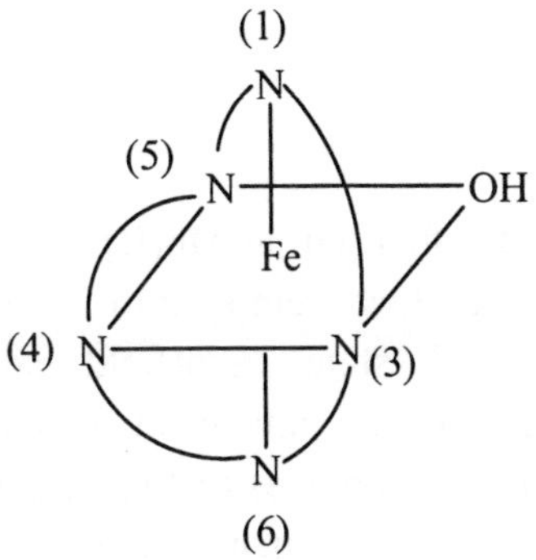

Fig. 3.1

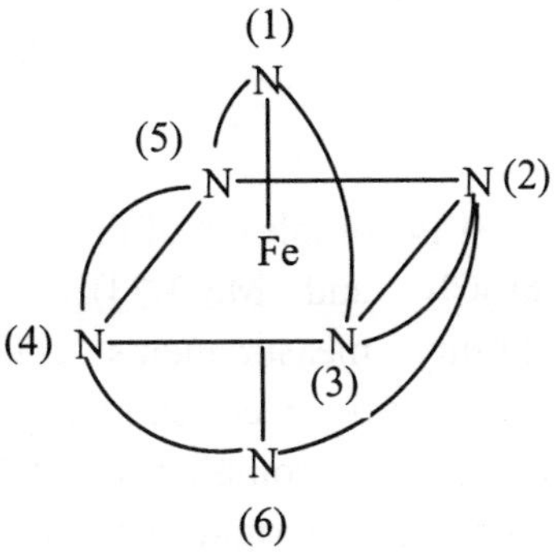

Fig. 3.2

Complexes of ligand **3.73** have also been prepared with manganese(II)[86,87] Zn(II), iron(II) and iron(III),[88] cadmium(II) and mercury(II)[89,90] as template and in some cases structural determinations have been reported. The complexes of these metals in different oxidation states have also been prepared by electrochemical means. All the complexes have been characterized as seven-coordinated ones, with pentadentate macrocyclic ligand lying in a plane and with two monodentate anions occupying axial positions. Comparing the Fe(III) and

Fe(II) complexes, a small increase in the average Fe-N equatorial bond length has been observed on going from the Fe(III) to the corresponding Fe(II) complexes. This fact was expected based on the radii of the metal ions. Work on iron complexes includes studies of charge-transfer photochemistry,[91] and low temperature magnetic susceptibility[92] which are largely dependent on the nature of the axial ligands. For example, unusual magnetic behaviour of FeL(CN)$_2$, L= **3.73**, complexes was observed and explained considering a structure in which the macrocycle acts as a quadridentate ligand with one of the secondary amino groups not coordinated.

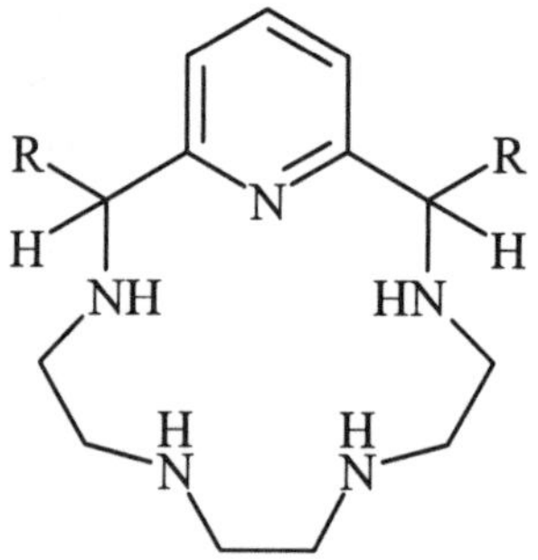

3.74

Interest in complexes of this ligand has focused on Mn(II) complexes Mn(**3.69**)Cl$_2^+$ and Mn(**3.73**)Br$_2^+$ whose low temperature magnetic susceptibility measurements and ground state zero-field splitting parameters have been shown perspectives as advanced materials.[93] Hydrogenation of complexes of **3.72** gives complexes of ligand **3.73**[94, 95] which is much less rigid than **3.69** and as a consequence is able to form complexes with a variety of coordination geometry.

Reaction of diacetylpyridine with the tetraamines 1,9-diamine-3,7-di azanonane(2,3,2-tet) and 1,10-diamine-4,7-diazadecane(3,2,3-tet) in the presence of metal ions gives complexes of the macrocyclic ligands [16]pydieneN$_5$, **3.75**, and [17]pydieneN$_5$, **3.76**. Complexes of **3.75** are formed by using the following metal templates: Fe(II), Fe(III), Cd(II), La(III), Co(II)], Ag(I), Zn(II), Hg(II), and Mn(II) and structural data have been reported for Ag(I), Mn(II)[96], Fe(II), Fe(III), Cd(II). It has been observed that the increase in the macrocyclic ring size results in structural changes.

3.75   **3.76**

For example, the pentagonal bipyramidal geometry of the iron(II) complex is still maintained but, the macrocycle is much more severely distorted than in the corresponding 15-membered complex. Thus, the deviation of a contributing atom from the FeN_5 least-square plane is 0.29 Å and some of the Fe-N bond lengths are increased to fit the six-membered ring into the girdle. Also, Mn(II), Fe(III), and Cd(II) complexes are still seven-coordinate but the plane containing the five nitrogen atoms and the metal ion is more distorted compared with complexes of **3.73**, so that the coordination geometry is pentagonal pyramidal for the Ag(I) complex with Ag(I) bound to the five nitrogen atoms of the ligand in a plane and there is a sixth nitrogen from an adjacent ion at a distance just longer than the equatorial Ag-N distances.

Complexes of **3.76** are formed using following metal templates: Cd(II), Ag(I), Zn(II), Hg(II), and Mn(II). It has been established that only $Mn(3.76)(NCS)_2$ is seven-coordinated although its structure in not a regular pentagonal bipyramid. Thus, the maximum deviation of a contributing atom from the least-square plane of MnN_5 is 0.46 Å. Actually four nitrogen atoms are approximately coplanar together with the metal atom and the pyridine nitrogen atom is 0.92 Å away from the plane. On the other hand, the unsaturated part of the macrocycle forms an approximate plane which intersects this MnN_5 plane at an angle of 41.8° and the metal ion is 0.77 Å above this plane. The macrocycle distortion produces a crowding which leads to the displacement of the axial ligands. The geometry of the complex is actually distorted from the pentagonal bipyramid towards a capped trigonal prism (Figure 3.3). It has been considered that the folding of the ligand is due to the its being too large for the manganese(II) ion when in the planar arrangement. This argument can explain the failure to prepare the template method complexes of this ligand with metal ions smaller than Mn^{2+}.

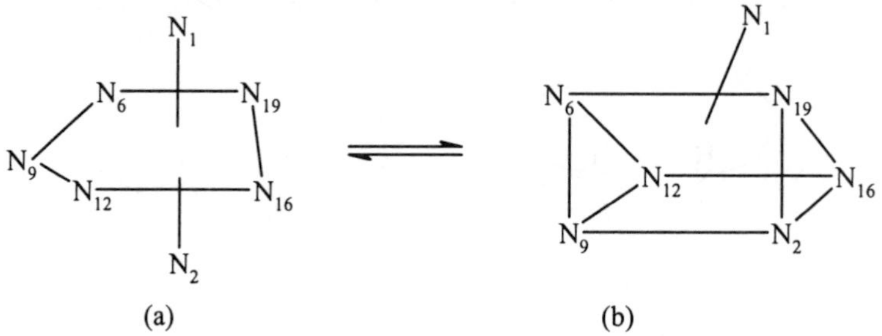

Fig 3.3. The coordination sphere of Mn(3.77)(NCS)₂:
(a) pentagonal bipyramid; (b) capped trigonal prism

Structural data for the Ag^+ complex with **3.76** shows that the macrocycle adopts a conformation with approximately C_2 symmetry in which the pyridine nitrogen lies closest to the AgN_5 plane. The Hg(II) and Cd(II) complexes of **3.72** are both six-coordinated. Structures of $[Cd(3.72)(Br)]Br \cdot 0.5H_2O$ and $[Hg(3.72)(Br)]_2[Hg_2Br_6]$ have been resolved. Both cations are six coordinated with geometry described as pentagonal pyramidal with the metal atom bonded to five nitrogen atoms of the macrocycle and the bromine atom in an axial position.

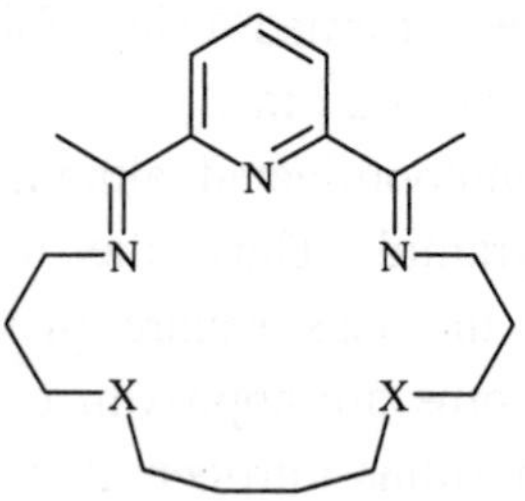

	X
3.77	NH
3.78	O

However, the sterically crowded axial site can be occupied by a unidentate ligand of small steric demands and good coordinating ability such as NCS. Indeed, this is the case of [Cd(**3.72**)(NCS)$_2$] which is seven-coordinated.

The template reactions of 2,6-diacetylpyridine with spermine, 4,9-diazadodecane-1,12-diamine) in the presence of trivalent yttrium, lanthanum, praseodymium, neodymium, samarium, gadolinium, dysprosium, holmium, erbium and ytterbium perchlorates in a 1:1:1 molar ratio resulted in a [1+1] Schiff base condensation. The 19-membered macrocyclic **3.77** formed as ligand with an N$_5$ set of donor atoms[97] has a cavity sufficiently large to enclose the metal ions. The coordination number of these ions is achieved by the incorporation of water molecules and counterions. In the absence of metal ions an amorphous material with indefinite compositions has been obtained.

The more rigid macrocycle **3.79** has been obtained[98] as Mn(II) and Zn(II) complexes. These metal ions with small preference for any particular coordination geometry control the stereochemistry reactions of precursors possessing rigid stereochemical features to form bipyramidal complexes in which **3.79** defines the equatorial plane.

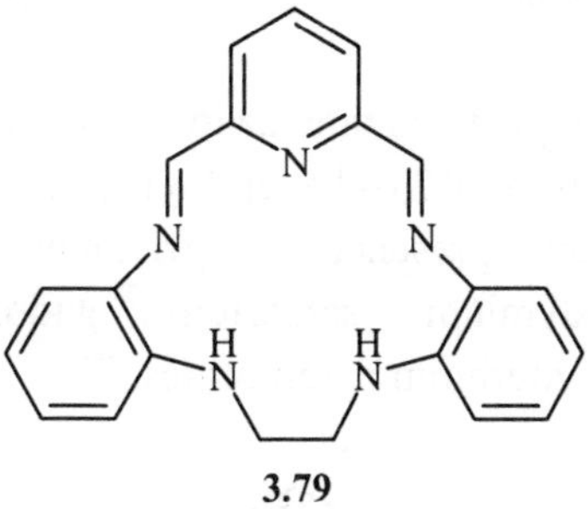

3.79

Diimine along with O, S and P donors. Macrocycles bearing beside nitrogen atoms, oxygen, sulfur or phosphor as donor sites in the macrocyclic ring system have been reported.[99] If terminal diamines with other potential donor groups are used in metal directed Schiff base condensation reactions, mixed-donor macrocycles of the type N$_3$X$_2$ are formed, where X = O or S. The aza-crowns have complexation properties that are intermediate between those of all oxygen crowns and of all nitrogen cyclames. For example, the macrocycle **3.80** has been obtained as ligand in the complex [Fe(**3.80**)(X)$_2$]·H$_2$O, X = CN or NCS, when

Fe(II) acted as template in an O_2-free solution of methanol. The IR spectra confirms that the complexes contain the coordinated macrocycle and electrical conductance proves the coordination of the counterion. The complex [Fe(**3.80**)(CN)$_2$]·H$_2$O has μ_{eff} = 5.09 B.M. at 293 K and Mössbauer spectrum at the same temperature shows a single quadrupole split doublet with an isomer shift (δ) of 0.84 mm s^{-1} and a quadrupole splitting (ΔE_Q) of 3,18 mm s^{-1}. These values show the existence of iron(II) as having S =2 configuration. For these reasons it was concluded that iron(II) is six-coordinate with one uncoordinated ether oxygen.

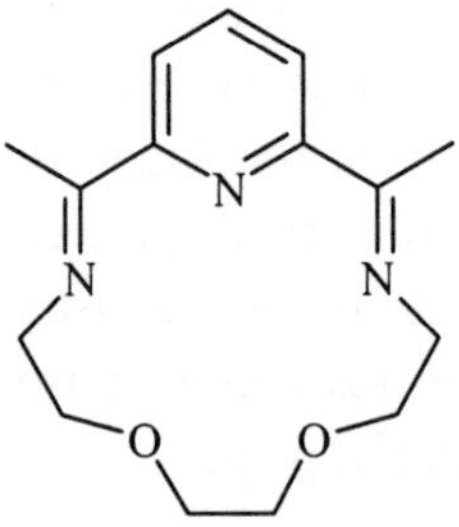

3.80 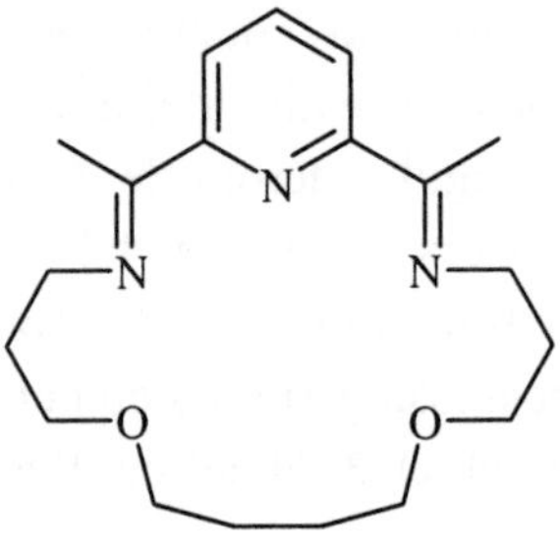   **3.81**

The 19-membered pentadentate azaoxa macrocycle **3.81** has been obtained by condensation of 2,6-diacetylpyridine with 1,12-diamine-4,9-dioxadodecane in the presence of trivalent yttrium, lanthanum, praseodymium, neodymium, samarium, gadolinium, dysprosium, holmium, erbium and ytterbium perchlorates.[100]

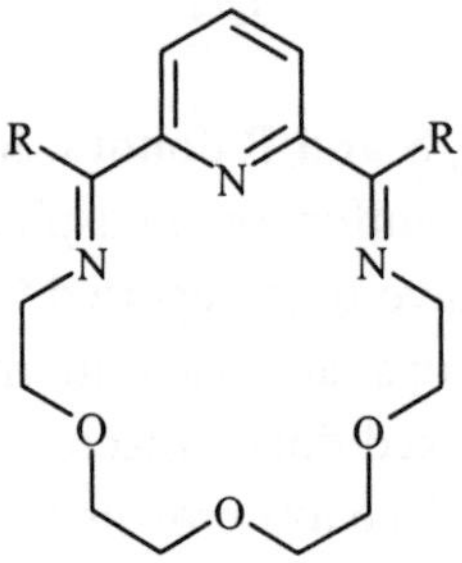

	R
3.82	CH$_3$
3.83	H

Condensation of 2,6-diacetylpyridine or 2,6-diformylpyridine with a long chain diamine $H_2N-(CH_2)_2-O-(CH_2)_2-O-(CH_2)_2-O-(CH_2)_2-NH_2$, resulted in and **3.83** rather than complexes of the analogous [2+2] ligand. The template preparation of M(**3.82**)(NO$_3$)$_3$ proceeds satisfactorily for M = La^{3+} or Ce^{3+}.[101] For heavier lanthanides, hydration of an imido linkage occurs. The stability of the complexes is different. The complexes of **3.82** are considerable less stable in water than the analogous complexes of the hexaimine **3.126** which are stable even in aqueous alkali, M(NO$_3$)$_3$(**3.82**) deposit metal hydroxide within a few seconds. Thus these trimide-triether complexes do not fall midway between the ready dissociating hexaether complexes (such that of 18-crown-6) and the very stable hexaimide complexes.

By using aromatic head units more rigid macrocycles can be prepared. Thus, rigid macrocycles **3.84**, **3.75** and **3.76** have been obtained using lanthanide ions as template.[102, 103]

The reactions between precursors in the absence of the metal ions resulted in an unidentifiable oil instead of **3.84** and no reaction occured in the 2,6-diacethylpyridine or 2,6-diformylpyridine and 1,2-bis(2-amino-phenoxy)-ethane mixture. The high thermal stabile complexes [Ln(**3.84**)][NO$_3$]$_3$·xH$_2$O (Ln = La^{3+}, Ce^{3+}, Pr^{3+}, Nd^{3+}, Sm^{3+}, Eu^{3+}, Gd^{3+}, Tb^{3+}, or Dy^{3+}) and [Ln(**3.86**)][ClO$_4$]$_3$·xEtOH (Ln = La^{3+} or Ce^{3+}) contain at least one nitrate, respectively, perchlorate group coordinated. Their nature has been established on the f.a.b. mass spectral data and their magnetic moments suggest that the 4f electrons are not involved in bond formation. The extension of bridge on going from **3.86** to **3.84** is accompaned by the increase of stability of the complexes, which can be attributed to the more flexible ligand which reduce the strain in the macrocycle.

Template effect of Mn^{2+} and Zn^{2+} - ions which also have a small preference for any particular coordination geometry, has been applied to obtain pentadentate macrocyclic ligands resulted through condensation of 2,6-diformyl-pyridine with primary diamines **3.85** and **3.87**.[104] For all the complexes the macrocyclic ligands define the equatorial plane in the pentagonal bipyramidal complexes.

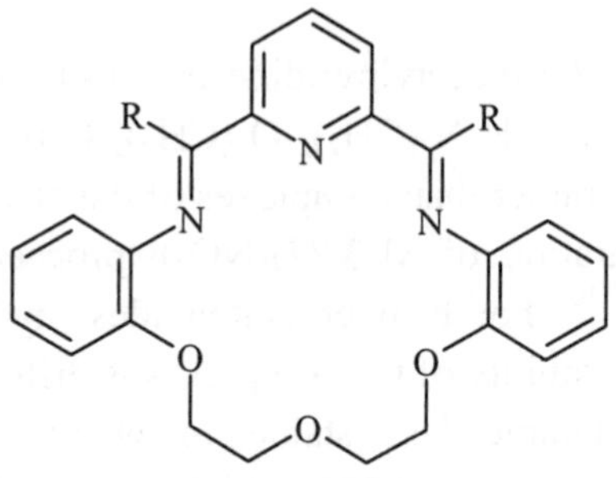

3.84

Metal complexes of the pentadentate macrocycle **3.85** have been isolated from the Schiff base condensation of 2,6-diformyl-pyridine and 1,2-bis(2-aminophenoxy)-ethane in methanol in the presence of perchlorates or nitrates of Mn(II) and Zn(II) and the crystal structure of the mangenese perchlorate complex has been reported. The complexes of **3.88 - 3.91** of the general type $[Cu(L)(X_2.)] \cdot xH_2O$ have been obtained as crystals in the same way thus resulting macrocycles of 15 to 19 members in their inner rings.[105] The f.a.b. mass spectral results confirm the monomeric[1+1] nature of complexes. They are very stable: they have been reduced using sodium tetrahydroborate to the corresponding amine macrocycles.

A range of transitional metal ions and trivalent lanthanides have been proved to be effective template agents in the formation of the macrocyclic Schiff-base ligand with more oxygen atoms as potential donors, like **3.92** and **3.93**. Binuclear complexes of the formula $[M_2(\mathbf{3.92})](X_4)_n \cdot xH_2O$, M = Co(II) or Ag(I) ; X = ClO_4 and/or X = NO_3 as well as mononuclear complexes of formula $[M(\mathbf{3.92})](ClO_4)_2 \cdot xH_2O$, M = Ni(II) or Zn(II) have been prepared recently by condensation of 2,6-bis(2-formylphenoxy-methyl)pyridine and 1,5-bis(2-aminophenoxy)-2-oxapentane in the presence of the appropriate metal salt.[106] Although no single crystal structure exist yet, information on the complexes have been obtained from other physical-chemical data.

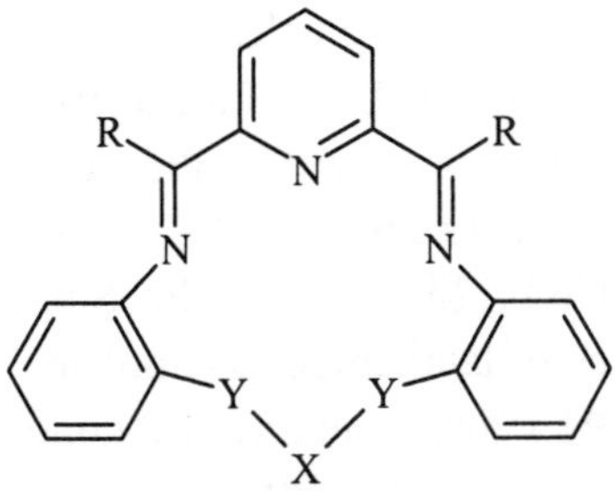

	Y	X	Ring size
3.85	O	CH2-CH2-	15
3.86	O	CH2-CH2-	15
3.87	S	CH2-CH2-	15
3.88	O	$-CH_2-CH_2-CH_2-$	16
3.89	O	$-CH_2-(CH_2)_2-CH_2-$	17
3.90	O	$-CH_2-(CH_2)_4-CH_2-$	19
3.91	O	$-CH_2-CH(OH)-CH_2-$	16

The mass spectra of the complexes confirmed the cyclic nature of the ligands and molar conductivity data show their ionic nature in dmf and acetonitrile. The ^1H NMR spectra of the diamagnetic Zn(II) perchlorate complex of **3.92** show a peak at approximately 8.65 ppm attributed to the imine protons. This peak together with the absence of those assignable to carbonyl and amine groups confirm that cyclization occurred and thereby the template effect of Zn(II).

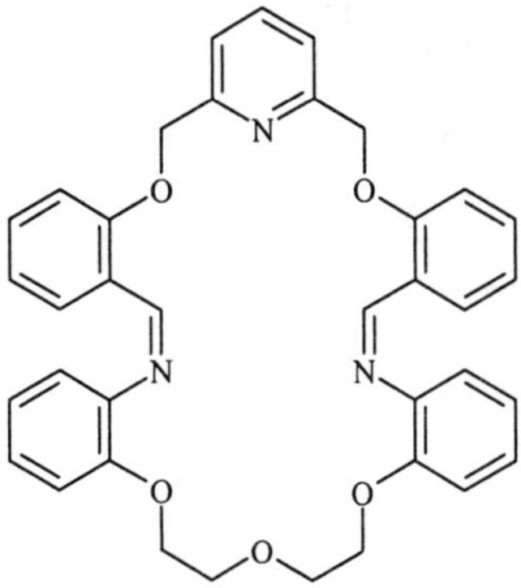

3.92 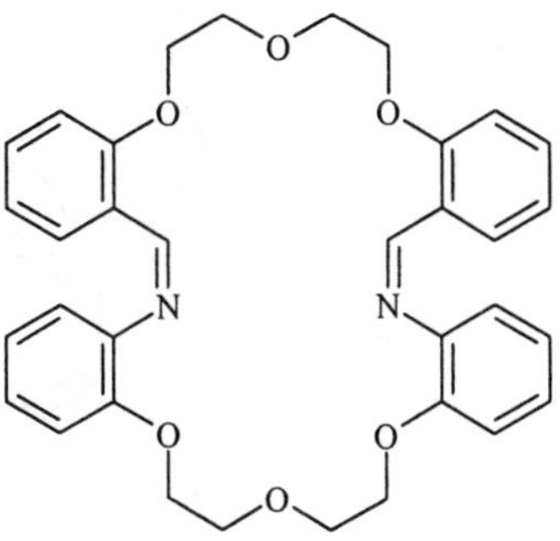   **3.93**

Alternatively, the mononuclear air stable complexes [Ln(**3.92**)][NO$_3$]$_3$·xH$_2$O· yEt$_2$O, Ln = La-Lu except Pm and Dy,[107] and [Ln (**3.92**)][ClO$_4$]$_3$·xH$_2$O, (Ln = La, Ce, Pr, Nd, Sm. Tb, Dy, Ho or Er) and [M(**3.93**)]X$_3$·xH$_2$O, where for X = NO$_3^-$, M = Y, La → Yb except Pm and Dy, and for X = ClO$_4^-$, M = La, Ce, Pr, Sm, Gd or Eu[108] have been obtained in the presence of lanthanide as template.

The macrocycle **3.94** which contains N$_3$P donor set has been obtained as ligand in the complex [Ni(**3.94**)X]PF$_6$, where X= Br or I, by refluxing a mixture of 2,6-diacetylpyridine, bis(3-amino-propyl)phenylphosphine and NiBr$_2$ hydrate in ethanol.[109] Diamagnetic behaviour and spectral properties support the trigonal-bipyramidal arrangement of donor atoms around nickel(II).

Oxygen or sulfur may belong to the head unit as it result in Scheme 3.13. Macrocycles in which furanyl fragment is part of the macrocyclic framework have been reported, using template techniques.[110] The slight soluble in water macrocyclic complexes [ML]X$_2$ where L = **3.95** or **3.96** and M = Co, Ni, Cu, and X = Cl, Br, NO$_3$ have been obtained[111] from template condensation of *o*- aminothiophenol or β-mercaptoethylamine, 2,5-diformylfuran and 1,3-dibrompropene in 2:1 molar ratio. The attempt to prepare the metal-free macrocyclic ligand was unsuccessful, thereby signifying the important role of the metal ions in cyclization.

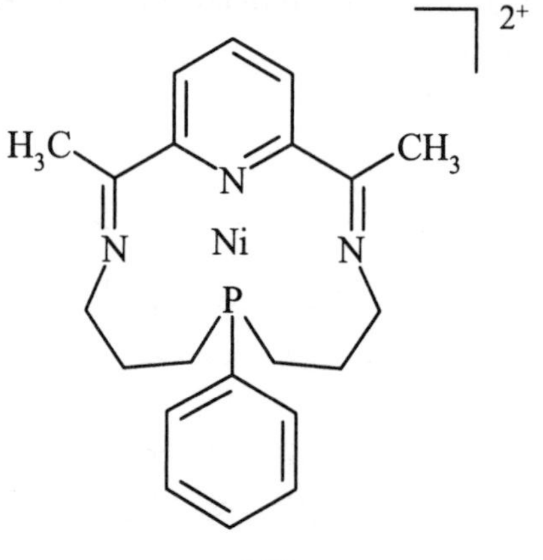

3.94

The copper(II) and cobalt(II) complexes show five-coordination with a square pyramidal geometry and a magnetic moment expected for a single unpaired electron and, respectively, four, whereas the diamagnetic nickel(II) complexes present a square planar geometry.

A copper(II) and copper(I) complex that contains a macrocylic ligand with an N_2S_2 donor atom set has been synthesized from template condensation of 2,2'-diaminobiphenyl, 1,4-bis(2-formylphenyl)-dithiabutane and copper(II) tetrafluoroborate. Crystal structure of [Cu(**3.97**)]$^{2+}$ shows copper(II) in an approximately square-planar metal center with a weak axial interaction with BF_4^- counterion.[112]

3.3.2. Tetraimine macrocycles

3.3.2.1. Aliphatic carbonylic precursors

Complexes of tetraimine macrocycles which can include metal ions in a square planar ligand field are traditionally prepared by template syntheses involving diketones and diamines in 2:2 mole ratios. The extend of degree of unsaturation reduce the flexibility of the ligand and further increases the preference for a planar structure. On going from diimine to tetraimine macrocycles the ligand field strength increases as the imines are stronger donors than amine groups. Also the macrocyclic cavity become smaller which forces the metal ions that are too large to lie out of the plane. This is the case of highly unsaturated rings like porphyrin. Several examples will be further presented. The condensation of 2,3-butanedione with 1,3-diaminopropane in 2:2 mole ratio in the presence of nickel(II), copper(II), cobalt(II)[113] or iron(II)[114] as template resulted in the tetramethyl-substituted 1,4,8,11-tetraazacyclotetradeca-1,3,8,10-tetraene, **3.98**. The reaction course is strongly dependent on the presence of H^+ ion. Actually, experiments show that a specific acid-induced intermediate or even the organic macrocycle is formed prior to the point where the metal ion plays a role. However, the attempt to obtain the organic ligand in the absence of metal ions failed. **3.98** gives rise to only one stereoisomeric complex and shows little tendency to adopt a folded configuration of the metal coordination sphere. For this reason, their complexes provide appropriate systems for the study of substitution reactions which involve the monodentate ligands in the apical positions.

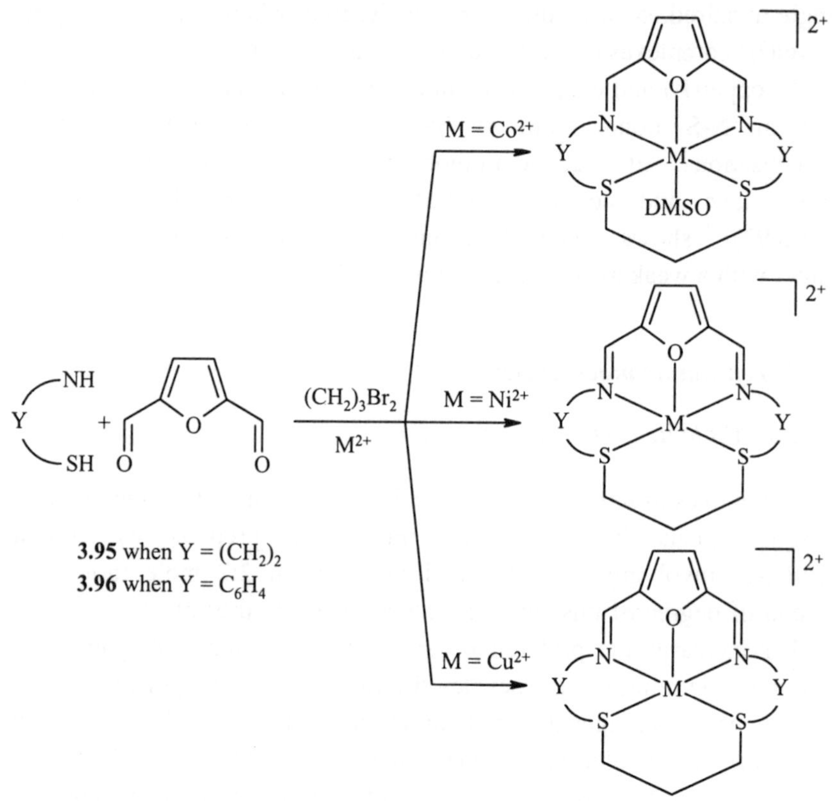

Scheme 3.13

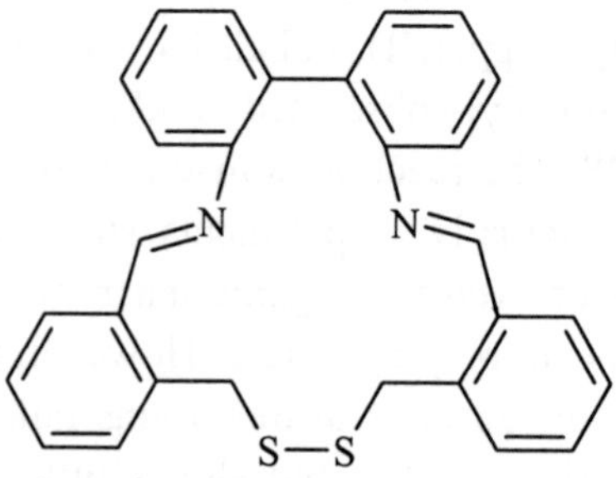

3.95 when Y = $(CH_2)_2$
3.96 when Y = C_6H_4

3.97

For example, in acetonitrile solutions $Fe(\textbf{3.98})(CH_3CN)_2](PF_6)_2$ undergoes reversible substitution reaction of acetonitrile with both imidazole and carbon monoxide. Single-crystal X-ray diffraction studies established that the four nitrogen atoms of **3.98** form a planar arrangement and the monodentate ligands are disposed *trans* to each other as in [Co(**3.98**)Cl$_2$](PF$_6$). The infrared spectra allows to make a difference between the electronic state of the ligand **3.98** in these complexes. Indeed, the strong band at ~1600 cm^{-1} which is assigned to the C=N stretching vibration is present in the spectra of nickel(II) and cobalt(III) complexes but is absent or very weak in all the Fe(**3.98**) derivatives. Further, a sharp medium-intensity band appears at about 980 cm^{-1} in the spectra of iron complexes. These differences have been attributed to delocalization of d electron density onto the ligands.

	R1	R2	R3	R4	X	Y
3.98	CH$_3$	CH$_3$	CH$_3$	CH$_3$	H	H
3.99	CH$_3$	C$_6$H$_5$	C$_6$H$_5$	CH$_3$	H	H
3.100	CH$_3$	Me	CH$_3$	CH$_3$	H	H
3.101	C$_6$H$_5$	C$_6$H$_5$	C$_6$H$_5$	C$_6$H$_5$	H	H
3.102	CH$_3$	CH$_3$	CH$_3$	CH$_3$	OH	H
3.103	CH$_3$	CH$_3$	CH$_3$	CH$_3$	OH	OH
3.104	C$_6$H$_5$	C$_6$H$_5$	C$_6$H$_5$	C$_6$H$_5$	OH	H
3.105	C$_6$H$_5$	C$_6$H$_5$	C$_6$H$_5$	C$_6$H$_5$	OH	OH

Iron, cobalt, nickel and copper[115] ions act as template in the condensation reaction between 1-phenyl-1,2-propanedione and 1,3-diaminopropane in methanol resulting in 2,9-dimethyl-3,10-diphenyl-1,4,8,11-tetraazacyclotetradeca-1,3,8,10-tetraene, **3.99**, as ligand. The free macrocycle **3.99** has some degree of stability in solution as [Cu(**3.99**)]$^{2+}$ undergoes transmetallation to [Zn(**3.99**)Cl]$^+$ by metallic zinc reduction.

A series of macrocycles based on **3.98** framework containing attached pendant coordinating groups offer the possibility to occupy one or both the axial coordination positions of the metal ions. With this idea, macrocycles **3.102** - **3.105** containing one or two hydroxyl groups were obtained as ligands. The cobalt(III) complexes of the type [Co(L)Br$_2$]Br where L stands for **3.98** or **3.103**[116] have been obtained and it has been established that the hydroxyl functions on complexed ligands are unreactive toward a variety of common reagents including acetyl chloride, benzenesulfonyl chloride and benzenesulfonyl isocyanate although the coordination of the OH groups was not proved.

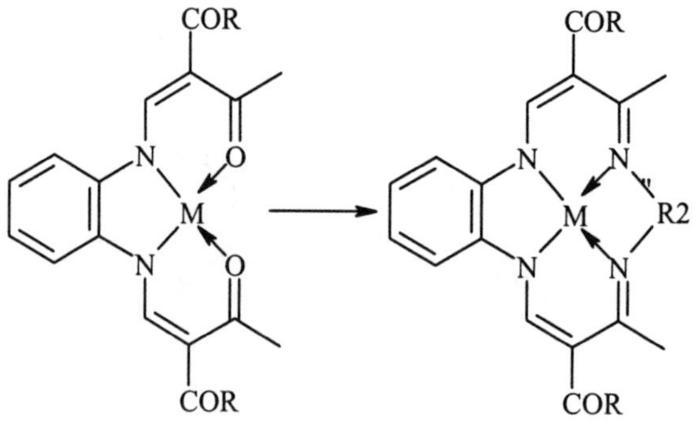

3.106   **3.107**

Scheme 3.14

The synthesis of unsymmetrical tetraaza macrocycles which can include metal ions in a square planar ligand field, is generally achieved by template synthesis using β-keto imines. The preparation of 14-membered macrocycles **3.107**, described by Jäger,[117] where R^2 = *o*-phenylene or –CH$_2$-CH$_2$- have been realized by condensation of complexes of type **3.106** with *o*-phenylenediamine or ethylenediamine. Bamfield[118] studied the importance of the position of the carbonyl group attached to the *meso*-position of the six-membered chelate ring for the conversion of Ni(II) or Cu(II) complexes with N-phenylaminomethylenecyclohexa-1,3-dione, **3.108**, into macrocycles **3.109** by interaction with ethylenediamine in boiling methanol (Scheme 3.15). Mass spectrum of the product showed a parent ion at *m/e* 384 thereby confirming that the cyclization occurred and that also amine exchange had taken place.

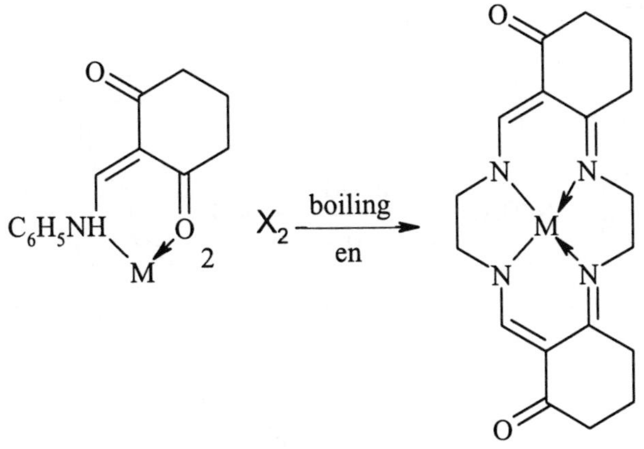

3.108   **3.109**

Scheme 3.15

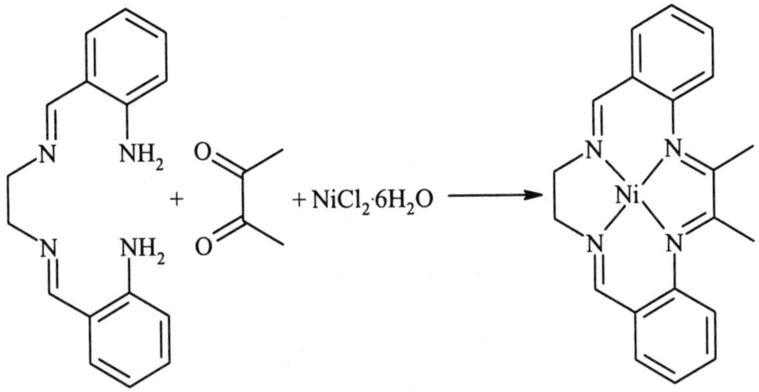

3.110

Scheme 3.16

Macrocycles which include aromatic rings have been obtained as ligands using nickel(II) as template in condensation of Pfeifer's amine (N,N'-bis(2-aminobenzilidene)-ethylenediamine and biacethyl generalize to Scheme 3.16[119] but, the attempt to generalize this type of reaction failed.[120]

Asymmetrical nickel(II) complexes with 14- (n=2) or 15-membered ($n = 3$) tetraaza[X]benzomacrocycles have been obtained by the reaction of a 1:1 mixture of the appropriate phenylenediamine and alkanediamine with 2,4-pentanedione in the presence of nickel(II) salts according to the Scheme 3.17.[121]

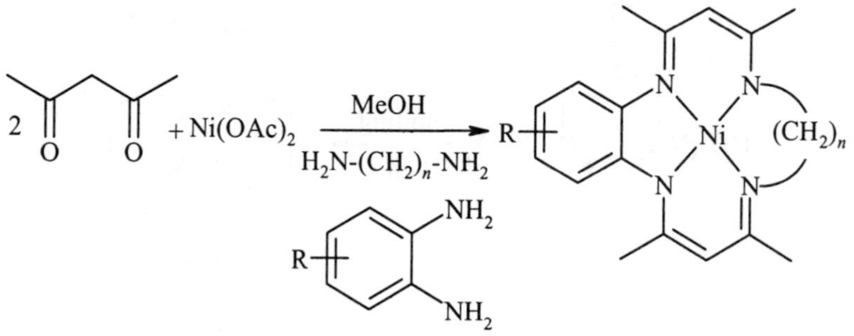

	R
3.111	CH_3
3.112	H
3.113	Cl
3.114	NO_2

Scheme 3.17

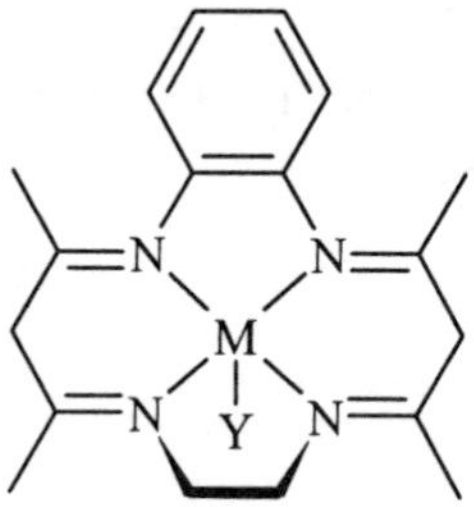

3.115

The template reaction between a divalent transition metal, *o*-phenylenediamine and bisacetylacetone-ethylenediamine yields 14-membered tetraaza macrocyclic six-coordinated complexes of the type **3.115**,[122] where M = Fe(II), Co(II), Cu(II), and X = Y = Cl⁻ or X = H_2O , Y = SO_4^2. The metal ions are coordinated by four azomethine nitrogen atoms bridged by acetylacetone moieties. These complexes are characterized by IR, electronic and magnetic measurements.

Investigation on 14-membered highly conjugated systems have essentially deal with imine type complexes. Some metal-directed template condensation reactions lead directly to conjugated metal complexes. Examples are metal directed condensation of propargylaldehyde or 2,4-pentadione with *o*-phenylenediamine.

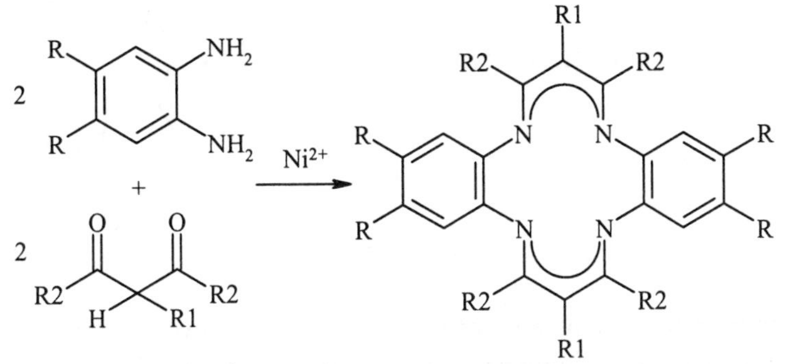

Scheme 3.18

The macrocyclic ligand obtained by nickel(II) template condensation of *o*-phenylene-diamine with 2,4-pentadione (**3.116**), first reported by Jäger, can be stripped from nickel(II) with anhydrous HCl in ethanol and isolated as hydrochloride salt.[123] Similar macrocyclic Ni(II) and Cu(II) complexes derived from acetylacetone and *m*- phenylenediamine or 3,4-toluenediamine have been prepared by template method but the free ligand could not be isolated.

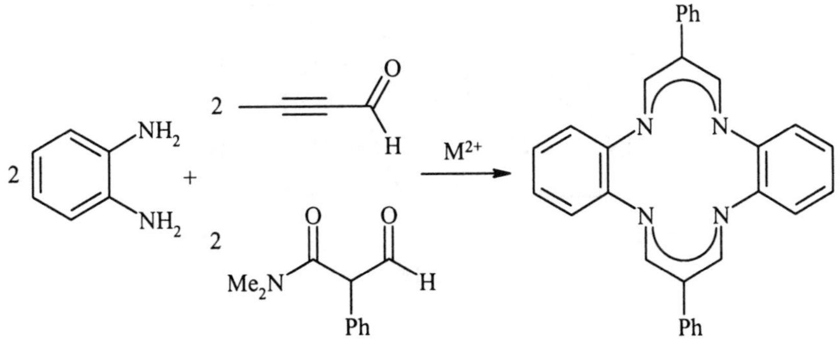

Scheme 3.19

Also Co^{2+} and Cu^{2+} can be used to effect the direct cyclisation of *o*-phenylene diamines with 1,3-dicarbonyl compounds according to Scheme 3.19.[124]

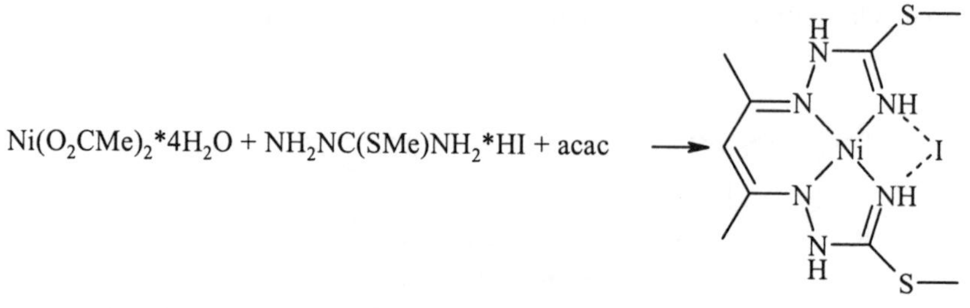

$$Ni(O_2CMe)_2 \cdot 4H_2O + NH_2NC(SMe)NH_2 \cdot HI + acac \longrightarrow$$

3.118

Scheme 3.20

Although metal ions are not essential for macrocycle formation from *o*-phenylene diamines with propynal or substituted acroleins, carrying out these reaction in the presence of Co, Ni, Cu results the corresponding macrocyclic complexes **3.117**.[125] These macrocycles are related to the porphyrins. The cavity size is however smaller than that of the N_4 cavity hole size of porphyrins.

The mononuclear complex **3.118** has been obtained by template synthesis according to reaction in Scheme 3.20.[126] Further, dinickel(II) bis(macrocyclic) complex **3.119** has been obtained from template condensation of **3.118**.

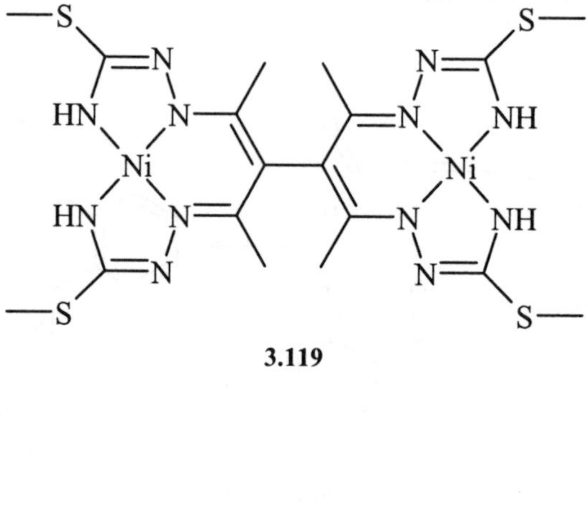

3.119

3.3.2.2. Unsaturated and aromatic precursors

The self condensation of *o*-aminobezaldehyde (AB) in the presence of metal ions was also one of the first example of metal template reaction. AB undergoes self-condensation forming two main products: a bis-anhydrotrimer and a tris-anhydrotetramer. In the presence of nickel(II) or copper(II) ions, self-condensation of *o*-aminobezaldehyde resulted in the tetradentate macrocycle **3.120** as a ligand. In the presence only of nickel(II) a second product was isolated and formulated [Ni(**3.121**)(H$_2$O)(ClO$_4$)$_2$].[127]

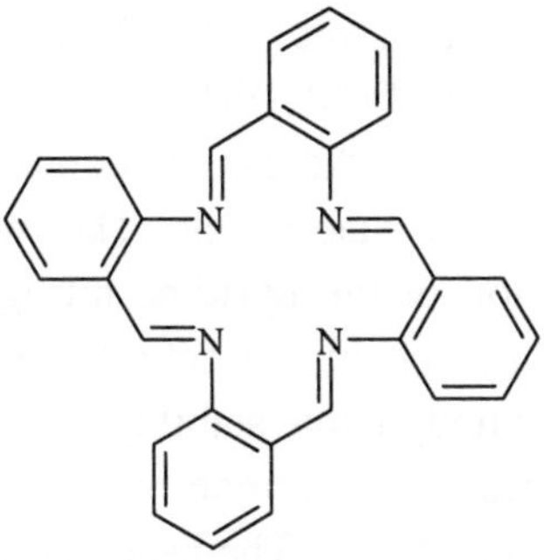

3.120 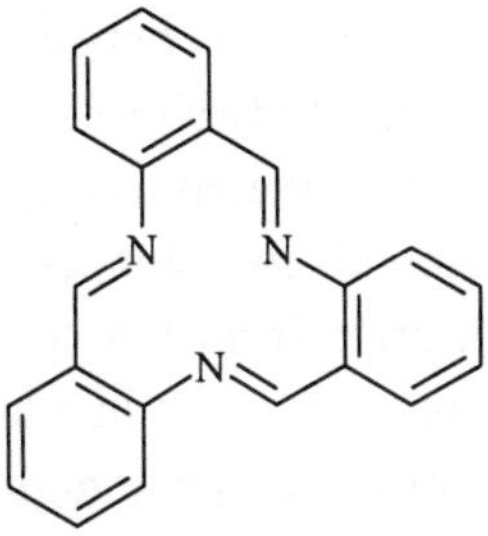   **3.121**

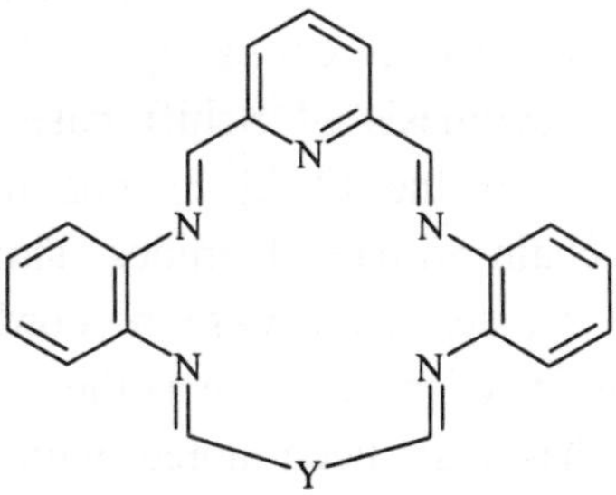

	Y	Ring size
3.122	-CH$_2$-CH$_2$-	15
3.123	OH 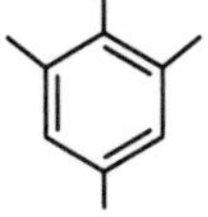 (aromatic ring)	18

The X-ray analyses confirmed the closed ring structure of the ligand and the octahedral environment of nickel(II).[128] Numerous reactions of AB have been run using various first row transition metal(II) nitrates. In each case, either **3.120** or/and **3.121** complexes have been obtained for metal ions from Fe(II) to Zn(II). The large size of metal ions (like Mn(II)) as compared to the above mentioned was considered to be the reason of the failure of the condensation in the presence of the other first-row metal ions.

The complexes $[M(L)(X_2)]\cdot xH_2O$, where L = **3.122** or **3.123,** and for M = Cu, n = 7 and X = NO_3 , ClO_4, for M = Ni, n = 7-8 and Co, n = 7 and X = NO_3, have been obtained as crystals in a metal ion directed 1+1 condensation reaction between the precursors. The X-ray structure analysis of $[Ni(\textbf{3.122})(NO_3)_2]$ and $[Cu(\textbf{3.123})(NO_3)](NO_3)$ shows that the nickel(II) is coordinated to only four atoms whereas the copper(II) atom accommodates all five donors of the same macrocycle. The two chelate rings which arise from the coordination of the pyridine nitrogen in the copper(II) complex result in a higher macrocyclic effect relative to that in nickel(II) complex and in a higher thermodynamic stability.

Formation of [2+2] macrocycles is common in Schiff base condensation reactions directed by metal ions.[129] The head and the lateral units can be varied, with the consequent formation of macrocycles with different donor atoms and/or different cavity sizes.[130] Lanthanide cations act as template in the synthesis of Schiff base macrocycles to yield mononuclear complexes of the [2+2] ligands derived from pyridine dicarbonyls and short chain primary diamines. Thus, for the [2+2] hexa-azamacrocycles, there is a steady decrease in coordination number of the metal within a given series of macrocycle as the series moves from La to Lu. A similar effect has also been noted with complexes of cyclic polyethers.

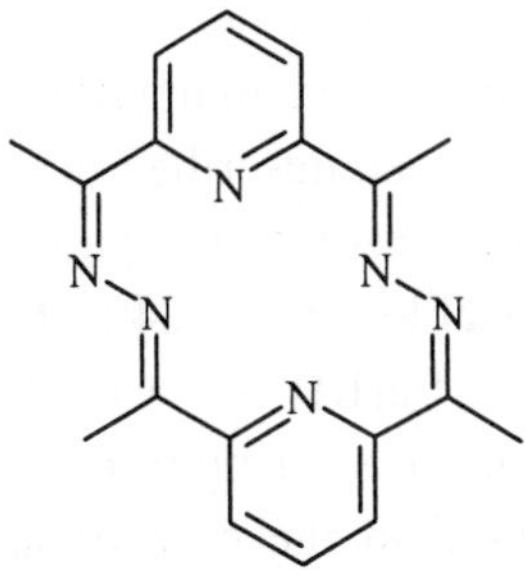

3.124 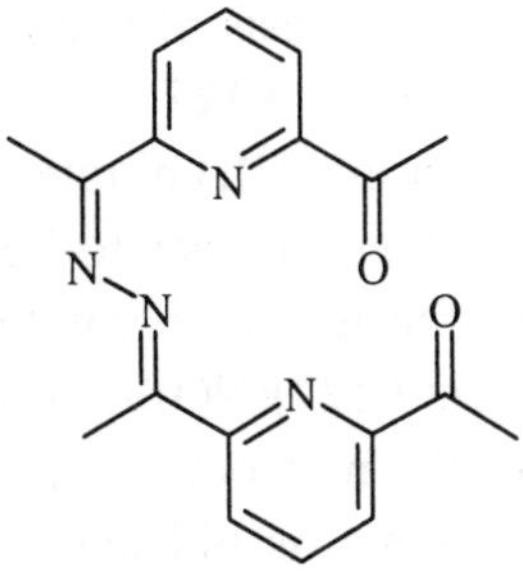   3.125

The relationship between the size of the cavity and coordination number on the one hand and the diameter of the metal ions, on the other hand, is most evident for this class of metal ions. For example, the ratio of cation diameter to ligand cavity size is the most important factor in the synthesis of 14-membered macrocyclic ligands in the presence of a lanthanide ion.

Thus, the heavier lanthanides (Tb - Lu) have been used as template in the synthesis of 14-membered tetradentate hexaaza macrocyclic ligands,[131] 3.124, whereas the lighter lanthanides (La-Gd) have been found to be ineffective for this type of ligands (in the same conditions). The nature of the counter ion is also important. For example, 14-membered hexaaza macrocyclic ligand 3.124 has been obtained via template condensation of 2,6-diacetylpyridine with hydrazine in the presence of $Y(NO_3)_3 \cdot 6H_2O$. Attempts to prepare this NNNN-donor macrocyclic compound in the presence of yttrium chloride or perchlorate resulted in the ring-opened complexes of (6-acetyl-pyrid-2-yl)methyl ketone azine 3.125. This was attributed to the greater flexibility of the podants than coronands and to the different complexing abilities of the counterions involved in the template synthesis. It is well known that the nitrate ions are very good complexing agents towards lanthanides/rare earth elements. The interaction between the metal ion and the counterion play the decisive role in the stabilization of the macrocycle. Based on the spectral data and thermogravimetric analysis, the coordination number of 10 was assigned for the yttrium ion in nitrate complex arising from the four nitrogen atoms of the hexaaza macrocycle and oxygen atoms of the three bidentate nitrate groups.

The template synthesis of lanthanide complexes of the 18-membered hexa-aza-macrocycle **3.127** only for lanthanum(III) nitrate and perchlorate, and cerium(III) nitrate have been described in a first stage.[132] The complexes show inertnes to ligand substitution regarding **3.127** and the anhydrous forms are thermally stable up to 240 °C. The X-ray structure of Ln(**3.127**)(NO$_3$)$_3$, Ln = La(III) or Ce(III), show the coordination of the macrocycle through the six ring nitrogen atoms. The three nitrate ions act as bidentate ligand in the case of lanthanum(III) complex which thus, reach the coordination number 12. The cerium(III) complex, formulated [Ce(**3.127**)(NO$_3$)$_2$(H$_2$O)]$^+$, shows the unusual coordination number 11. In this case, only one nitrate group is disposed on one side of the macrocycle act as bidentate ligand. The remained nitrate acts as monodentate ligand and together with the water molecule they occupy the opposite side of the macrocycle. Appropriate combination of counterions and experimental conditions resulted in metal – templated macrocycle synthesis of **3.127** for every lanthanide(III) ion except the radioactive Pm using 1,2-diaminoethane, 2,6-diacethylpyridine and a lanthanide acetate. The complexes of the general formula LnL(CH$_3$COO)$_2$·nH$_2$O, n varied between 3 and 6 depending on the metal ion has been obtained. Perchlorates LnL(OH)(ClO$_4$)$_2$·nH$_2$O (n=0-2) have been obtained in very low yields.

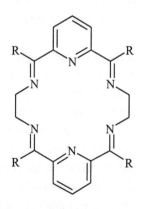

	R
3.126	H
3.127	CH$_3$

The authors concluded that the ability of the lanthanide (III) ions to act as templates agrees for the formation of the 18-membered macrocycle and is less affected by the contraction in the ionic radius that occurs in the 4f series and more by the nature of the counterion: the acetate favours the reaction much more effectively than chloride and perchlorate do.

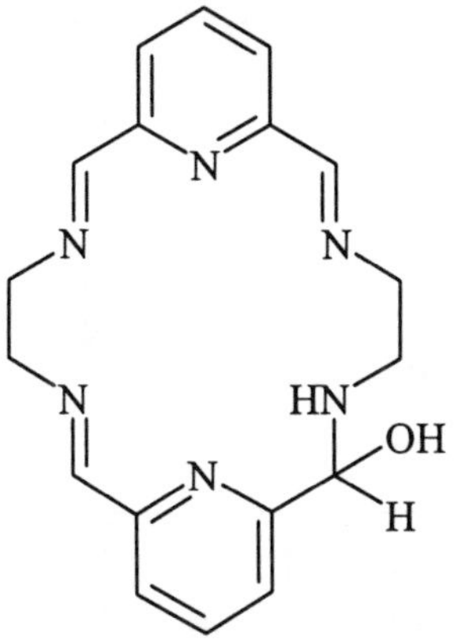

3.128

The template condensation of 2,6-pyridinecarbaldehyde and ethylenediamine in the presence of lanthanide nitrates have been reported. Two types of complexes have been obtained whose differences have been put into evidence by IR spectroscopy, although in both cases either carbonyl or primary amino groups were absent. The former group - complexes of the heavier lanthanide, Nd → Lu except Eu and Pm, show a distinctive sharp band at ca. 3220 cm^{-1} which proved the presence of a secondary amine group. The second group refers to the complexes of the lighter lanthanide, La→ Pr and Eu. The structure of Sm(III) complex show that it contains discrete [Sm(**3.126**)(NO$_3$)(OH)(H$_2$O)]$^+$ complex cations in which the Sm^{3+} ion is ten-coordinate being directly bonded to the heteroatoms of the sexidentate ligand, to one chelating nitrate group, and the oxygen atoms of the OH$^-$ ion and of one H$_2$O molecule. It was considered that the differences between the two types of complexes result from the differences in ionic radius. Thus, for the lighter and larger radius lanthanides, an addition of the solvent molecule occurs in order to release strain in macrocyclic ligand and so, to produce a more flexible species.

The second group refers to complexes containing macrocycle **3.128** as ligand. The lanthanide complexes of macrocycle **3.126** have been shown to be inert to release of metal ions in solution.

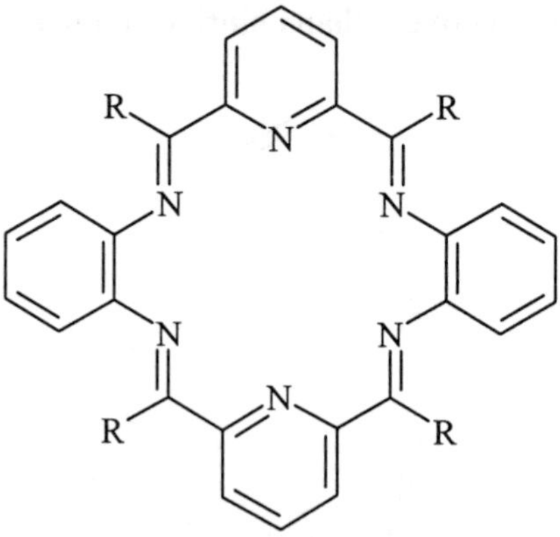

	R	Z
3.129	CH_3	H
3.130	H	H

Schiff-base macrocycles derived from 2,6-diformyl-[133] or 2,6-diacethyl pyridines[134] and aromatic diamines would exhibit more rigidity due to extended conjugation and consequently their lanthanide complexes would be thermodinamically more stable. The 18-membered macrocyclic ligand **3.129** has been obtained by condensation of 2,6-diacethylpyridine with and without the presence of the copper(II) template ion. The complex has been formulated as $[Cu_2(\textbf{3.129})](NO_3)_4$ on the basis of mass spectroscopy, analytical and spectral properties. It has been established that pyridyl groups are coordinated in an equivalent manner. The molecular model shows that the ligand is sufficiently too large to circumscribe two metal ions with a distance of 2.66 Å between them, which explains the magnetic moment of 1.70 BM.

The template synthesis and properties of complexes of Ce, Pr and Nd with the ligand **3.128** has been reported. The different geometrical requirements of metal ions lead to complexes of varying stoichiometry. For the complexes the 18-membered hexaazamacrocycles **3.129** results from the template condensation of pirydylformaldehyde and aromatic fenylene-diamine as lateral units. Typically the lanthanide ion coordinates to the six nitrogen donor atoms of the macrocycles, leaving the metal ion exposed for potential ligands to coordinate on either side of the macrocycle.

Schiff-base condensation of 2,6-diacetylpyridine and 2,6-diaminopyridine in the presence of hydrated lanthanide(III) nitrate in 2:2:1 mole ratio yields the complexes $[Ln(L)(NO_3)(H_2O)_3][NO_3]_2 \cdot 2H_2O$, $[Ln(L)(NO_3)(H_2O)_2][NO_3]_2 \cdot 2H_2O$, Ln = Nd, Sm, Eu, Tb, Ho, or Er) and, $[Ln(L)(NO_3)(H_2O)_2][NO_3]_2 \cdot 3H_2O$, Ln = Gd, Dy or Y, where L stands for **3.130**. The perchlorato complexes, $[Ln(L)(ClO_4)_2 (H_2O)]ClO_4 \cdot 3H_2O$, (Ln = Eu or Tb) and acetato $[Ln(L)(CH_3CO_2)_3(H_2O)] \cdot 3H_2O$, (Ln = Eu or Tb) complexes were synthesised by using the corresponding lanthanide (III) perchlorate or acetate as metal template. The (perchlorato)(nitrato) and (nitrato)(isothiocyanato) dysprosium complexes have been obtained by the above mentioned complexes by the anion metathesis with $NaClO_4$ and KSCN, respectively. The formation of the lanthanide(III) complexes of **3.130** demonstrates the template potential of these metal ions in the assembly of Schiff-base macrocycles having pyridine head and lateral units. In the (nitrato)(aqua) complexes the metal ion is coordinated to the macrocycle, to one bidentate nitrate and to two water molecules. In the (perchlorato)(aqua) and (acetato)(aqua) complexes, the metal ion is coordinated to the macrocycle, to two perchlorates or three acetates and to one water molecule.

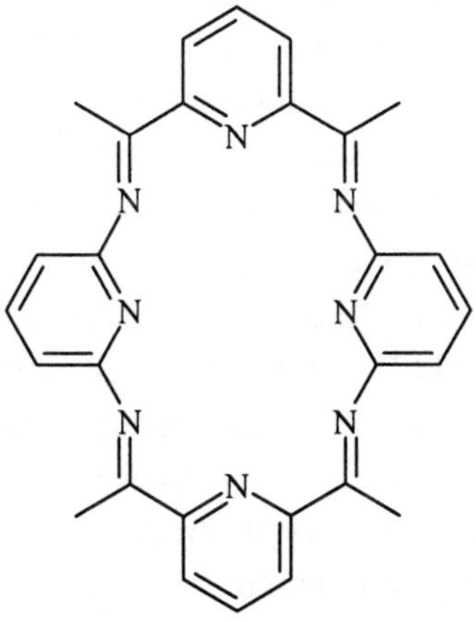

3.131

Lisowski[135] *et al.* have used lanthanide(III) ions as template in condensation of *R,R'-1,2*-diaminocyclohexane and 2,6-diformylpyridine to obtain the chiral macrocycle 4(R),9(R),19(R),24(R)-3,10,18,25,31,35-hexaaxapentacyclo-[25.3.1.1^{12,24}.0^{4,9}0^{19,24}]-dotriaconta-1(31),2,10,12,14,16 (32),17,25,27,29-decaene, **3.131**, as ligand in the complexes [LnL](NO$_3$)$_3$.nH$_2$O. It has been shown that the Tb(III) and Eu(III) complexes exhibit circulary polarized luminescence.

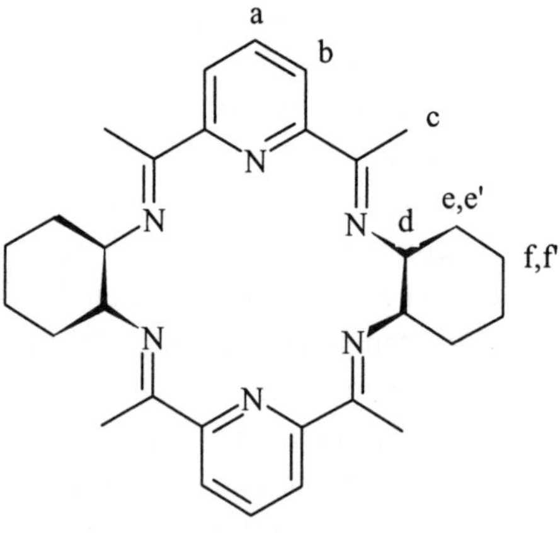

3.132

The complexes of the paramagnetic Ce(III), Eu(III) and Yb(III) have been studied based on ^{1}H and 13C NMR spectra in methanol-chloroform solution which indicate D_2 symmetry with a helical, twist conformation of the ligand and symmetrical coordination of the nitrate and/or solvent ligands above and below the macrocycle plane. These complexes discriminate between the chiral molecules which can replace the nitrate and/or solvent molecules. Indeed, they form diastereoisomeric complexes with D- and L-aminoacids. Racemic mixtures [Ln*rac*L]Cl$_3$·nH$_2$O, Ln = La, Ce, Pr Eu,[136] Gd,[137] Nd, Tm, or enantiopure compounds [LnL]Cl$_3$·nH$_2$O have been obtained using either racemic *trans*-1,2-diaminocyclohexane   or   *trans-R,R'*-1,2-diaminocyclo-hexane, respectively. The solid-state crystal structure of some complexes have shown lower symmetry compared to that found in solution. The lanthanide(III) ions have characteristic nine-coordination geometry composed of six nitrogen atoms belonging to the macrocycle. In all complexes, the macrocycles adopt a twist-bent conformation characteristic for complexes of hexaazatetraimine macrocycles. The nature of the counterions affects the structure of the complexes. Thus, whereas [Gd*rac*L](NO$_3$)$_3$·nH$_2$O exhibit D_2 symmetry, [La*rac*L]Cl$_3$·nH$_2$O shows lower symmetry due to the unsymmetrical coordination of chloride anions above and below the macrocycle plane. Changes can be noticed as a result of the changing of the anion La(III) complex.

The symetrical macrocycle **3.132** has been obtained by template action of Ln^{3+} (La-Dy except Ce and Pm) and Y^{3+} in the Schiff base condensation of 2,6-diaminopyridine and phtalic dicarboxaldehyde. Experimental works show that the nature of the solvent is very important. Thus, in super dry ethanol or methanol, in the presence of hydrated lanthanide(III) nitrate, compounds of undefinite compositions are obtained. In dry acetonitrile, discrete mononuclear complexes are formed.

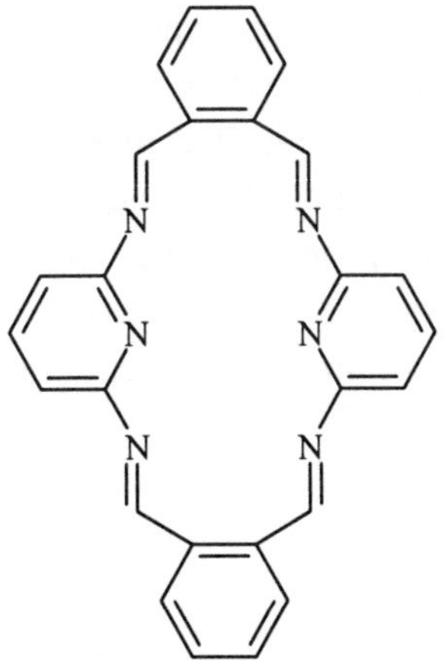

3.133

This behavior was explained by the more negative reaction enthalpy in acetonitrile. Four types of complexes have been obtained in this case: bis-(nitrato)(aqua) complexes [La(**3.132**)(NO$_3$)$_2$(H$_2$O)](NO$_3$)·2H$_2$O a. [Dy(**3.132**)(NO$_3$)$_2$(H$_2$O)$_2$](NO$_3$)· H$_2$O; nitrato(hydroxo) complexes of the type [Ln(**3.133**)(NO$_3$)(HO)] (NO$_3$)·nH$_2$O (Ln = Pr^{3+}, Nd^{3+}, Eu^{3+}; n = 2 for Pr^{3+}, Nd^{3+}, 3 for Eu^{3+}); (nitrato)(aqua)(hydroxo) complexes [Ln(**3.133**)(NO$_3$)(HO)(H$_2$O)](NO$_3$)· nH$_2$O (Ln = Sm^{3+}, Gd^{3+},); and, bis(nitrato) complexes [Ln(**3.133**)(NO$_3$)$_2$] (NO$_3$)·4H$_2$O (Ln = Tb^{3+}, Y^{3+}).

When the Schiff condensation occurs in ethanol in the presence of lanthanum(III) chloride, the complex [La(**3.133**)Cl$_3$(C$_2$H$_5$OH)]·H$_2$O is formed. The composition of the complexes was established based on f.a.b. mass spectra and the structure is formulated on the basis of IR and electronic spectra. They exhibit high thermal stability, the macrocycle remains intact with metal ions up to 388 °C. Magnetic moments of the Gd^{3+}, Nd^{3+} and Dy^{3+} complexes show that they are paramagnetic with μ_{eff} very close to the Van Vleck values of the free metal ions.

The 20-membered macrocycle **3.137** has been obtained in the presence of copper(II)[138] or cobalt(II)[139] as template from the corresponding precursors in ethanol as binuclear complexes. The spectral properties of Cu$_2$(**3.137**)(OH)$_2$(ClO$_4$)$_2$ do not show that the sulfur atom is bonded to any copper(II) ion which is in a square-planar environment being bonded to two nitrogen atoms from the macrocycle and two bridging hydroxo anions.

When 1,3-diamino-2-hydroxypropane (dahp) is used as lateral unit the product is a [3+3] macrocycle **3.134** which includes three lanthanum ions. The form and size of the resulted macrocycle is determined by the metal used. In presence of La(NO$_3$)$_3$ the yellow product contains the macrocycle deprotonated on each OH site with three lanthanide ions each having two nitrate ions to balance the charge. The crystal structure data indicated an arrangement of three lanthanum ions in an isoscel triangle.[140]

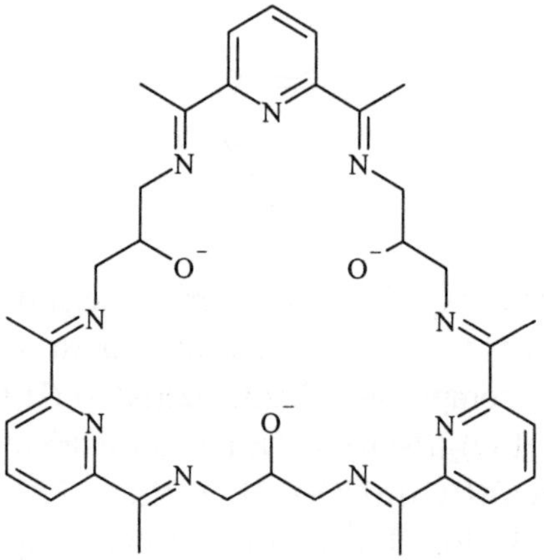

3.134

A series of binuclear hexaazamacrocyclic complexes [M$_2$(**3.135**)X$_4$], M= Co, Ni or Zn, and [Cu$_2$(**3.136**)X$_4$], where X = Cl or NO$_3$, have been prepared by template condensation reaction of phtalaldehyde with diethylenetriamine in methanol.[141] The yields were low in all cases but the attempt to obtain the metal-free ligands was unsuccessful. The macrocyclic donors fix the metal ions, and thereby determine, within the limits of the macrocycles flexibility, the internuclear distance, the coordination geometry and hence, the disposition of the bridging ligands.

Mutual influence between the metal centres was put into evidence by the magnetic moments. For the copper(II)complexes, the EPR spectra was interpreted in terms of the strong dipolar and exchange interactions between the copper(II) ions in the unit cell.

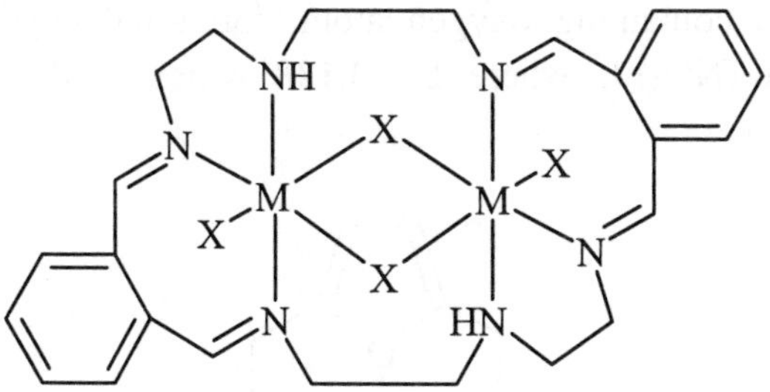

3.135

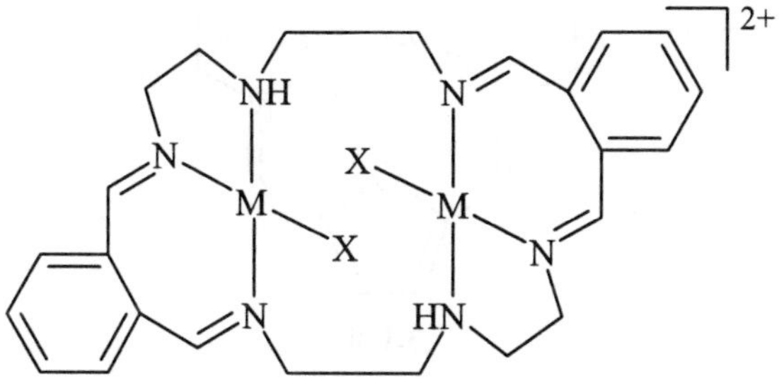

3.136

Magnetic moments at room temperature of 1.36 BM show some kind of antiferromagnetic interaction but no quantitative appreciation of this interaction could be estimated. It has been postulated – and experiments confirmed, that the small cavity size makes it possible, that only the presence of a single atom-bridge between the metal centres is allowed, like hydroxo or alkoxo.

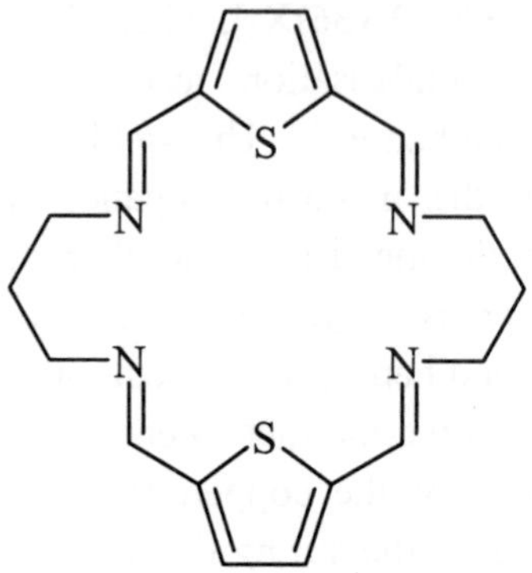

3.137

Head units containing oxygen atom, furan-dialdehyde lead to the complexes M (NO$_3$)$_3$L where L= **3.138** which readily dissociate in water.[142,143]

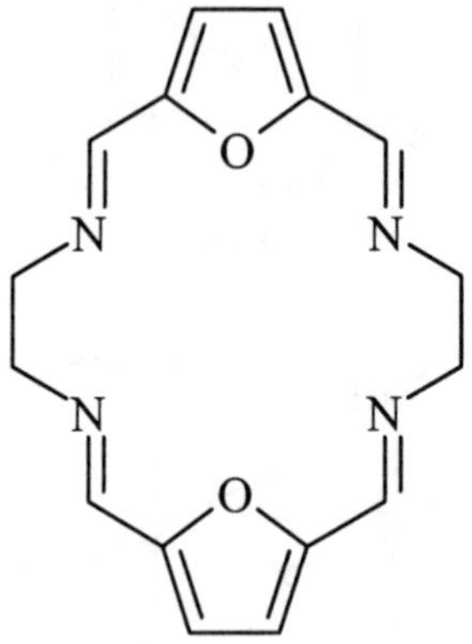

3.138

The condensation of 3,4-diethylpyrole-2,5-dicarbaldehyde with 4,5-diamino-1,2-dimetoxybenzene in the presence of uranyl nitrate as template was performed with the formation of complex **3.139** in 75% yield.[144]

Single crystal X-ray analysis of the complex shows that uranyl ion is coordinated to all six nitrogen atoms in a planar fashion and the coordination geometry about it is hexagonal pyramidal. The complex is thermal stable, it decompose around 300°C. The use of 2,6-diformyl-4-substituted phenols as head units and aromatic diamines leads to symmetric [2+2] Schiff - base macrocycles with a larger cavity size compared with those arising from pyridine derivatives as head unit. Additionally, the increased denticity makes them more suitable for larger lanthanides as the coordination number of the lanthanides increases as the size increases. The complexes of the 18-membered tetraaza macrocycle **3.140** its chloro **3.141** and bromo analogues **3.142** have been obtained for all lanthanide(III) ions except lanthanum and cerium, in the same experimental conditions. The yields increase along the lanthanide series denoting that the ring size is more suitable for the small ions. The macrocycle **3.140** behaves as a neutral ligand with undeprotonated OH-groups. The complexes are stable in the solid state as well as in DMF or methanol solutions. However they decompose in the presence of other competing ligands.

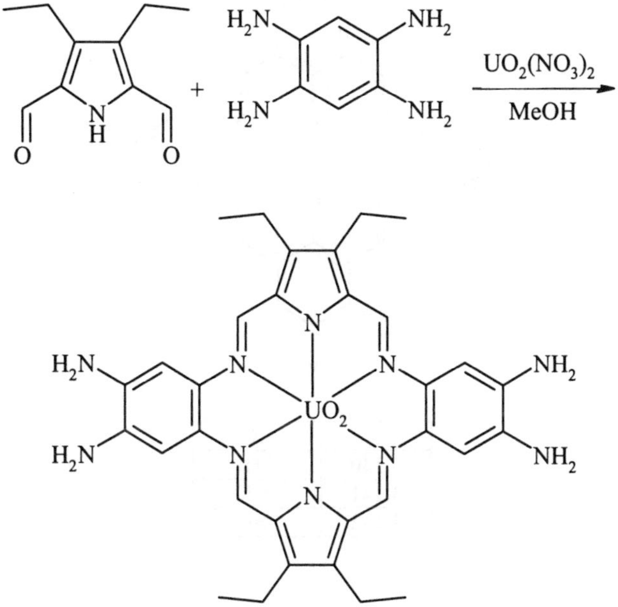

3.139

Scheme 3.21

Pyridine groups have been incorporated in the macrocycles **3.143**[145] and **3.144**[146] when increases in the stereochemical rigidity and binding ability towards lanthanide actions would be expected. Additionally, pyridines containing chelate rings have lower ring strain than analogous chelate rings formed from alkylamines. These properties are often associated with an increase in the thermodynamic stability of the complexes. The Schiff-base condensation of 2,6-diaminopyridine and 2,6-diformyl-4-methylphenol in acetonitrile is promoted by lanthanide(III) metal ions Ln = Y^{3+}, La^{3+}, Pr^{3+}, Nd^{3+}, Sm^{3+}, Eu^{3+}, Gd^{3+}, Tb^{3+}, Dy^{3+}, Ho^{3+}, Er^{3+}, Tm^{3+}, resulting in the complexes of the symmetric macrocycle **3.143**. The yields of the complexes decrease with decreasing the ionic radii denoting that the "best fit" between the sizes of the metal ions and the macrocyclic cavity controls the synthesis.

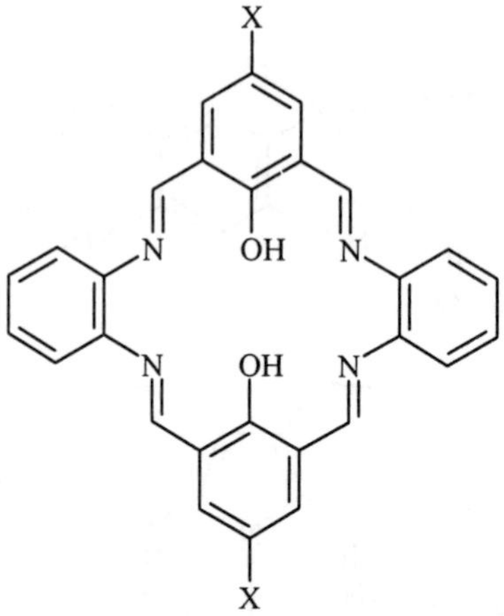

	X
3.140	CH_3
3.141	Cl
3.142	Br

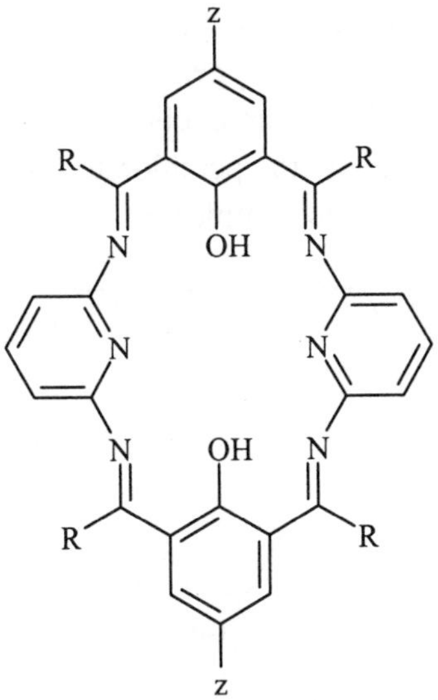

	R	Z
3.143	H	CH$_3$
3.144	H	Cl

The complexes have been formulated [Ln(**3.143**)(NO$_3$)(H$_2$O)$_3$]NO$_3$. nH$_2$O except Pr^{3+} and Eu^{3+}, for which the formula [Pr(**3.143**)(NO$_3$)$_2$(H$_2$O)$_3$]NO$_3$·3H$_2$O and [Eu(**3.143**)(NO$_3$)(H$_2$O)$_2$]NO$_3$· 5H$_2$O, respectively, were proposed. It results that at least one phenolic OH is deprotonated due to the basicity induced by pyridine fragment. The complexes are stable under air in the solid state. Infrared spectra support the presence of the coordinated nitrate ion as bidentate ligand. The complexes undergo anion metathesis when treated with pechlorates or SCN$^-$.

The formation of the complexes [Ln(L)(NO$_3$)$_2$(H$_2$O)$_2$] NO$_3$·3H$_2$O, (Ln = La, Pr, Nd, Gd, Dy, Ho, Er or Y), [Ln(L)(NO$_3$)$_2$(H$_2$O)$_2$]NO$_3$·4H$_2$O, Ln = Sm, Eu or Tb, and the perchlorates [Ln(L) (H$_2$O)$_2$]ClO$_4$·H$_2$O, Ln =Eu or Tb) where L stands for **3.144** demonstrates the template potential of these metal ions in the assembly of Schiff-base macrocycles having phenol head and pyridine lateral units. The difference in the size of the lanthanide actions does not affect their template potential in the formation of the complexes of L^6 and **3.144**. This indicates the adaptability of these macrocycles to fold according to the geometric requirements of the metal ions. The formation of these macrocycles along with coordinated anions or water shows the ability of oxygen donor ligands in stabilizing the lanthanide(III) ions in the macrocyclic frameworks. The complexes undergo anion exchange with perchlorate and thiocyanat to give perchlorato and isothiocyanato complexes.

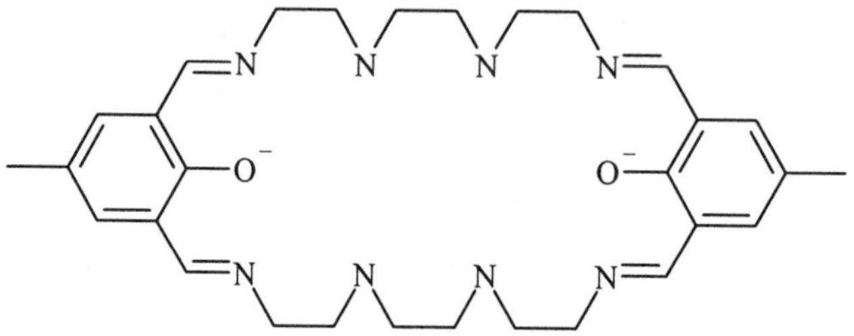

3.145

The off-white complexes $Ln_2L(NO_3)_4 \cdot nH_2O$ where L = **3.145**, $n = 0$ for Ln = La, Ce and 2 for Ln = Pr → Gd or orange $Ln_2L_7(NO_3)_{4-x}(OH)_x$, x = 1 or 2 for Ln = La, Gd, Eu, Ce ones have been obtained from diluted solutions. Their composition has been established based on f.a.b. mass spectrometry and complete cyclization was proved by IR spectroscopy.[147] Several experiments underline the role of the lanthanide metal ion in this condensation. Thus, the condensation of the precursors in the presence of divalent cations Ca^{2+}, Sr^{2+}, Ba^{2+} or Pb^{2+} leads to intractable solids and direct condensation in absence of any metal ions leads to intractable oils. It results that the lanthanide ions have the size and charge that are just right for the formation of the cyclic Schiff base **3.145** and they serve to organize reactants towards an exclusive formation of this ligand.

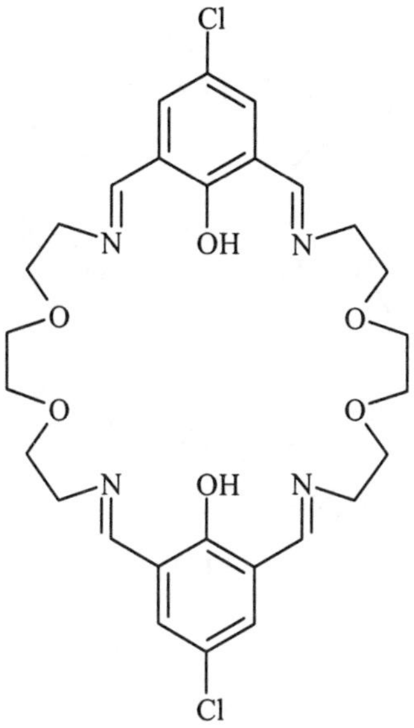

3.146

Similar complexes have been obtained with the ligand **3.146**, Ln_2 $L(NO_3)_4 \cdot nH_2O$, Ln = La, Tb, Dy; n = 2-4 through the condensation of the keto-precursor 2,6-4-chlorophenol and the diamine H_2N-$(CH_2$-CH_2-$O)_3$-$(CH_2)_2$-NH_2 in the presence of $La(NO)_3$.[148] The magnetic moment of the terbium complex (13.42 BM, 9.5 per terbium atom) shows that no metal-metal interaction occurs and further, that the ligand behaves as a binucleating but not as a compartmental one.

3.3.2.3 Pendant-arm macrocycles

Arms bearing additional potential ligating groups have been introduced at both carbon and nitrogen atoms of macrocycles which have generally been based on polyaza-donor sets.

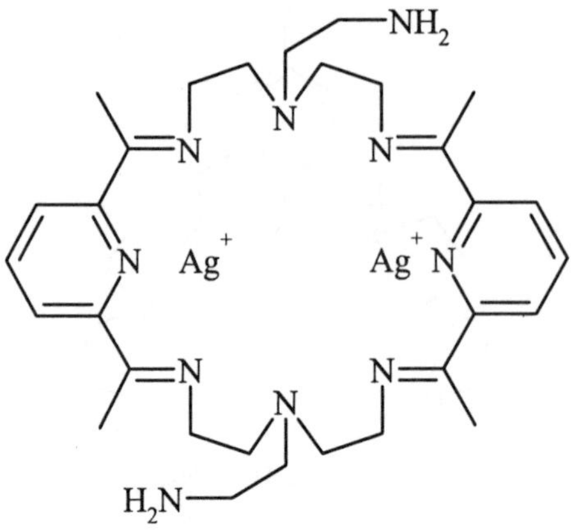

3.147

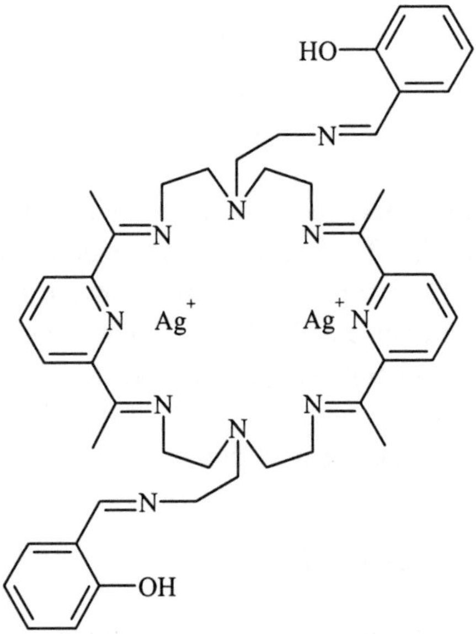

3.148

One potential in this area derives from the concept that the presence of two pendant arms, bearing ligating groups, attached at appropriate positions on a macrocyclic framework, would result in an "opened" cryptand, thus leading to modified complexation properties relative to the corresponding clathrochelates or simple macrocyclic precursors.[149] For example the binuclear silver(I) complex **3.147** was prepared by silver(I) templated [2+2] cyclocondensation of tris-(2-aminoethyl)amine (tren) and 2,6-diacetylpyridine.[150]

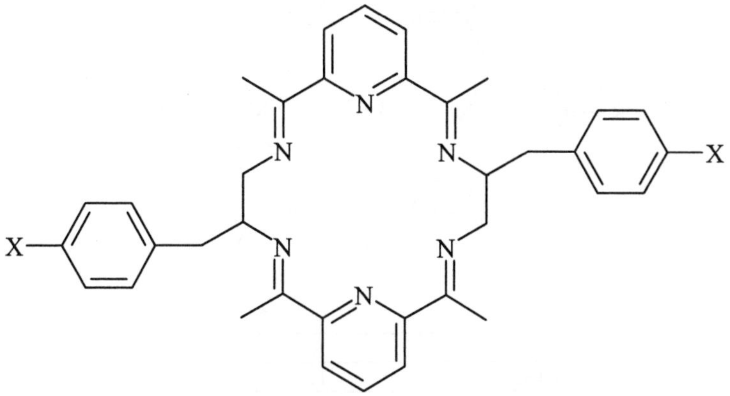

	X
3.149	H
3.150	OH
3.151	NH₂

The structure shows that the silver ions are bound in the diimino pyridyl head units of the macrocycle separated by 3.14 Å. The two pendant arms can be functionalized by condensation with salicylaldehyde leading to **3.148** thus proving the high stability of the macrocycle. Transmetallation using copper(II) salts resulted in the isolation of dark green crystals of the tricopper(II) hydroxospecies $Cu_3(OH)(L)(ClO_4)_3 \cdot H_2O$.

The lanthanide-templated synthesis shows the preference for the isomer with the side-chain in *cis* positions relative to the metal macrocycle moiety. For example, the 18-membered hexaaza macrocycles carrying a benzyl-type group as pendant at each di-imine side chain has been synthesized by metal-templated cyclic Schiff-base condensation of 2,6-diacethylpyridine with substituted *S*- or *R,S*-1,2-diaminopropanes.[151] The use of (*S*)-diamines resulted in the complexes [LnL](CH₃COO)₂Cl·nH₂O where Ln = La(III), Eu(III) or Gd(III) and L stand for **3.149**, **3.150**, and **3.151**, and *n* = 0.25 – 2.75. These complexes were obtained as mixtures of the two possible isomers, containing, respectively the (*S,S*)-5.14- and (*S,S*)-5.15-disubstituted metal-macrocycle moieties.

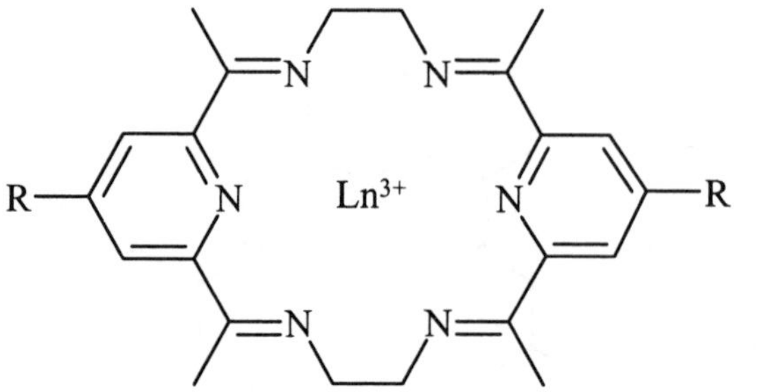

	R
3.152	H
3.153	-O-(CH$_2$)$_2$-NH-CH$_3$
3.154	-O-(CH$_2$)$_2$-NH$_2$
3.155	-O-(CH$_2$)$_2$-N(CH$_3$)COO-C(CH$_3$)$_3$
3.156	--HN(CH$_3$)-(CH$_2$)$_2$-NH-CH$_3$
3.157	-HNCH$_3$
3.158	-(CH$_3$)N-(CH$_2$)$_2$-OH

Using *R,S*-diamines, a mixture of the complexes [Ln(**3.151**)](CH$_3$COO)$_3$·nH$_2$O where Ln = La(III), Eu(III) or Gd(III) and $n = 1.5 - 4$, were obtained, three of them being identified as the (*R,R,SS*)-5.14, (*R,R,SS*)-5.15 and (*R,S*)-5.15 isomers. The crystal structure of the lanthan(III) complex with the ligand containing non-functionalized benzyl substituent, **3.151**, established unambiguously the *S* configuration of both stereocenters. This proves that no racemization or isomerization occurred during the complex formation. The lanthanide-macrocycle moieties are inert to metal release or exchange in the presence of competing ligands or competing metal ions, whereas the exocyclic anionic ligands are labile and can be easily exchanged. The pendant amino-benzyl or hydroxy-benzyl groups can be derivativised without the destruction of the complex. Lanthanide tris-acetates act as template in a [2+2] condensation to obtain tetra-imine complexes **3.152 - 3.158**. Here, the 18-membered macrocycles contain methylamino functions directly graphed onto the pyridine ring.[152]

3.4. Cage Ligands

Schiff condensation is adequate to obtain macropolycyclic molecules containing various binding subunits. They may act as receptors binding more than one guest metal ion in close proximity and this fact may allow magnetic interaction or electron transfer studies. According to the number of connecting bridges used for their constructions and to nature of the subunits used as building blocks, a variety of macrocyclic structures may be envisaged: bis-macrocycles, axial macrobicycles, lateral macrobicycles, cylindrical macrotricyclic systems and so on.[153]

Using 2,2',2"-triaminotriethylamine (tren) the macrobicyclic cages are formed in one-step condensation reactions. The amine is well suited for occupying the unique donor site and the top three octahedral sites in a C_{3v} model (capped octahedron). Also, tren is known to adopt a similar configuration in that its tertiary amine nitrogen serves as an apical donor in a trigonal bipyramid, whereas the three primary amine nitrogens are located in equatorial sites. For this reason, tren has been selected as appropriate component for the design of the cage ligands. Using 2-pyridinecarboxaldehyde as second component in a Schiff condensation, a heptadentate ligand has been obtained in the presence of nickel(II) as template in water. It should be noted that the role of nickel(II) is to remove the formed ligand in solution so that, according to Lindoy classification, this process can be described as a kinetic template one.[154]

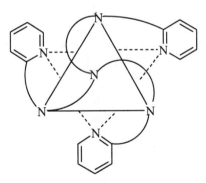

Fig. 3.4

The single-crystal X-ray structure of $[Ni(tren-py)]^{2+}$ salts, Fig. 3.4, shows that the Ni-N_{apical} distance of 3.25 Å precludes any perturbation of the octahedral ligand field by the lone pair of electrons on that nitrogen atom, although such interactions have been proved for some tren complexes. The magnetic moments of the $[Ni(tren-py)]^{2+}$ salts are all in the range expected for a high-spin d^8 configuration. However the lower values compared with that of related bipy complexes, for example, might suggest a lower effective symmetry.

The condensation of a tris(3-aminopropyl)amine with benzene-1,3-dicarbaldehyde in the presence of $AgNO_3$ produces the dinuclear silver(I) complex of the octaazahexadentate Schiff base macrocyclic ligand **3.159**[155] which is converted to a Cu(I) cryptate by transmetallation.[156]

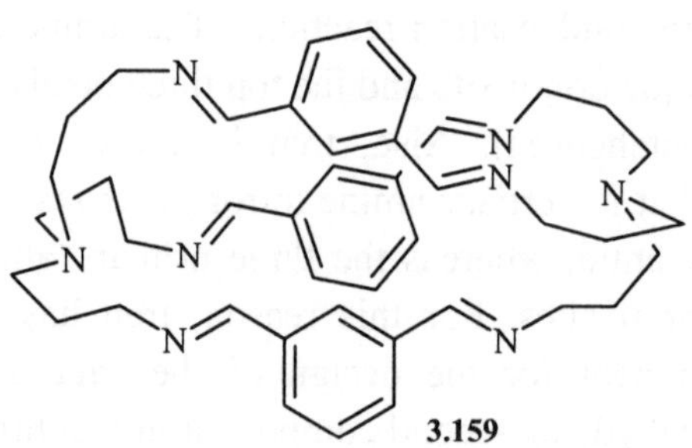

3.159

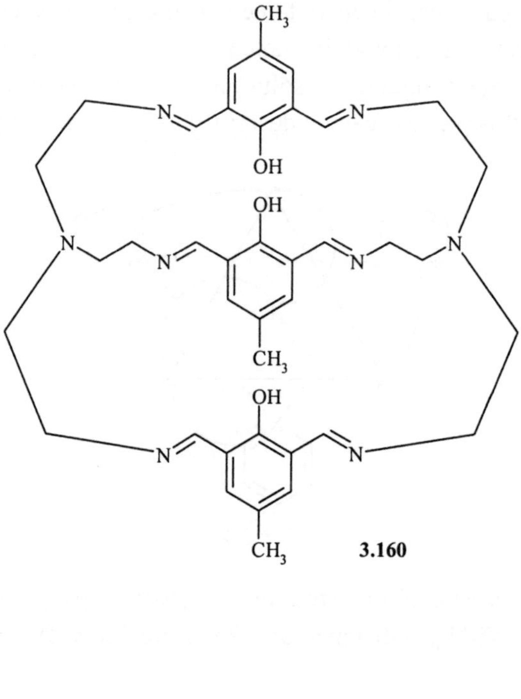

3.160

The condensation of 2,6-diformyl-4-methylphenol and tris(2-aminoethyl)amine in the presence of lanthanide nitrate Ln = La – Lu, Y resulted in mononuclear complexes of the general formula $LnL(NO_3)_3 \cdot xH_2O$, where L = **3.160**.[157] The X-ray analyses of Ce(III), Nd(III) and Eu(III) and Y(III) compounds show that they all contain the complex cation $[LnL(NO_3)]^{2+}$, **3.161,** and that the metal ions are placed at one end of the cavity of the cryptand, which is nine coordinated through the three imino-nitrogen atoms, three phenolic oxigen atoms, one of the bridgehead nitrogen atom and two oxygen atoms of a bidentate nitrate anion.

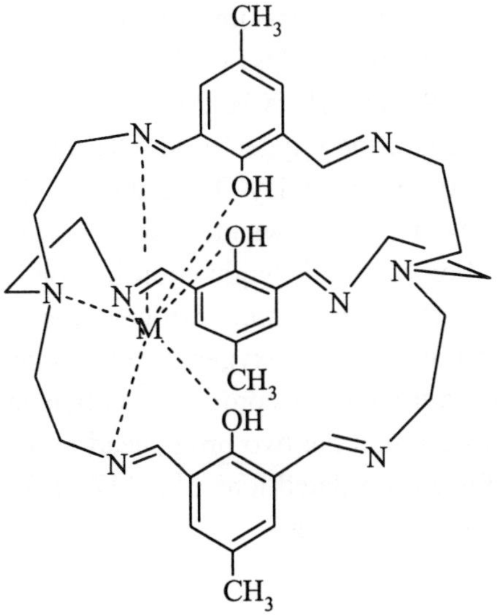

3.161

The coordination of the polyhedron is described as a monocapped square antiprism in which the coordinated bridgehead nitrogen atom is the cap. It has been noted that the Ln - bridgehead nitrogen atom distance is longer than other lanthanide ion – imine nitrogen atoms denoting a weak interaction. A progressive decrease of the Ln-donor atom distances was observed upon decreasing the ionic radii of the metal ions. All complexes adopt the same pseudo-triple-helix conformation around the metal ion, although the helicity and cell dimensions are different. It has been proved that both in solution and in solid state the complexes have a similar structure.[158] Thus, proton NMR spectra of the diamagnetic La, Lu and Y complexes in D_2O solution indicates C_3 symmetry, with the metal ion in a noncentred position of the ligand cavity. These studies also reveal the increase of the rigidity of the en moieties of the ligand as the ionic radius of the metal ion decrease.

The mononuclear complexes [LnL(NO$_3$)](NO$_3$)$_2$·xH$_2$O, (Ln = Y, La, Ce, Pr, Nd, Eu, Gd, Tb, Dy, Ho, Er. Tm or Yb) and the binuclear ones [Ln$_2$L(NO$_3$)$_2$](NO$_3$)·xH$_2$O·xEtOH with de-protonated form (L-3H)$^{3-}$, were synthesised by a [2+3] condensatiom of tris(2-aminoethyl)amine with 2,6-diformyl-4-methyl-phenolate in the presence of hydrated lanthanide nitrates as template, in ethanol under very dilute conditions.[159] The X-ray structure of the bimetallic dysprosium complex has been resolved. The complex shows the composition [DyL(NO$_3$)][DyL(NO$_3$)$_5$]·2MeCN. In the anion [DyL(NO$_3$)$_5$]$^{2-}$, the Dy(III) is 10-coordinated, being bonded to five bidentate nitrate ions and the oxygen atoms define a distorted bicapped dodecahedron with the average Dy-O distance of 2.43 Å. The crystal structure of lutetium complex shows identical coordination polyhedron, a distorted dodecahedron, where the eight coordination is completed by one bridgehead nitrogen atom, three imino-nitrogen atoms and to the three μ-phenolate oxygen atoms and one oxygen of a monodentate nitrate ion.[160] The cryptand adopts a triple helix conformation twisting around a pseudo C_3 –axis which runs through the two bridgehead nitrogen atoms with the distance between the encapsulated ions of 3.447(1) Å. A slight deviation of metal ions from this axis occurs.

3.5 Compartmental Ligands

Schiff bases compartmental ligands are derivatives of 2,6-disubstituted phenols, 1,3,5-triketones and β-ketophenols.[161, 162] These molecules form the "head" of the new ligand and suffer condensations with a wide range of amines including, dialkylalkanediamines, amino-acids, 2-aminoalkylpyridine, aminophenols or aminothiophenols which make the lateral units. These types of compartmental ligands have been designed according to Scheme 3.22. The first type, a macrocyclic one, results from a [2+2] condensation of heads and lateral units. The other two types are acyclic ligands: one of them, an "end-off", form B, in which one donor bridge is removed, and the second one, a "side-off", form C, in which one non-donor bridge is removed.

A large variety of complexes with these shape have been synthesized. The two coordination sites can be occupied by identical or different metal ions, thus resulting homo- or heterobinuclear complexes. For a given head unit, the lateral chain can be identical or different, thus creating different coordination sites. Also, compartments with different unsaturation degree and lateral chain containing additionally donor groups enrich this class of complexes.[163]

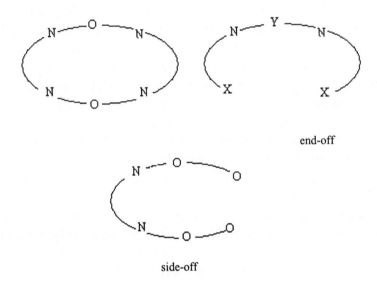

end-off

side-off

Scheme 3.22

Complexes with macrocyclic compartmental ligands are thermodinamically stabilized and kinetically retarded with regard to metal dissociation and metal substitution compared with acyclic ligands (macrocyclic effect).

3.5.1. Closed-chain ligands

3.5.1.1. N_2O donor precursors

Robson first reported the synthesis of the binucleating ligand capable of securing two metal ions in close proximity with important implication for metal-metal interactions. Using metal-directed template condensation of two molecules of 2,6-diformyl-4-methylphenol and two molecules of 1,3-diaminopropane, the planar binucleating macrocycle **3.163** has been obtained as ligand in the insoluble crystalline complexes $Cu_2(L)Cl_2 \cdot 6H_2O$, $Cu_2(L)(ClO_4)_2 \cdot 2H_2O$, $Ni_2(L)Cl_2 \cdot 2H_2O$, $Ni(H_2L)(ClO_4)_2 \cdot 2H_2O$, $Co_2(L)Cl_2 \cdot 2CH_3OH$, $Fe_2(L)Cl_2 \cdot 2CH_3OH$, $Mn_2(L)Cl_2 \cdot 2H_2O$, $Zn_2(L)Cl_2 \cdot 2H_2O$.[164] The binucleating ability of these ligands stems from the readiness of the phenol to deprotonate and bridge two metal ions. Efforts to isolate the metal-free ligands, from condensation of 2,6-diformyl-4-methylphenol with 1,3-diaminopropane under a wide variety of conditions yielded polymeric products.

Evidence for the presence of the binucleating ligand has been obtained from infrared spectra. The disappearance of the aldehydic C=O stretching band at 1680 cm^{-1} and the appearance of a strong band at ~ 1635 cm^{-1} assigned to $\nu_{C=N}$, in the infrared spectra of all the complexes proved that aldehyde groups have been completely converted into Schiff base residue. The binuclear nature of the complexes was confirmed by a single-crystal X-ray diffraction study of $Cu_2(\textbf{3.163})Cl_2 \cdot 6H_2O$ and $Co_2(\textbf{3.163})Br_2 \cdot CH_3O$[165]. It has been shown that the compound contains the copper(II)/cobalt(II) binuclear entity with the phenolic oxygen atoms of the macrocycle as bridges. The metal ion coordination geometry is square pyramidal, the basal plane is described by two nitrogen and two oxygen atoms whereas the apical positions are occupied by halide ions. The copper(II) and cobalt(II) ions are 0.21 and 0.30 Å, respectively, out of the ligand plane. These structural parameters as well as the number of unpaired electrons explain the variation in J value across the series copper, nickel (-27 cm^{-1}), cobalt (-9.3 cm^{-1}), iron (-4.2 cm^{-1}), and manganese (+0.2 cm^{-1}) complexes, the last one, actually, is becoming ferromagnetic. It has been established that the bicopper(II) complex is reduced in successive, quasi-reversible, one-electron steps at -0.52 and -0.91 V (vs. NHE). It was possible to isolate and characterise the mixed valence species $Cu^{II}Cu^{I}$(**3.163**)ClO_4 complex. It is stable in solutions containing no oxygen and reacts with CO to form the adduct $Cu^{II}Cu^{I}$(**3.163**)$(CO)(ClO_4)$.[166] The ability of the macrocycle **3.163** to keep the metal ions in its plane makes it appropriate to study metal-metal interactions in a system with a high degree of symmetry, to evaluate the specific contribution to intramolecular coupling of a metal centre and electrical properties both for hetero-binuclear and mixed-valence complexes. With a view to this, six-coordinated, binuclear complexes $[M_2$(**3.163**)$(base)_4](BF_4)_2$, where $M = Ni^{2+}$, Co^{2+}, Fe^{2+}, and *base* stand for pyridine(py), imidazole (Im) and methylimidazole (MeIm) have been obtained.[167] Indeed, the X-ray crystal structure of $[Fe_2$(**3.163**)$(Im)](BF_4)_2$, confirms the six-coordination for the iron ions and that they are only 0.01 Å out of the plane of the ligand **3.163** with a Fe-Fe distance of 3.117(2) Å. These structural parameters assert the variation of the exchange interaction parameters of -7.5, -4.1 and -23 for Fe, Co and Ni complexes, respectively, in the series $[M_2$(**3.163**)$(py)_4](BF_4)_2$ are controlled by the number and distribution of unpaired electons.

Since the works of Robson, the use of 2,6-diformyl-4-substituted phenols as precursors in the preparation of macrocyclic and /or macroacyclic Schiff bases compartmental ligands and of the related homo- and heterodinuclear complexes containing d- and/or f-metal ions were extended. The analogous ability of the 2,6-diacethyl-4-methylphenol was also explored, but the presence of a methyl group seems to reduce the ability to create imine due to both steric requirement of the bigger methyl group and its positive inductive effect. Considering the Robson's complex as an archetype, a variety of possibilities involve the change of the 4 position substituents with potentially electron-donor or electron-attractor units. Similar complexes containing halide atoms or an electron-attractor like CF_3 instead the methyl group were obtained.

The structure of Cu(**3.165**)$(ClO_4)_2$[168] was resolved. In the binuclear copper(II) entity, each copper ion has the [4+2] coordination with two nitrogen and two oxygen atoms in the basal plane. These atoms belong to the macrocycle and the phenolic oxygen atoms act as bridges between the two metal centres. The two oxygen atoms which belong to perchlorate groups, occupy the apical positions.

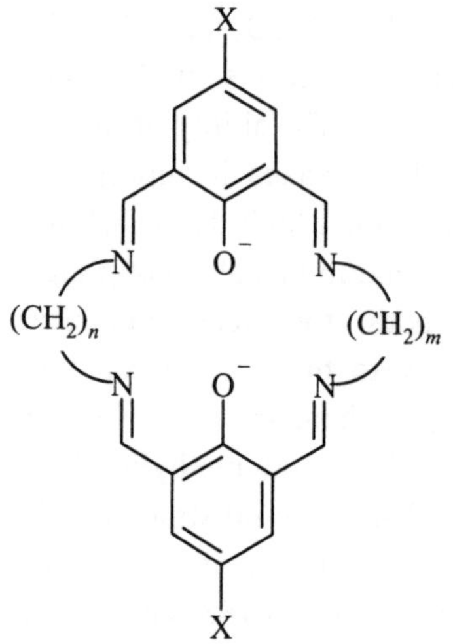

	m	n	X
3.162	2	2	CH_3
3.163	3	3	CH_3
3.164	4	4	CH_3
3.165	3	3	CF_3
3.166	4	4	CF_3
3.167	2	3	CH_3
3.168	2	4	CH_3
3.169	2	5	CH_3
3.170	3	4	CH_3

The complexes show reversible redox in two steps. Each step involves one electron process corresponding to the $Cu^{II}Cu^{II}/ Cu^{II}Cu^{I}$ and $Cu^{II}Cu^{I}/Cu^{I}Cu^{I}$ couples. It has been concluded that the replacement of the CH_3 by electron-attracting group CF_3, the dinuclear species become more reducible as the two reduction potentials are −0.53 and −0.91 V for **3.163** and −0.39 and −0.74 for **3.165**, vs. NHE, respectively. It has been noticed that the replacement of the CH_3 by electron-attracting group CF_3, does not modify the magnetic properties of the complexes. Thus, the energy gap between singlet and triplet states was found to be around −700 cm^{-1} for both complexes.

The changes in the length and nature of lateral chain were considered in order to modulate the flexibility of the ligands and further, the electrochemical and magnetic properties of the complexes. The binuclear complexes were severely strained when $x = 2$. The longer lateral chain means a more flexible coordination sphere that allows a tetrahedral distortion favouring the Cu(II) $\rightarrow$ Cu(I) reduction. This fact was proved by the couples $Cu^{II}Cu^{II}/Cu^{II}Cu^{I}$ and $Cu^{II}Cu^{I}/Cu^{I}Cu^{I}$ which are in the order $-0.53 > -0.32$ and $-0.91 > -0.81$, for the series of **3.163**, and **3.162**. More reductible copper(II) binuclear cations have been obtained when buthylene groups were the lateral chains as in **3.164**. The binuclear complexes $[Mn_2(\textbf{3.164})(O_2CCH_3)_2]^{169}$ and $[Mn_2(\textbf{3.163})(O_2CCH_3)_2]$ have been obtained by template condensation around manganese(II). The configuration about each Mn is described as a distorted octahedron with N_2O_2 donor set of the macrocycle describing a plane and acetate or dmf in the apical positions. Structural analyses have shown different core structures of the two complexes as a result of the mismatch between the cavity size and the ionic radius of the manganese(II). Thus, the centrosymmetric binuclear core is nearly coplanar for **3.164** whereas Mn deviates from he basal plane by 0.75 Å for **3.163**.

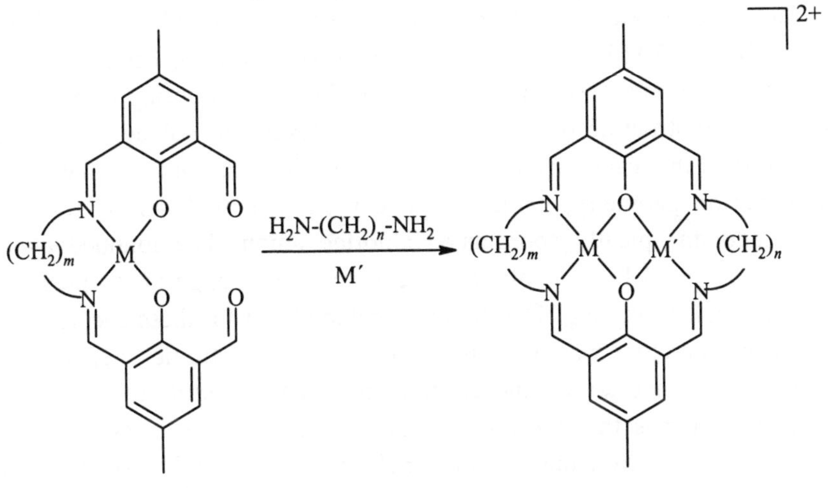

Scheme 3.23

The structural differences explain the magnetic behaviour of the two homobinuclear manganese complexes. $[Mn_2(\textbf{3.164})(O_2CCH_3)_2]$ shows strong antiferromagnetic interaction with the overall exchange integral $J = -5$ cm^{-1}. In this case, the $2p_x$ and $2p_y$ orbitals of the phenolic oxygen have their lobes spread on the coplane allowing an efficient overlapping between 3d orbital of manganese and phenolic 2p orbital. Such an overlapping is not possible for $[Mn_2(\mathbf{3.163})(O_2CCH_3)_2]$ because of the large deviation of the two manganese from basal plane.

Macrocycles with different lateral chains $n \neq m$ can be obtained by a stepwise template reaction. Two strategies were adopted. The first one involves the formation of an open mononuclear complex which further reacts with a diamine to form a mononuclear macrocyclic complex. The second metal ion can be incorporated in the formed cavity. The second strategy may involve a different metal ion as template and thus, heterobinuclear complexes can be obtained (Scheme 3.23) in one pot reaction. Both the strategies refer to Cu(II) or Ni(II) as template and benefit by the fact that these ions bind preferentially to the N_2O_2 site in the precursors.[170] The complexes $[Cu_2(\mathbf{3.167})]Cl_2 \cdot 2H_2O$ and $[Ni_2(\mathbf{3.167})]Cl_2 \cdot 2H_2O$ as well as those of symmetrical **3.162** and **3.163** were obtained by stating from corresponding mononuclear species following the first strategy.

Homo- and heterobinuclear complexes of the type $[M_1M_2L]^{2+}$ where L = **3.162** or **3.163,** and M_1 = Cu(II) and M_2 = Cu(II), Fe(II), Mn(II), Co(II), Ni(II) and Zn(II) have also been obtained according to Scheme 3.24.[171] The structure of some of this complexes was solved and it has been shown that each metal ion in the given binuclear unit, has the same coordination geometry, namely, square pyramidal with an N_2O_2 basal plane and an apically coordinated chloride anion. The founded close proximity ($\sim$ 3.14 Å) of the metal centres suggests a possible intramolecular coupling. The studies concluded that for these species, the molecular structure was less important than the electronic one. Indeed, for the homobinuclear complexes, an increased net of antiferromagnetism interaction across the transition series from manganese to copper was observed.[172] The exchange parameter J, of -30, -71, and -101 cm^{-1}, for Mn(II), Fe(II), and Ni(II), respectively, shows an increasing antiferromagnetic exchange interaction along this series. For all the complexes reversible to quasi-reversible Cu(II)Cu(I) electrochemistry was observed with the Cu(II)Cu(I) reduction potential, $E_{fCu(II)Cu(I)}$ = -1.068 V (vs. ferrocene/ferrocinium(+1)), almost invariant across the series M_2 = Cu(II), Fe(II), Mn(II), Co(II), Ni(II) and Zn(II).

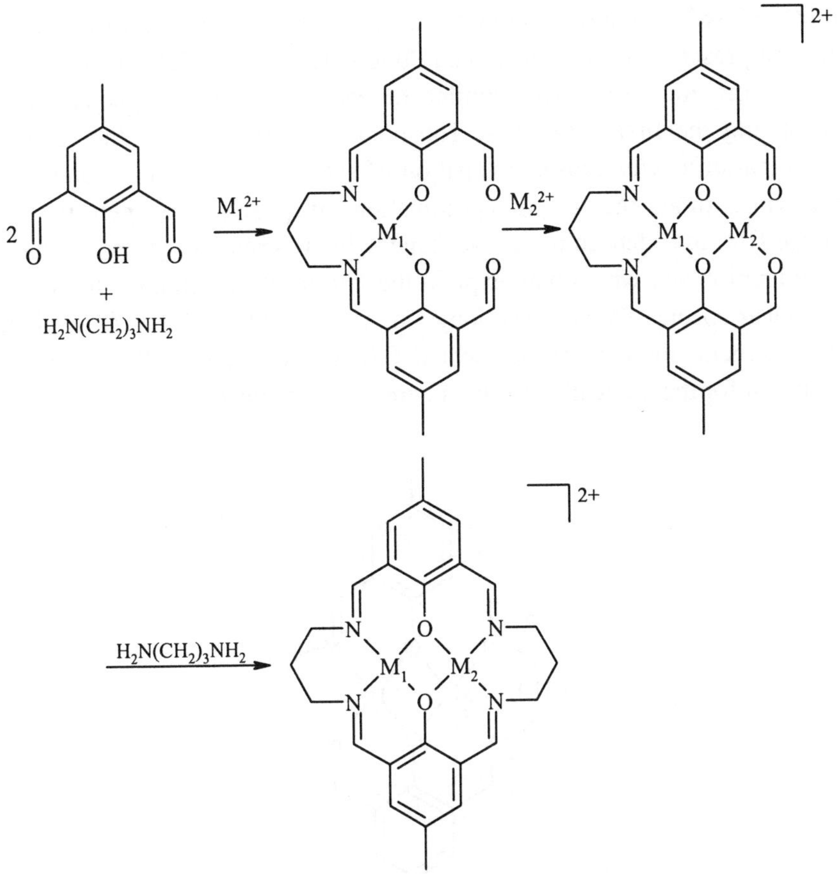

Scheme 3.24

Hetero-three-nuclear complexes containing unsymmetrical macrocycles $n \neq m$, **3.167**, **3.168**, **3.169**, **3.170**, of the general formula Pb(ML)$_2$]X$_2$, (M= Cu^{2+}, or Ni^{2+}; X = ClO$_4$, PF$_6$, BPh$_4$ or BF$_4$) have been obtained following the last strategy[173] in which Cu^{2+} (or Ni^{2+}) and Pb^{2+} ions act as the first and the second template ions, respectively. The high yields of 40-80 % for Ni$_2$Pb and 55-95 % for Cu$_2$Pb complexes were explained by the high flexibility of the lead(II) ion in coordination as well as the flexibility of the second coordination site. It should be noticed that the macrocycle **3.167** could not be obtained in a trinuclear complex because the second N$_2$O$_2$ coordination cavity is too small and rigid to incorporate a large lead(II) ion. The X-ray crystallographic analysis of [Pb(Cu(**3.163**))$_2$](ClO$_4$)$_2$, demonstrate the formation of the macrocycle and the presence of the sandwiched Pb ion. The two Cu(**3.163**) entities in the cation are structurally similar to each other, the copper(II) ion occupying the first N$_2$O$_2$ sites. The bound of the Pb(II) in the second coordination cavity causes elongation of the Cu-N and Cu-O bonds. In the same time, the strong coordination of phenolic oxygen to the copper(II) ion determine a weakened field strength of the second coordination site which may explain the failure of the tentative to obtain binuclear species Pb ML. The long Cu...Cu separation ($\sim$ 6.0 Å) explain the magnetic moment values which fall in the range 1.78 - 1.90 μ_B, common for magnetically isolated copper(II) complexes.

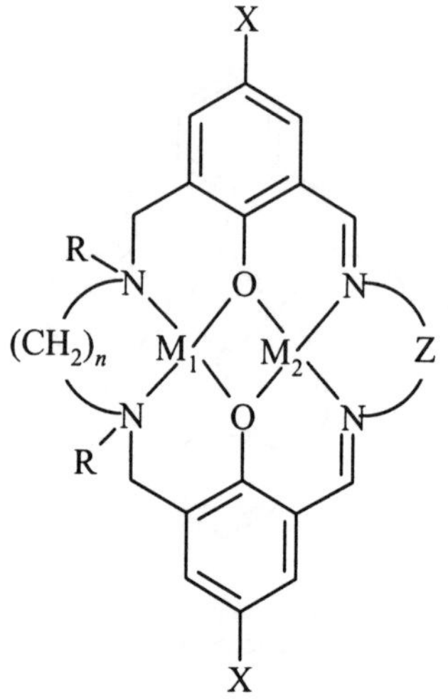

	n	X	R	Z
3.171	2	CH$_3$	H	CH$_2$-CH$_2$
3.172	2	Br	CH$_3$	CH$_2$-CH$_2$
3.173	2	Br	CH$_3$	CH$_2$- CH$_2$-CH$_2$
3.174	2	CH$_3$	CH$_2$-Py	CH$_2$- CH$_2$-CH$_2$
3.175	3	CH$_3$	CH$_2$-Py	CH$_2$- CH$_2$-CH$_2$
3.176	3	CH$_3$	CH$_3$	CH$_2$-CH$_2$
3.177	3	CH$_3$	CH$_3$	CH$_2$- CH$_2$-CH$_2$
3.178	3	CH$_3$	CH$_3$	CH$_2$-CH$_2$-CH$_2$-CH$_2$

Trinuclear complexes [Pb(ML^{m,n})$_2$](ClO$_4$)$_2$, where M = Ni, Cu, (m,n) = (2,3), (2,4) or (3,3) were used as precursors to obtain CuMn and NiMn species by methatesis. The Mn(II) in [NiMn(Cu(**3.168**)L^{2,4}(dmf)$_2$](ClO$_4$)$_2$ is 0.966 Å away from the basal N$_2$O$_2$ least-squares plane and the two dmf molecules occupy the open space. This is explained by the preference of nickel(II) for a planar geometry which limits the flexibility of the tetramethylene lateral chain and hinder the accommodation of Mn(II) in the formed cavity.[174]

Macrocyclic complexes **3.172** – **3.178**, with a reduced degree of unsaturation can be obtained when an acyclic complex precursor like **3.179**, possessing two dissimilar coordination sites an amine, N(amine)$_2$O$_2$, and an O(formyl)$_2$O$_2$ one, suffer template cyclization with an aliphatic or aromatic diamine, according to Scheme 3.25.[175] The resulted macrocycle contains two different coordination sites with respect to the saturation or unsaturation of the donor nitrogen.

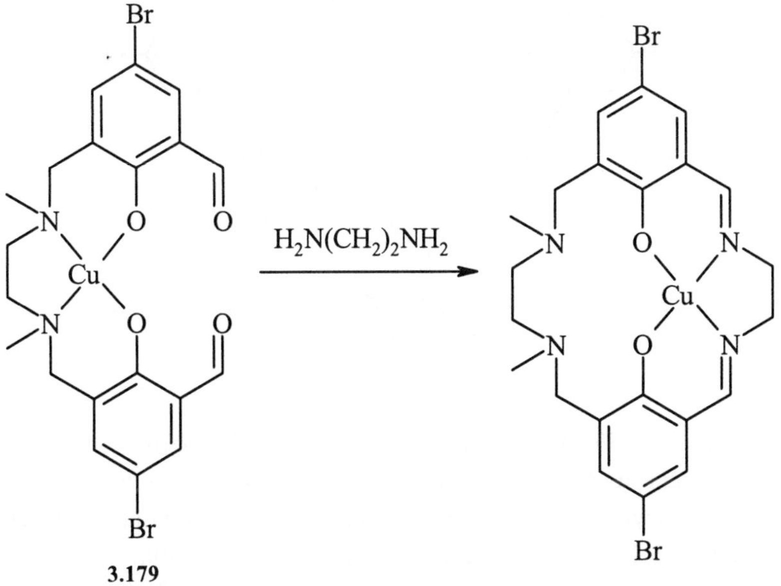

3.179

Scheme 3.25

The crystal structure of the mononuclear [Cu(**3.172**)]·2PrOH shows that Cu(II) occupys the N(imine)$_2$O$_2$ coordination site and thus it is demonstrated that the copper(II) ion migrates from the N(amine)$_2$O$_2$ site to the N(imine)$_2$O$_2$ one during cyclization process. Similar migration was observed when the cyclization step occurs in the presence of Pb(II)[176] when the complex [CuPb(**3.173**)(BzO)(dmf)](ClO$_4$) is formed. In DMF, with excess of perchloric acid, it resulted in the cyclic condensation of the constituents, **3.180**.

Macrocycles **3.181** – **3.185** containing different constituents can be obtained following also a stepwise template strategy.

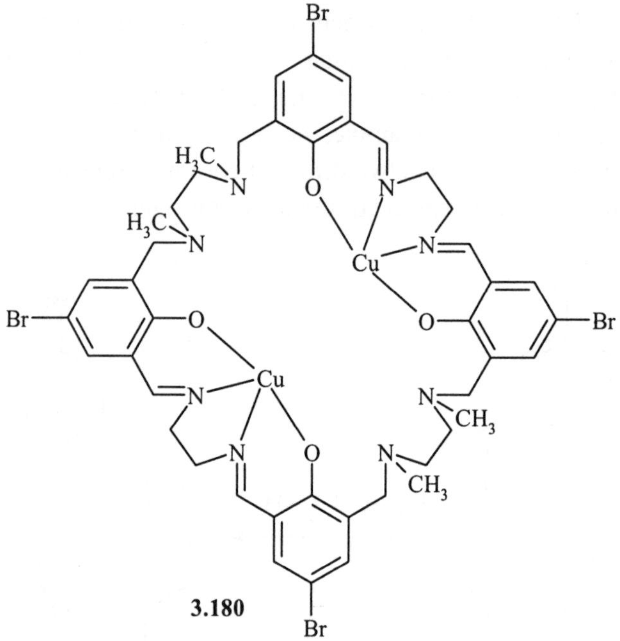

3.180

Using 1,3-diformyl-1,3-diacetyl- or 1,3-dibenzoyl-2-hydroxy-5-methylbenzene as head unit and, 1,3-diaminopropane as lateral bridge, copper macrocyclic complexes have been obtained following the reaction in Scheme 3.26.[177] The complexes undergo sequential one-electron transfers at two different potentials and except for **3.181**, reversible reduction steps have been observed. The potentials of the first step remain almost invariant in the range -0.459 – -0.515 V, whereas that of the second reduction step depends on the alkyl or aryl substituents.

The open chain copper(II) complex [Cu$_2$L(OH)](ClO$_4$)$_2$·nH$_2$O, **3.185**, reacts with appropriate 1,3-diketophenols to form the macrocyclic complexes **3.186** – **3.189**.

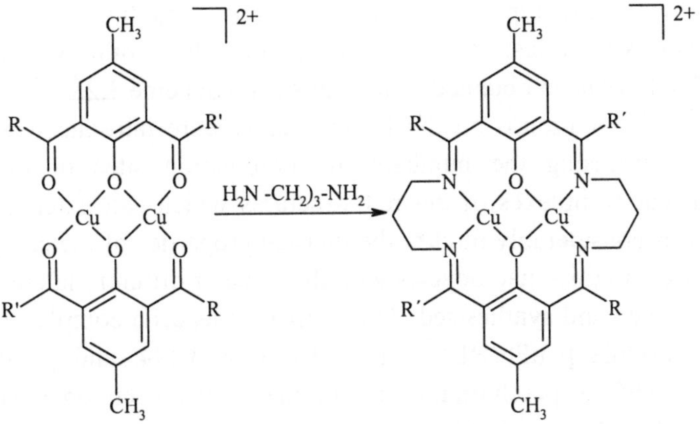

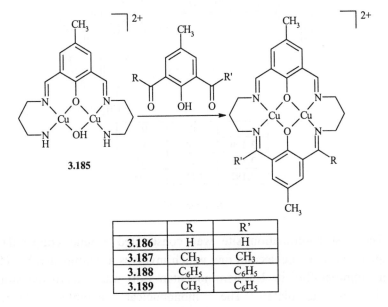

	R	R'
3.181	H	H
3.182	CH₃	CH₃
3.183	C₃H₇	C₃H₇
3.184	C₆H₅	C₆H₅
3.185	CH₃	C₆H₅

Scheme 3.26

	R	R'
3.186	H	H
3.187	CH₃	CH₃
3.188	C₆H₅	C₆H₅
3.189	CH₃	C₆H₅

Scheme 3.27

Following this type of reaction, various substituted macrocyclic complexes with a less degree of saturation which otherwise are not accessible have been obtained as it results from Scheme 3.28.

Potent donor groups on the lateral chains were introduced with the hope of increasing the numbers of coordination sites or to obtain multinuclear complexes of desired structures which can extend, and in some cases considerably modify the unusual properties of the complexes. With a view to this, macrocycles with the same or different lateral chains were designed and synthesised. The copper(II)-lead(II) complexes of the general formula $[CuPb(HL)]^{2+}$, L = **3.193** or **3.194**, and $[CuPb(L)]^{+}$, **3.193** - **3.195** as perchlorates or chlorides, have been obtained by a "stepwise template reaction".

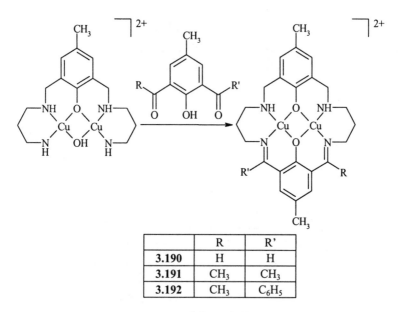

	R	R'
3.190	H	H
3.191	CH₃	CH₃
3.192	CH₃	C₆H₅

Scheme 3.28

The first coordination site was constructed around copper(II) as template whereas the second one which involves a diaminoalcohol (1,3-diaminopropan-2-ol or 1,5-diaminopentan-3-ol) was constructed around Pb(II) as template. The mononuclear complex [HCu(H-**3.196**)](ClO₄)·0.5H₂O was obtained in the presence of Ba(II) but the product shows that the second coordination site, the five-coordination one, is occupied by a proton instead of Ba(II). The single-crystal X-ray analysis[178] of the complex [CuPb(**3.193**)](ClO$_4$)$_2$·dmf shows that the copper(II) ion is bound to the first N$_2$O$_2$ four-coordination site creating a nearly planar structure. In the second coordination site, Pb(II) shows a six-coordinate geometry.

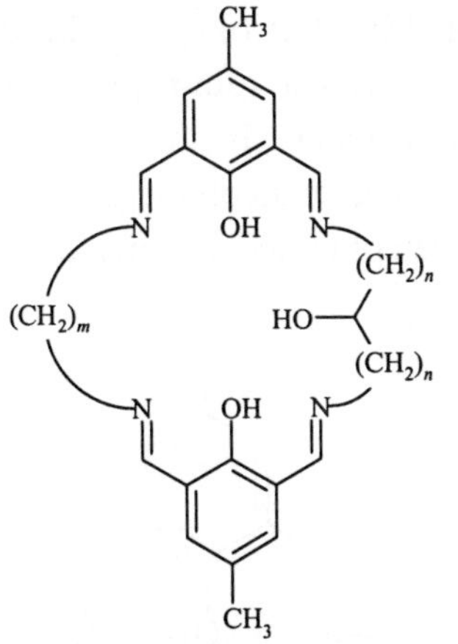

	m	n
3.193	2	1
3.194	3	1
3.195	2	2
3.196	3	2

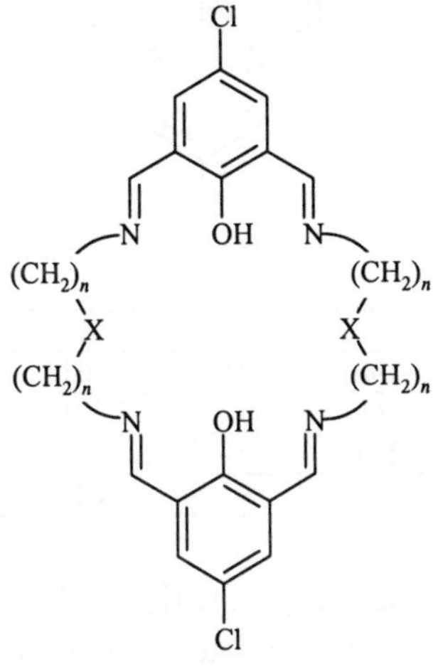

3.197

The two phenolic oxygens and the two imino nitrogens coordinated to Pb(II) describes a square plane with a considerable deviation of the metal ion from the least-square N_2O_2 plane due to the coordination of the alcoholic oxygen on the lateral chain and, on the other side to the mismatch between the size of the cavity and the radius of Pb(II). The sixth position is occupied by dmf oxygen. The ESR spectrum measured in frozen dmf solution shows a spin-triplet state of a dimeric species produced by the out-of-plane bonding between CuN_2O_2 entities. Macrocycle **3.194** containing an alcoholic group on one lateral chain as a potent donor has also been obtained as bimanganese(II) complex.[179]

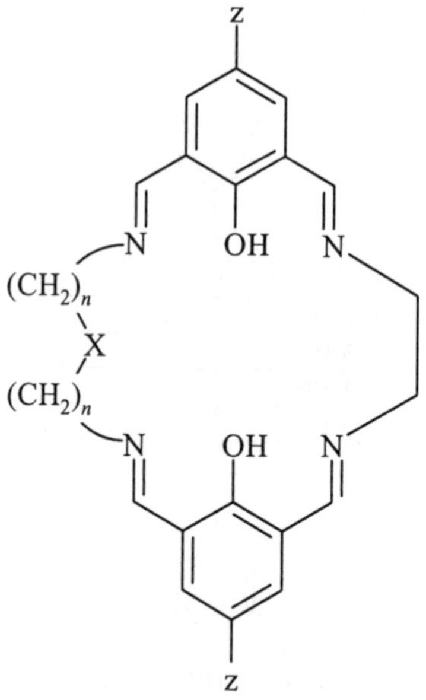

	X	Z
3.198	NH	Cl
3.199	S	Cl
3.200	NH	CH$_3$
3.201	S	CH$_3$

Mononuclear and binuclear Cu(II) and Ni(II) complexes with macrocycles **3.197** in which the lateral diamine briges were replaced with: 1,5-diamino-3-azapentane, 1,5-diamino-3-thiapentane,[180] 1,5-diamino-3oxapentane and 1,5-diamino-2-hydroxy-propane have been obtained. The mononuclear complexes have been obtained both by direct metal – ligand synthesis and also by metal template one, whereas, the binuclear complexes were obtained only when a metal ion is used as template.

Binuclear complexes $NiCuL(OAc)_2 \cdot 3H_2O$ which contain asymmetric cyclic ligands L = **3.198** or **3.199** were obtained and the different spin states for the central metal ion has been interpreted as a prove that the complexes are a mixture of positional isomers. The macrocycle **3.200** possessing two different metal coordination sites with N_2O_2 and N_3O_2 donor sets has been obtained as ligand following the strategy Scheme 3.23, in two series of binuclear $Cu^{II}M^{II}$ and $Ni^{II}M^{II}$ (M^{II} = Mn, Fe, Co, Ni, Co or Zn) complexes. It has been established that the copper(II) or nickel(II) occupying the first compartment as in the precursors. The similar thioether analogue has been obtained as mononuclear [Cu(**3.200**)](ClO$_4$)$_2$ complex with lead(II) as template for the cyclization reaction. Further, metal methatesis leads to the binuclear complexes [CuM(**3.200**)(NCS)$_2$]·H$_2$O, M = Co(II), Ni, Zn and mixed valence Cu(II)Cu(I).[181] The crystal structure of dmf adduct [CuZn(**3.200**)(NCS)$_2$]·dmf show that copper(II) has been moved in the N_2O_2S coordination site.

2,6-diformylpyridine N-oxide was used as precursor to obtain symmetrical ligand using Fe(II), Co(II) or Ni(II) ions as template when homo- and heterobinuclear complexes 3.202 were obtained.[182]

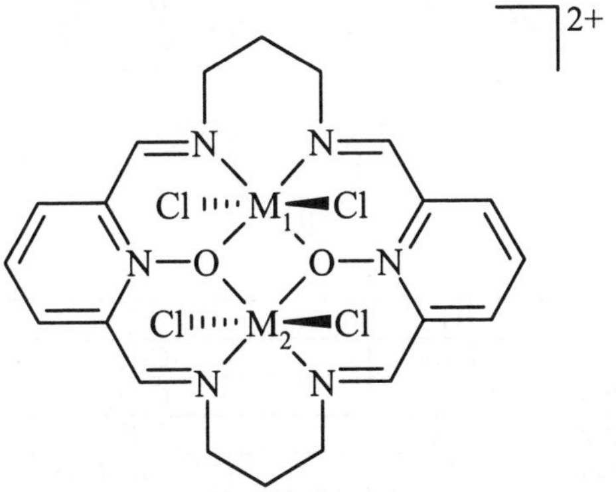

3.202

3.5.1.2. N_2S donor precursors

Binucleating macrocyclic ligands containing sulfur donor atoms were studied in the hope that they might show enhanced affinity for "soft" cations, in particular those of second and third transition series, compared with analogous ligands containing nitrogen and oxygen. These complexes are models for metalloproteins containing thiolate such as Fe-S cluster, nickel-iron hydrogenase and Cu_A site of cytochrome C oxidase. Schiff-base complexes have been obtained by [2+2] condensation of a thiolate "head units" like 2,6-diformyl-4-methylthiophenolate or 2,6-diformyl-4-*tert*-butylthiophenolate, with polyamine as lateral units, in the presence of a metal salt.[183] Like phenolate analogues, the thiophenolate group allows the communication between the metal centres through thiolate bridges. However, some differences are noticed due to the specific character of the sulphide group. Thus, the thiolate complexes have the tendency to aggregate *via* thiolate bridges, which further avoid formation of disulfides. But, the bonding of thiolate group to the aromatic ring reduces drastically this tendency and thus, the nuclearity of the complexes can be controlled. The enhanced polarizability and the pyramidal geometry adopted by the sulfur atoms lead to different magnetic and redox properties.

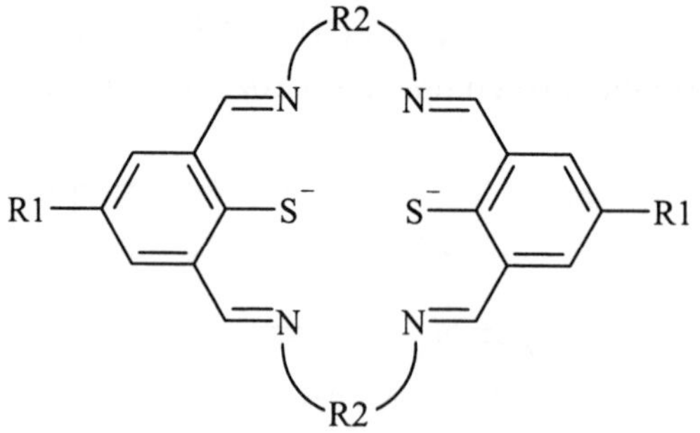

	R1	R2
3.203	-CH$_3$	-C$_3$H$_6$-
3.204	-C(CH$_3$)	-C$_3$H$_6$-
3.205	- CH$_3$	-C$_4$H$_8$-
3.206	-CH$_3$	-C$_2$H$_4$-NH-C$_2$H$_4$-
3.207	-CH$_3$	-C$_2$H$_6$-NH-C$_2$H$_6$-
3.208	-CH$_3$	-CH$_2$-CH(OH)-CH$_2$-
3.209	-CH$_3$	- C$_2$H$_4$-CH(OH)- C$_2$H$_4$-

Macrocycles **3.203 – 3.209** are mostly templated by nickel(II) or zinc(II). For example, complexes of the formula [Ni$_2$L]X$_2$, where X = ClO$_4$, NCS or CF$_3$SO$_3$, were obtained from the corresponding precursors and, some of them, were structurally characterized by Brooker's group. The crystal structures show that the macrocycle **3.203** is enough flexible to adopt two different conformations in order to provide the incorporated metal ions with a suitable geometry. Thus, Zn(II) ions have a trigonal-bipyramidal environments whereas nickel(II) lie in a square-planar N$_2$S$_2$ environment.[184] With a longer lateral chain, the folding of the macrocyle is possible. For example, the ligand **3.205** is folded in a way that the two phenyl rings are almost perpendicular to each other, thus favouring π-π stacking interactions. As a consequence, nickel(II) ions suffer a tetrahedral distortion away from the square planar arrangement.[185] As expected, the change in the macrocyclic cavity affects the properties of the complexes of **3.205** and **3.203** which show four one-electron redox processes and are considered to be predominantly metal centered. The [Ni$_2$L]$^{2+/+}$ and [Ni$_2$L]$^{+/0}$ processes seems to be most affected by the macrocyclic cavity. Thus, although [Ni$_2$LSj]$^{2+/+}$ with j molecules of solvents, S, processes are reversible for both the complexes, it occurs at -1.01 V for **3.203** while the complex with the larger macrocyclic cavity is easier reduced, by 90 mV. The [Ni$_2$LSj]$^{+/0}$ process is reversible for **3.205** and irreversible for **3.203** although they occur at a similar potentials, -1.46 and -1.45 V, respectively.

Schiff base macrocycles **3.207**, **3.208**, and **3.209**, have been prepared by using zinc(II) ions to template reaction of 2,6-diformyl-4-methylthiophenolate with diethylenetriamine, dipropylenetriamine or triethylenetetramine as lateral units. X-ray crystal structure determinations of [Zn$_2$(**3.207**)]$^{2+}$ and [Zn$_2$ (**3.208**)]$^{2+}$ show a tetrahedral geometry similar to that reported for a related copper(II) complex containing acyclic thiophenolate ligand.

3.5.1.3. Multicompartmental ligands

The chemistry of Robson-type complexes has been expanded by increasing the number of donor sites thus allowing accommodating more than two metal ions in specific compartments. The number of coordination sites can be increased either by increasing of the incorporated phenol groups, by introducing additional alcohol groups in the lateral units or both of these procedures. In general, the pairs of metal ions are present: four, six, eight and twelve.[186, 187, 188,189]

Planar tetra- and octanuclear copper(II) and nickel(II) complexes of the [2+2] macrocyclic ligands **3.210** and **3.211** which include the bridging abilities of both phenol and alcohol groups have been obtained. The flexibility of the lateral chain allows the twist and fold of the macrocycle so that it become suitable for the formation of the mixed-valence $Mn^{II}_2M^{III}_2$ complexes when manganese(II) acts as template.[190]

Tetranuclear Ni_4 and Zn_4 complexes[191] have been obtained with tetradentate Schiff base ligand **3.212** but attempt to obtain a similar one with manganese as template in the condensation reaction of 2,6-bis(aminomethyl)-4-methylphenol with 2,6-diformyl-4-methylphenol resulted in the dimanganesse(III) complex $[Mn_2(\textbf{3.212})(\mu\text{-}O_2CCH_3)(\mu\text{-}OH)(CH_3OH)_2][ClO_4]_2 \cdot 2CH_3OH$.[192]

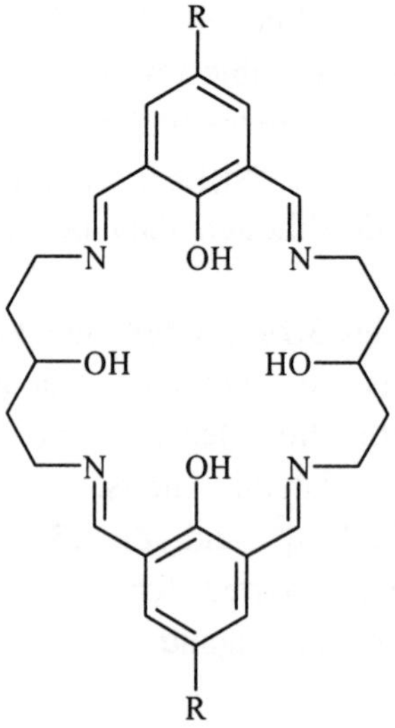

	R
3.210	CH_3
3.211	$-C(CH_3)_3$

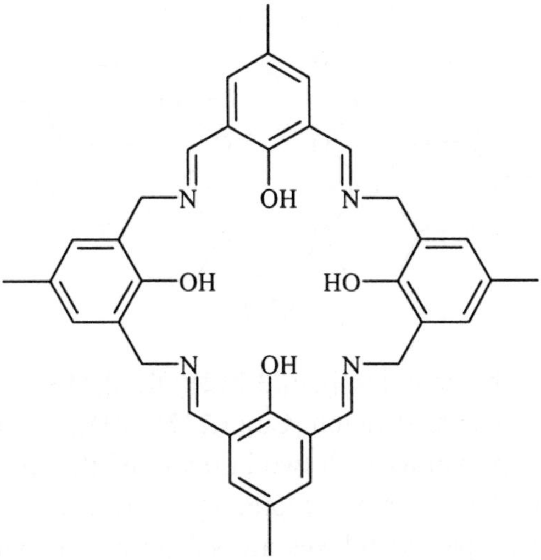

3.212

Tetranuclear complexes of the "dimer-of-dimer" type have been obtained by metal-directed template synthesis of Cu(II), Mn(II) or Ni(II). These complexes contain macrocycles **3.212 – 3.216** obtained by [2+2] condensation of the appropriate precursors. It should be noticed that the tetranucleating ligands have been obtained only when the linear tetramine precursor is alkylated at the secondary nitrogen atoms in the case of Mn(II) or Ni(II) as template.[193]

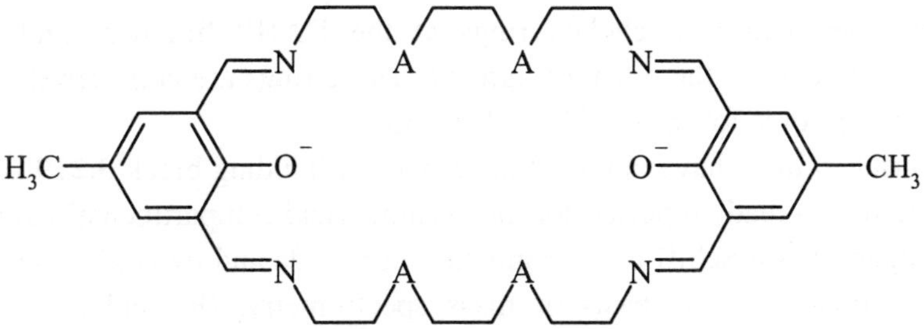

	A
3.213	NH
3.214	N-CH$_3$
3.215	N-CH$_2$- CH$_3$
3.216	S

3.217

For example, the complex [Cu$_4$(**3.212**)(OH)$_2$][MeCO$_2$]$_2$[PF$_6$]$_2$·MeOH contain the nearly planar skeleton [Cu$_4$(**3.212**)(OH)$_2$]$^{4+}$. The two dinuclear units are in close proximity with two bridges and the apical coordination site of the copper(II) ions are occupied by an acetate ion. The magnetic moments of the copper complexes are subnormal at room temperature, suggesting antiferomagnetic interactions within the complex molecules. The cryomagnetic studies show J values between -259 and -357 cm^{-1}.

Hexanuclear rings **3.218** have been obtained in the complexes of the general formula [Cu$_6$(**3.218**)(μ_2-OH)$_3$]$_2$(X)$_6$·nH$_2$O, where X = NO$_3$, ClO$_4$, or BF$_4$ and [Ni$_6$(**3.218**)(μ_2-OH)$_3$(H$_2$O)$_6$]$_2$(X)$_6$·nH$_2$O, where X = NO$_3$ or ClO$_4$.[194] In the syntheses of these complexes the metal salts have reacted with 2,6-diformyl-4-tert-buthylphenol(DTBP) in methanol followed by addition of NEt$_3$ and 1,3-diamino-2-hydroxypropane (DAHP) and refluxing. The use of NEt$_3$ facilitates the deprotonation of the secondary alcohol and causes it to act as a donor. The resulted macrocyclic ring forms five-membered chelate rings at the DAHP fragment and six-membered ones at the phenol fragment. These rings are considered to be the driving force for the [3+3] condensation.

Busch *et al.*[195] have obtained the binuclear building block **3.219** using the strategy already reported for the asymmetrical compartmental ligands. The ligand has a Schiff base fragment on one side and two oxime groups on the other. On the basis of mass spectrometry, IR- and electronic spectroscopy the tetranuclear complex was formulated as [Cu$_2$(**3.219**)]$_2$[Cu$_2$Cl$_4$]. Crystal structure of [Cu$_2$(**3.219**)]$_2$[Cu$_2$Cl$_4$] shows that two dicopper(II) cations are bound to each other forming a dimer of dimers. The structure of a dicopper(II) complex cation is generally similar to those of analogous complexes with Robson-type dicompartmental ligands, but affected by its unsymmetry.

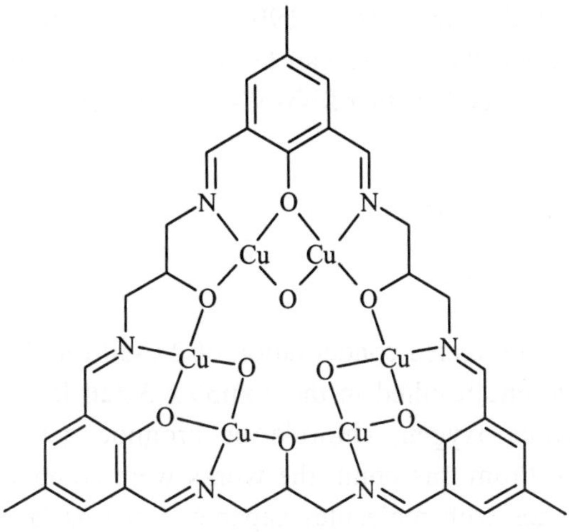

3.218

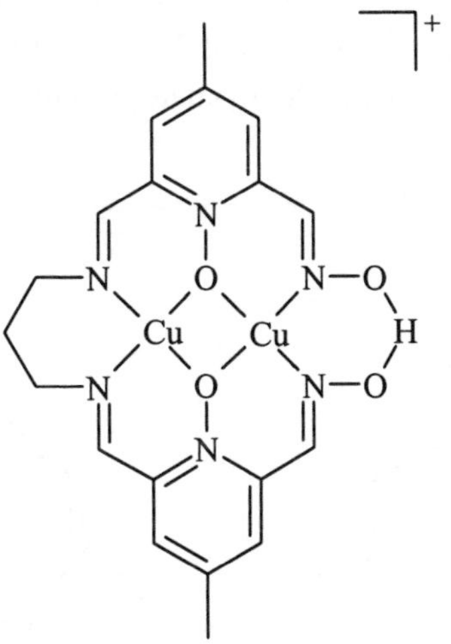

3.219

The metal ion is square-pyramidal, each forming an elongated fifth bond, either to an oxime oxygen of a different ligand or a chloride and the copper-copper separation of 3.07 Å. The $[Cu_2Cl_4]^{2-}$ anion bridges two

[Cu$_2$(**3.219**)]$^+$ cations and the oxime-copper link to the other copper, thus resulting a chain structure. The magnetic moment of 0.65 B.M. measured at room temperature indicates antiferromagnetic coupling and a cryomagnetic study gave $2J = 690 \pm 10$ cm^{-1} and no significant interaction between the dimers can be seen. The founded value for $2J$ falls in the usual range for other Robson-type dicopper(II) complexes.

3.5.2. Open chain ligands

3.5.2.1. End – off ligands

The copper(II) directed condensation of 2,6-diformyl-4-methylphenol with an alkyl amine resulted in the complex **3.220** in which beside the bridging phenolic oxygen, chloride or bromide acts as the second bridging group. From this point, the works were extended by replacing these monoamines with molecules capable to act as lateral chains in a system which lodges two metal ions in a close proximity. Thus, diamines, N-substituted alkyl- or aryldiamines,[196] aminoacids[197], 2-aminomethylpyridine[198],[199,] or o-aminophenol were used as lateral units. Also, a large variety of the second bridges were employed including hydroxo, alkoxo, halide, cyanide, cyanato, organic bases.

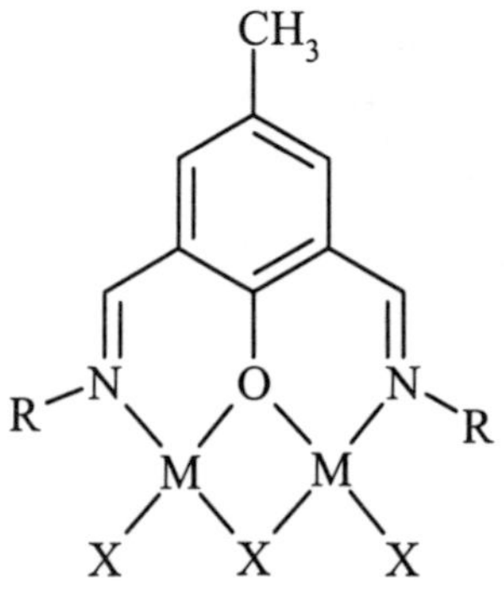

3.220

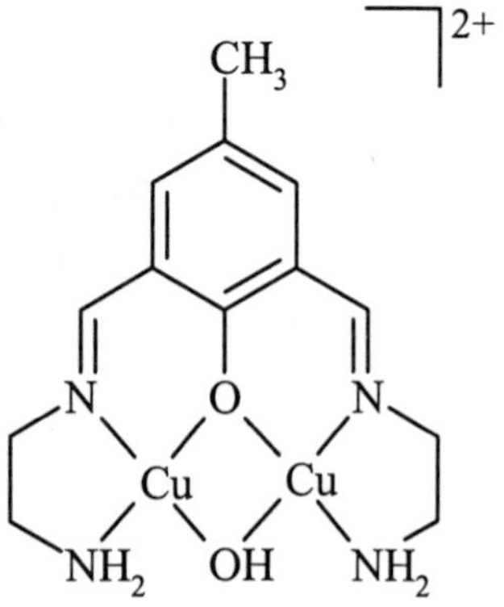

3.221

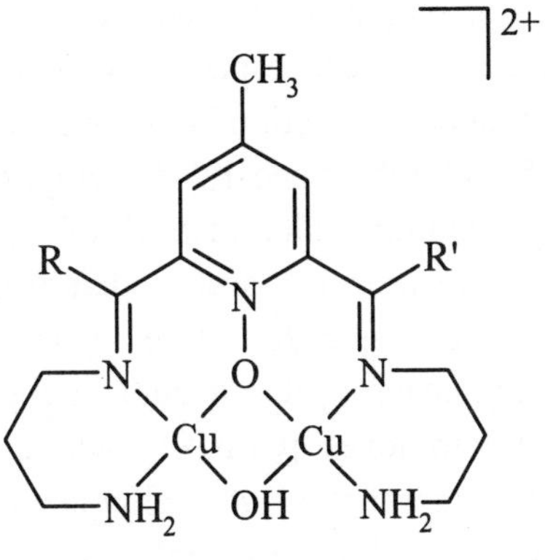

3.222 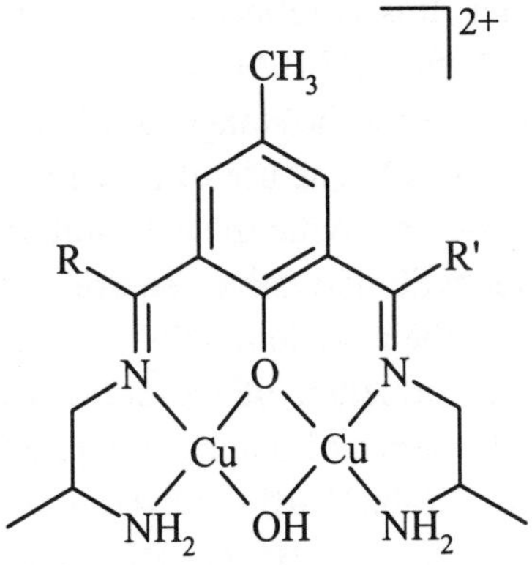   3.223

The reaction of the sodium salt of 2-hydroxy-5-methylbenzene-1,3-dicarbaldehyde with copper(II) perchlorate and diaminoalkanes in aqueous solution under high dilution resulted in the acyclic compound **3.221, 3.222** and **3.223** which contains the hydroxide group as exocyclic ligand.[200] The hydroxy group can be replaced with the weak organic acids like *p*-nitro-phenol, 2-hydroxy-5-methylaceto-phenone or pyrazol which also serve as exocyclic bridging ligands. The flexibility of the ligands can be modulated by the lateral chain. Thus, 1,3-diaminopropane in **3.222** offers the more flexibility in the metal-linkages which is reflected by the Cu-O-Cu angle. It suffers an argmentation compared with those in the similar other complexes.

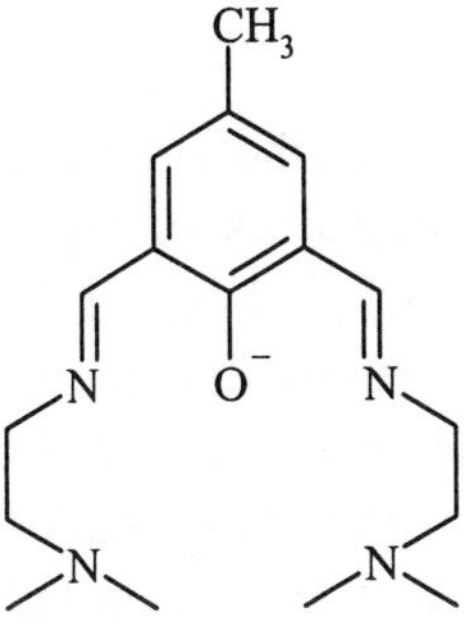

3.224 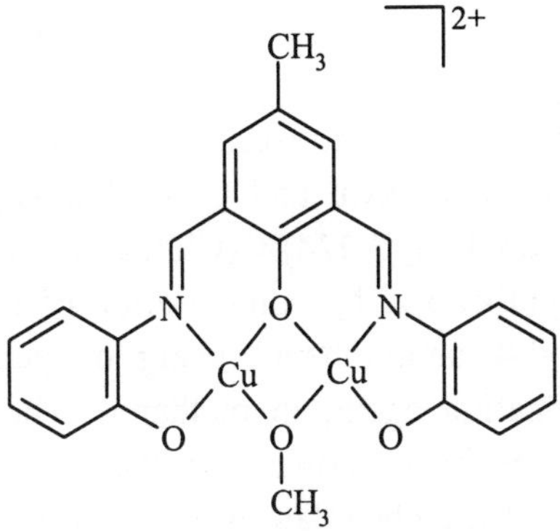   3.225

Further, this is reflected in the range of magnetic moment values of 0.67, 1.20 and 1.27 B.M. for **3.221, 3.222** and **3.223,** respectively, which are indicative for the stonger antiferromagnetic interaction between metal centres in **3.222**. The use of 2-hydroxy-5-methylbenzene-1,3-diacethyl or 2-hydroxy-5-methylbenzene-1,3-dibenzyl sodium salts instead of 1,3-dicarbaldehyde derivative requires a mixture of water and methanol as solvent as their sodium salts are less soluble in water. Also, in these cases, the formation of the corresponding macrocycles occurs in high yields. The series of binuclear copper(II) complexes, $[Cu_2LX]^{2+}$ where L stands for 2,6-bis[N-(β-dialkylaminoethyl)iminomethyl]-4-methylphenol, **3.224,** and X = Cl, Br, or OH, were prepared. All the complexes show demagnetization by a spin-spin-pairing process between two neighbour tetragonal pyramidal copper(II) ions connected with the phenolic oxygen and the bridge X. As the degree of the spin-exchange through the first bridging group is the same throughout the series of complexes, it has been considered that the energy separation between the singlet and triplet states reflects the effect of the second bridging group. Thus, the $-2J$ values obtained on the basis of Bleaney-Bowers equation arrange the second bridging groups in the order X = OH >> Br > Cl.

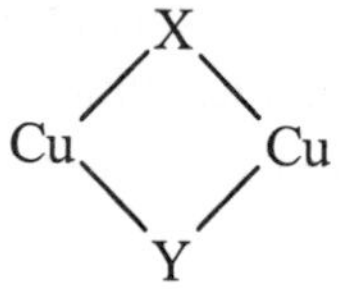

$$X \neq Y$$

Fig. 3.5

With the aim to discriminate between the contribution arising from each of the CuXCu and CuYCu linkages arranged as in Fig. 3.5, the complexes $[Cu_2(\textbf{3.224})X](ClO_4)_2$ where X = OH⁻, 1,1-N₃⁻ or 1,1-OCN⁻ were synthesized by condensation of the appropriate diamine with 2,6-diformyl-4-methylphenol using copper(II) as template and LiOH as deprotonating agent, in methanol. The X-ray difraction studies of the complexes have shown that the copper(II) ions are connected by an oxygen of the phenolato ligand and, in the same time, by a coordinating atom of the second bridging ligand. The copper(II) ions are in [4+2] surrounding with the CuOCuN networks roughly planar and the apical positions occupied by oxygen atoms of the perchlorate anions.

As it is shown in Table 3.1 the three complexes have the CuOCu angles of the same order of magnitude whereas the CuXCu one depends strongly on the nature of the X ligand. All the three complexes exhibit antiferomagnetic interaction which also strongly depend on the nature of the exogenous bridge X and the powder EPR spectra at 77 K for X = 1,1-N_3^- or 1,1-OCN$^-$ are typical of a triplet state. The singlet state is strongly stabilized for X = OH$^-$, whereas for X =1,1-OCN$^-$ singlet state is very weakly stabilized. Examining the nature of the bridging ligands and the bridging angle, it can be observed that the trend agrees well with the former observation that a singlet ground state corresponds to larger angles and a triplet ground state. This behaviour can be rationalized. The two CuOCu bridging angles are of the same order of magnitude in hydroxo-bridged complex, thereby a synergic effect in payring of electrons occurs. The end-off azido bridge exerts a ferromagnetic contribution and, additionally, the larger CuOCu bridging angles of 98.7°, bring a larger antiferromagnetic contribution from phenolato bridge than in X = OH$^-$. However, altogether the singlet state is less stabilized than in the case of X = OH$^-$. In the O-cyanato-bridged complex, the two CuOCu linkages exert almost opposite contribution as the OCN bridge is very efficient to favour the ferromagnetic interaction.

Table 3.1. Relevant angles and distances of some binuclear copper(II) complexes

X	CuOCu,	CuXCu	$Cu_1 ... Cu_1$	J, cm^{-1}
OH	97.4(2)	99.3(2)	2.924	- 367
N_3^-	98.7(3)	100.0(4)	2.972	-86.5
OCN$^-$	98.1(3)	99.3(4)	2.933	-3.8

Gagnee *et al.* have studied/reexamined the electrochemical properties of several complexes derived from this head and observed some mixed-valent Cu(II)-Cu(I) species from cyclic voltametric scans. The end-off binuclear complexes $[Cu_2(L)(X)](ClO_4)_2$, where L is the binucleating ligand 2,6-bis[N-(2-pyridylmethyl)formamidoil]-4-methylphenol, **3.226**, and X = N_3^- or OCN$^-$ OH$^-$, have been synthesized using copper(II) acetate as template in a mixture of tetrahydrofuran and acetonitrile as solvent. The compounds are isomorphous and, as in previous cases, the copper(II) ions lie in [4+2] environments, with nearly planar Cu(I)Cu(II)OX networks, the apical positions being occupied by the oxygen atoms of the perchlorate anion. The same trend as in $[Cu_2(\mathbf{3.224})(X)](ClO_4)_2$, can be observed for the angle $Cu(I)O_{phenolate}Cu(II)$ on changing the exogenous ligand X. So, the angles are 99.19, 97.5 and 100.5° for OH, OCN and N_3^- respectively.

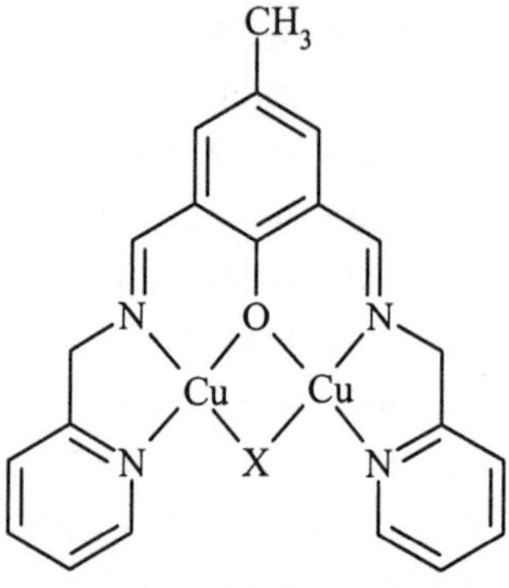

3.226

All the complexes exhibit antiferromagnetic interactions with a singlet ground state except $[Cu_2(\mathbf{3.226})(OCN)](ClO_4)_2$, which shows ferromagnetic behaviour with a rare triplet ground state. The singlet-triplet (S-T) energy gap varies as J_{OCN}(43 cm^{-1}) > J_{N3} (-161 cm^{-1}) > J_{OH} (-364 cm^{-1}).

No major differences can be observed between the energy gap values of $[Cu_2(\mathbf{3.224})(OH](ClO_4)_2$, and $[Cu_2(\mathbf{3.226})(OH](ClO_4)_2$, (-364 and -367, respectively). However, the effect of the lateral units on the magnetic behaviour can be observed for the two series when X = 1,1-N_3^- or 1,1-OCN$^-$ and the differences should be rationalized taking in mind structural aspects. In these cases, a more pronounced angle between the direction of these exogenous ligands and the plane of the bridging network has been observed for complexes of **3.226** compared with the similar complexes of **3.224.**

The reaction between the precursors with $Th(NO_3)_4 \cdot 4H_2O$ in the presence of stoichiometric quantity of LiOH and $Mg(CH_3COO)_2 \cdot 4H_2O$, resulted in the orange complex $Mg[Th_2L_3]_2 \cdot 6H_2O \cdot dmf$, where L = **3.227** whose structure has been determined by X-ray crystallography.[201] The ionic complex contains two formula units in the monoclinic unit cell. As shown in the formula, each anion is a binuclear unit in which two Th ions are bridged by three oxygen atoms from three different ligands. Each ligand, which is trianionic pentadentate, is coordinatd with two atoms to the first Th atom, with two atoms to the second Th atom, and with the central oxygen atom to both Th atoms.

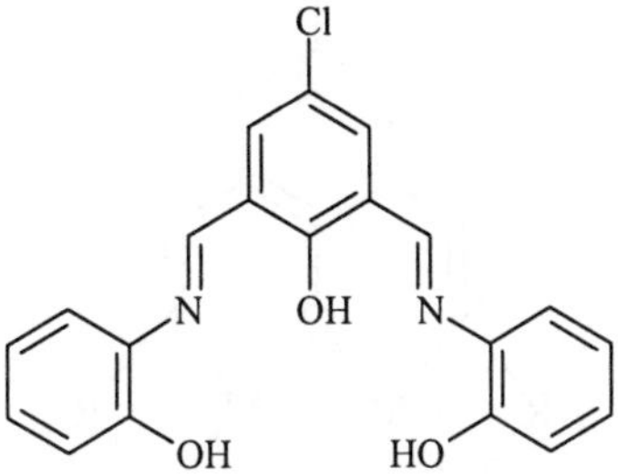

3.227

Binuclear complexes $Cu_2(L_A)X\cdot MeOH$, $Cu_2(L_B)X\cdot MeOH$, $Mn_2(L_A)(OH)\cdot MeOH$ and $Mn_2(L_B)(OH)\cdot MeOH$, where L_A stand for **3.229** - **3.231** and L_B for **3.232** - **3.233**, have been obtained when copper(II) or manganese(II) acted as template in the condensation of 2,6-diformyl-4-chlorophenol with aminomethane- or aminoethane-sulphonic acid.

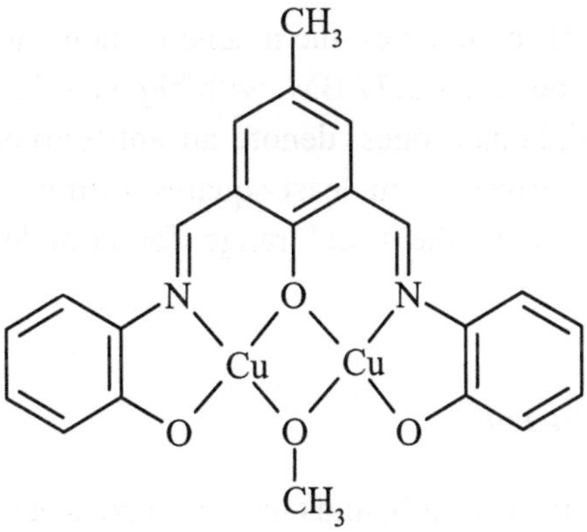

3.228

Similar copper(II) complexes have been obtained when (±)1-aminomethanephosphonic acid was used istead of corresponding aminosulphonic acid. The -OH or -Cl anions act as exogenous bridge thus enhancing the stability of the complexes. Magnetic moments of 1.33-1.77

BM per copper(II) atom are higher than for the phosphonate analogues than sulphonate ones but, criomagnetic studies show antiferromagnetic interactions between the metal ions for all the copper complexes. The longer side chain determin lower J value for **3.232** complexes.

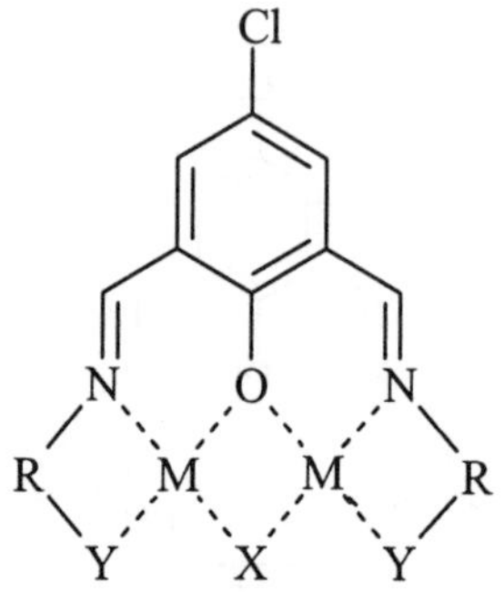

	R	Y
3.229	-CH$_2$-	-SO$_3$H;
3.230	-CH$_2$-	-PO$_3$H$_2$
3.231	-CH(CH$_3$)-	-SO$_3$H
3.232	-CH$_2$-CH$_2$-	-SO$_3$H;
3.233	-CH$_2$-CH$_2$-	-PO$_3$H$_2$

For the copper(II) complexes, the magnetic moment values per copper atom was found to be 1.33 - 1.77 BM, with higher values for phosphonate analogues than sulphonate ones, denote an antiferromagnetic behaviour and the g and J derived from least-squares fitting by using Bleaney-Bowers equation are in the usual range for complexes of analogous ligands.

3.5.2.2. Side – off ligands

The side-off binucleating ligands have two adjacent coordination sites situated in close proximity, the inner one containing N$_2$O$_2$ donor set and the outer one, the O$_2$O$_2$ coordination set. Synthesis of the side-off binucleating ligands occurs according to the procedure like that presented in Scheme 3.24 when the process is interrupted in the second stage. The ligands have been obtained as mononuclear or binuclear copper(II) complexes and as the heterobinuclear ones for M_1= Cu and M_2= Ni(II), Co(II), Fe(II), Mn(II) according to the method of Okawa and Kida (1972). Both the coordination sphere is obtained by metal template route.

Enlargement of the inner coordination site has been realized when an additional donor atom was introduced thus resulting a potential pentadentate coordination compartment. The mononuclear copper(II) **3.234** complex has been obtained when 3-formylsalicylic acid and diethylenetriamine were heated in the presence of copper(II) acetate. The complex can further undergo a transmetallation with UO_2^{2+} or the second coordination site can be occupied by UO_2^{2+}, Ni^{2+} or Cu^{2+} or ions.[202]

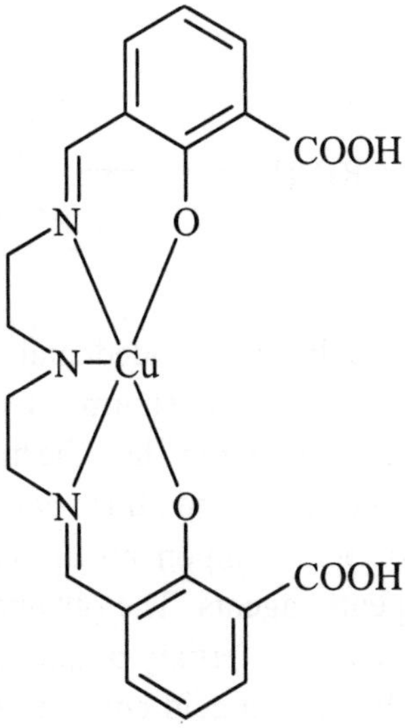

3.234

The ligands **3.235** and **3.236** were obtained as mononuclear UO_2^{2+}, Ni^{2+} or Cu^{2+} complexes. On the basis of IR data, it was proposed that the copper(II) ion is located in the inner coordination site of complexes of **3.236** whereas UO_2^{2+} and Ni^{2+} are located in the outer one.

Chapter 4

Mannich Condensation

Mannich and Kater discovered in 1912 the property of formaldehyde to bind an amine with a carbon acid via a methylene bridge.

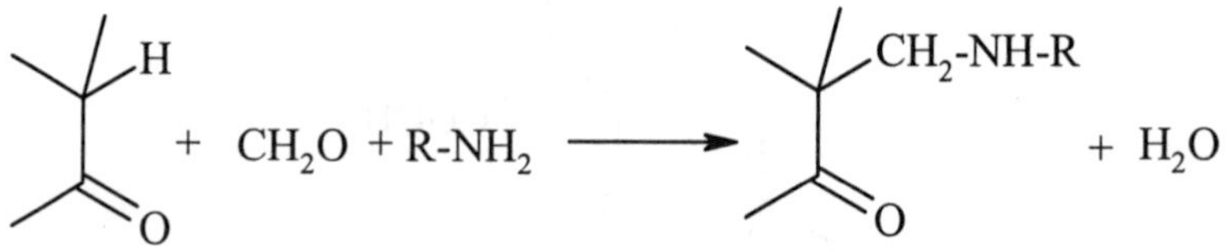

The method has since been extended and is increasingly used in preparative chemistry providing an enormous pool of information for synthetic chemists. Much research has been directed towards the production of pharmaceuticals and this has in turn, lead to the predominant selection of acid components among the substances which are recognised as therapeutic agents. The obtained molecule, a Mannich base, is a primary, secondary or tertiary amine, depending on the nature of the starting amine. The carbon acid contains an α activated hydrogen by a carbonyl or a hydroxyl group[203,204] so that beside the amine, the Mannich base contains another potential donor atom conveniently situated close to a chelate ring with a metal ion. The preorganising ability is increased when the Mannich base is obtained from a diamine or a diacid. In these cases a potential dinucleating ligand can be obtained. These kinds of ligands have been designed with the aim to obtain binuclear metal complexes which have special magnetic and redox properties.[205] Some complexes can be considered as models of iron or copper biosites[206] and Mannich bases with a polymer structure have been used as selective ion exchangers.[207]

In the last decades, metal ion directed Mannich condensation - developed by Sargeson's group[208] allows access to an easy synthesis of large rings which give rise to particularly effective and selective complexing agents. In this synthesis, a diprotic acid, AH_2, and two molecules of formaldehyde bridge two *cis*-coordinated amine groups by a new $-CH_2-A-CH_2-$ chain (capping reaction). Most of the ligands are saturated polyamines although they can incorporate heteroatoms like oxygen, sulfur, phosphorus or arsenic which may be involved in the coordination. The nature of the ligands depend strongly on the "organising" capacity of the metal ions and in this respect cyclic or cage ligands can be obtained. The reactions do not occur in the same way in the absence of metal ions. Thus didactically, the metal directed Mannich condensation is considered as a kinetic template reaction. The practical observations of the metal directed Mannich condensations underline its simplicity and efficiency as well as the fact that yields of the complexes obtained in this way are generally high. The high yields in a reaction mixture from which a multitude of alternating products are possible suggest that the reaction must be kinetically extraordinary favoured and the metal ion must strongly induce the reaction. Here, we are going to make a recension of the template Mannich bases along with a description of their properties compared to the parent precursors. We will describe some individual systems in more detail. The work is organised roughly by the ligand shape, starting with acyclic ligands and progressing to cyclic and cage macrobicycles (cryptands).

4.1. Mechanistic Aspects

The mechanism of the intramolecular condensation between a coordinated amine, formaldehyde and an acid component has been suggested on the basis of accumulated experimental results. The rate determining step is the formation of a coordinated imine. In this respect, the metal directed Mannich reaction has to be understood in terms of metal-ion – imino chemistry, namely the formation, the stability and reactivity of coordinated imines.

Curtis[209] studied the condensation of $[Me(en)_3]^{2+}$, Me=Ni, Cu with acetone and assumed as a common process that the coordinated nucleophiles attack the reactive centers in the metal complexes under the formation of the carbon-nitrogen double bond (imine). It was asserted that neither the coordinated nucleophile nor the ligand which is attacked ever leave the metal ion. The imine formation involves the formation of a carbinol-amine intermediate.

Intramolecular cyclization of chelated imines has been performed with Co(III)-amine complexes containing α-keto-carboxylato[210,211] or amino-ketone[212] as coligands in alkaline solutions. The imine complexes have then been isolated by the addition of concentrated acids as the respective salts (usually chlorides). However, the coordinated imine is sensitive to attack of a nucleophilic group like CN⁻, BH₄⁻ or carbanions. Such behaviour was explained by the fact that the metal ion occupies the position of a proton and, additionally, to the kinetic inertness of the Co(III)-N bond. In this respect, the coordinated imine can be regarded as having an intermediate character between imine and iminium ion. Remarkably, such condensations lead in purely organic chemical systems to polymers. Clearly the role of metal ion in these reactions appears to be the coordination of the amine groups to the cobalt(III), and to inhibit the intermolecular condensation between the amine and carbonyl groups, which is commonly observed in organic chemistry. The metal ion acts to direct the reaction via the small chelate rings resulting in a favourable outcome compared with alternatives which would not form a ring.

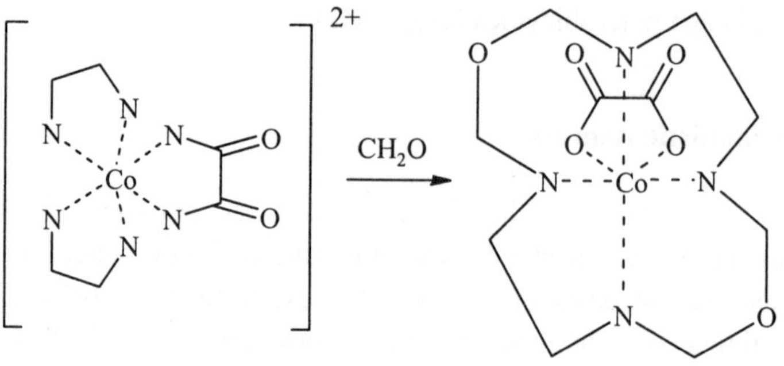

Scheme 4.1

Condensation of formaldehyde with the kinetically robust system [Co(en)₂gly]Cl₂ in basic medium[213] leads in the first instance to α-hydroximethylserine and, further to 1,4,8,11-tetraaza-6,13-dioxa-tetradecane (dioxacyclam). The latter type of condensation also takes place when bis(ethylenediamine)-oxalatocobalt(III) complex is treated with formaldehyde.[6] In this case [(oxalato)(dioxacyclam)-cobalt(III)]chloride has been obtained according to Scheme 4.1 and characterised. The reaction occurs with coordinated nucleophiles and without the rupture of the metal-ligand bonds.

Intramolecular reactions of coordinated amines like the one presented above, lead to the assertion[214] that the formation of a coordinated imine from a metal-amine complex, aqueous formaldehyde and an acid padlock must be the first step.

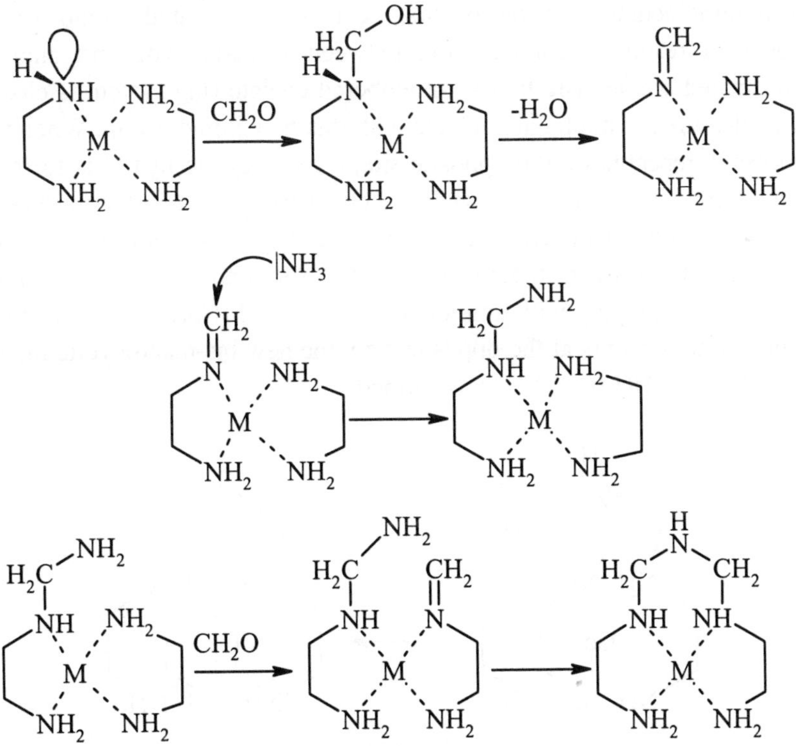

Scheme 4.2

It is well known that the coordinated amine has a smaller nucleophilicity compared with the free amine. Therefore, in some cases, it has been postulated that the disruption of the metal-nitrogen bond takes place before the nucleophilic attack and is followed by its restauration. The nucleophilic character of the coordinated amine can be increased when the nature of the reaction medium favours its deprotonated form.[215] The metal directed reactions take place in basic medium usually created by Li_2CO_3 (in water), or triethylamine (in a non-aqueous medium) so that deprotonation can occur. The nucleophilic attack of the formaldehyde carbon atom by the deprotonated amine gives the coordinated carbinolamine. Elimination of water then leads to a coordinated imine according to Scheme 4.2. In a further step, the addition of the anion of the acid molecule to the carbon-nitrogen double bond occurs. In a third step an intramolecular reaction occurs. A new coordinated imino group appears a result of condensation between formaldehyde and another coordinated amine. The first six-membered chelate ring system is closed when the already bound acid reacts with this new imino group. When the process is repeated on the opposite site, a macrocyclic ligand is formed. The same path is followed when tris(ethane-1,2-diamine)metal complexes, like $[Co(en)_3]^{3+}$, are capped along the C_3 axis of the ion. Scheme 4.3 illustrates the formation of the second chelate ring which results in the capping with ammonia at one side of $[Co(en)_3]^{3+}$. When the same reaction occurs at the opposite side, the new bis-macrocyclic ligand sepulchrate, **4.104** (see 4.5.1), is formed.

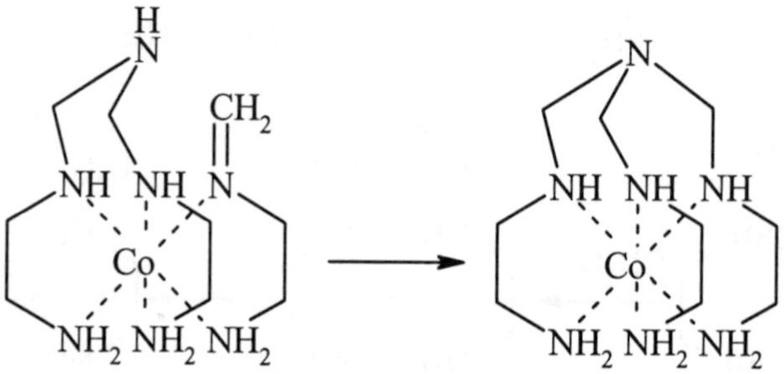

Scheme 4.3

The presence of the coordinated imine is shown by the successful isolation of some intermediates. Thus, during the formation of

[Co(diNOsar)]$^{3+}$ (diNOsar = **4.106**), the intermediates [Co(en)$_2$(NH$_2$CH$_2$CH$_2$N=CH$_2$]$^{3+}$ and [Co(N-Me-NOsen]$^{3+}$ have been isolated.[13] Imine intermediates were identified and characterized by ^{1}H and ^{13}C NMR spectra during the condensation of [Co(tame)$_2$]$^{3+}$, (tame = 1,1,1-tris(aminomethyl)ethane) with formaldehyde and nitromethane.[216]

Jackson *et al.*[217] have performed the capping reaction of {[Co(H$_2$N-H$_2$C-H$_2$C-S-)$_3$]$_2$Co}$^{3+}$ in two steps. As the imine complex was considered to be the key intermediates, the authors tried and succeeded to isolate and characterize it.

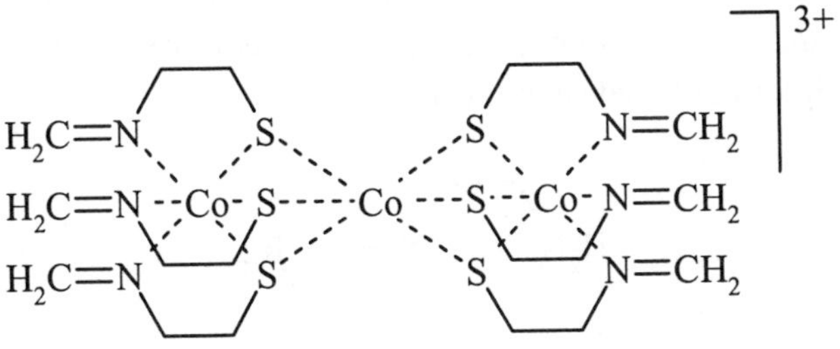

4.1

The hexaimine **4.1** has been obtained as triflate and perchlorate. These salts are stable in acidic aqueous or non aqueous solutions but decompose to the corresponding precursor hexathiolates in alkaline aqueous solution. In a second step, the hexamine was capped with NH$_3$ or nitromethane resulting in aza-, respectively, nitro-capped complexes (see **4.2.3**).

Macrocyclization occurs using other metal ions as templates, such as copper(II), nickel(II), and palladium(II). Other nucleophiles like aliphatic amines, diethyl malonic ester, phosphines or arsine might play the role of the acid component. When no other nucleophile is available, other coordinated amine groups within the same coordination sphere can act as nucleophile. In these cases, the short link -HN-CH$_2$-NH- between the two adjacent amine groups could not be stable in the absence of the metal ion.

Beside the electronic factors presented above, steric factors play a very important role in these reactions. These contributions will be presented for various classes of ligands in the following chapters.

4.2 Acyclic Ligands

4.2.1. Polyamine ligands

The metal directed condensation of amines with formaldehyde was often employed to obtain new macrocyclic ligands. However, with nickel(II) and copper(II) which form labile metal ion amines, polyazaacyclic ligands were obtained as unexpected products. Generally, most of the ligands of this class have been obtained when limited amounts of formaldehyde and capping agent, like nitroethane are employed, and the reaction can be intercepted at the "half-built" stage. Comba *et al.*[218] have isolated **4.2** as Cu(II) complex when $[Cu(en)]^{2+}$ reacted with formaldehyde and nitroethane.

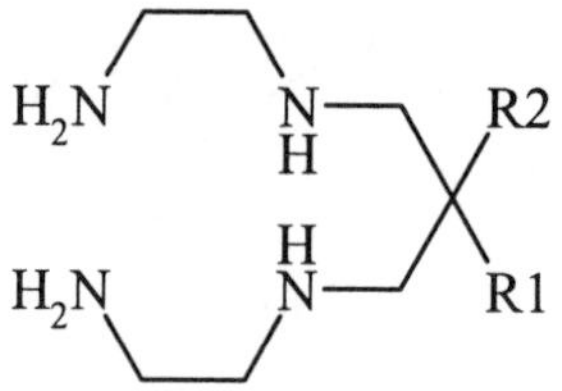

	R_1	R_2
4.2	NO$_2$	CH$_3$
4.3	NH$_2$	CH$_3$
4.4	CO$_2$Me	H
4.5	H	H
4.6	CO$_2$Et	CO$_2$Et

The nitro group can be easily reduced to the corresponding amine resulting in a copper(II) complex of **4.3**. When diethyl malonate is used as a capping agent, new amino acid ester complexes $[Cu(\mathbf{4.4})](ClO_4)_2$ and $[Cu(\mathbf{4.6})](ClO_4)_2$ were isolated from basic methanol solution. The formation of $[Cu(\mathbf{4.4})](ClO_4)_2$ implicates a dicarboxylate intermediate, **4.6**, which following decarboxylation, is esterified by the solvent. No coordination of the pendant groups has been noticed neither for amino- nor carboxilato-pendant ligands.

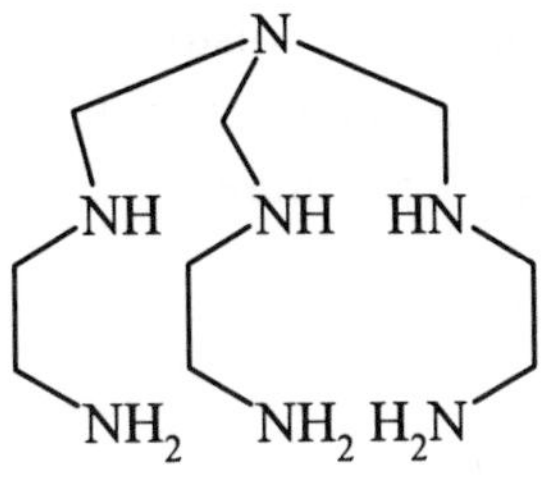

4.7

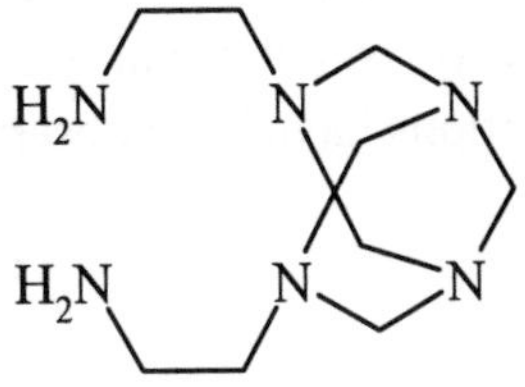

4.8

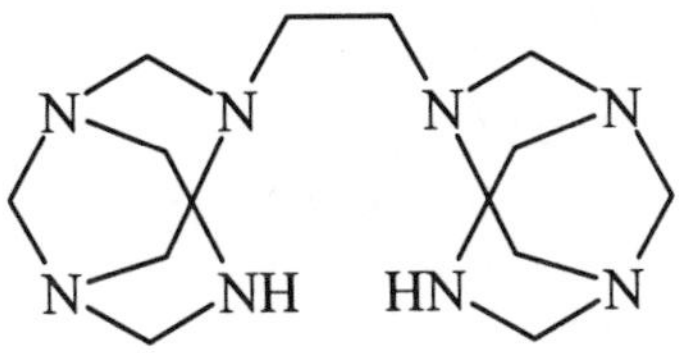

4.9

The hexadentate tripodal ligand **4.7** [tris(((aminoethyl)-amino)methyl)-amine (trivial name semispulchrate (semisep))] is formed with a good yield as Ni(II) or Cu(II) complexes by reaction of the corresponding $[(Me(en)_3]^{2+}$ formaldehyde and ammonia.[219,220,221] As minor products, **4.8** and **4.9** are found as Ni(II) or Cu(II) complexes, respectively. The ligands are formed by linking two respectively, one ethanediamine residues by a "cap" derived from five formaldehyde and two ammonia residues, and is structurally related to the hexamethylenetetraamine cage.[222] The ligands **4.8** and **4.9** are examples of sterically hindered linear polyamine and their complexes generally undergo relatively slow hydrolysis reaction, which involves attack on the bicyclic cap.

The above mentioned condensations may lead to a multitude of cyclic or acyclic products due to the similarity of the amino groups. The number of possible products is limited when the donor nitrogen is involved in a heterocycle. Comba *et al.* [223] have obtained in essentially quantitative yield the $[Cu(4.10)]^{2+}$ cation (**4.10** = N,N'-bis(2-pyridylmethylene)-1,3-diamino-2-methyl-2-nitro propane) using *cis*-bis(2-aminomethylpyridine)-copper(II) as precursor in basic methanol. The copper(II) is essentially four-coordinated and a small tetrahedral distortion from square-planar geometry, induced by enforced planarity of the coordinated pyridine ring, was observed. The distortion from planarity is not usual for a CuN$_4$ staturated chromophore but similar distortions have been observed in blue Cu proteins, in nature.

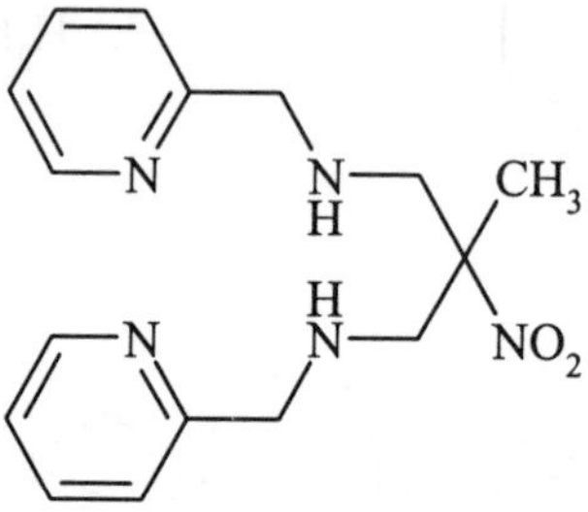

4.10

4.2.2. NO donor ligands

Reaction of the copper(II)-ethanolamine complex with formaldehyde and nitroethane leads to a noncyclic molecule **4.11** as ligand with N$_2$O$_2$ donor set and subsequently, to their amino derivative.[224] As for the precursor, the deprotonation of one alcohol group occurs providing strong hydrogen-bonding interactions between two units to form a pseudo-macrocycle **4.13**.

The Mannich type reaction in which an aminoacid acts as substrate was carried out and it was extended to amino-acid complexes of Cu(II),[225, 226, 227, 228] Ni(II)[229] and Co(III).[12] Despite the large amount of work, information concerning the effect of the nature of the metal ion, of its coordination sphere, the pH, etc. is rather lacking.

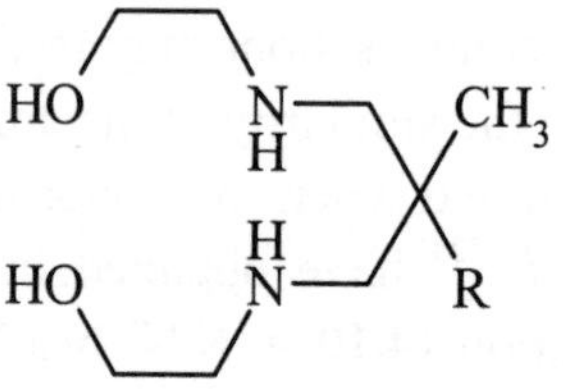

	R$_1$
4.11	NO$_2$
4.12	NH$_2$

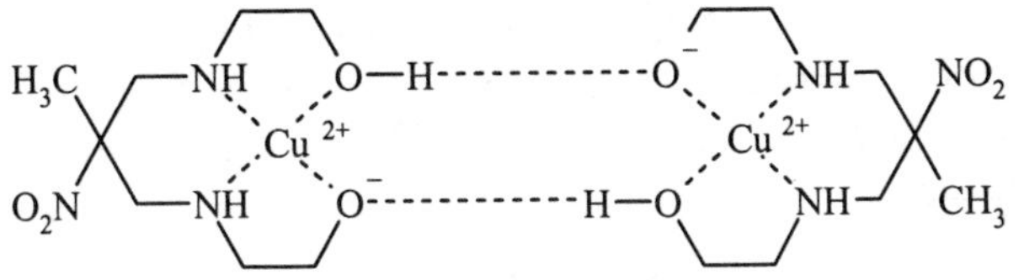

4.13

It is generally accepted that, in the base-catalyzed reaction, both the nitrogen and α-carbon atoms of the chelated amino acids are attacked by formaldehyde. In contrast, in the absence of base, formaldehyde reacts only with the amino-nitrogen and the α-carbon atoms remain unaffected. This is found in the case of *trans*-Ni(gly)$_2$·2H$_2$O. In the absence of base, the product is an N-substituted metaformaldehyde ring arising from the condensation of formaldehyde at the nitrogen atom of glycine chelate. When the pH was adjusted to 8.5 the two glycine rings were bridged by a pentamethylene diamine grouping akin to the structure of hexametylene tetramine. At intermediate pH values, a mixture of the two compounds is obtained. The two glycine rings in the resultant complex are *cis* with respect to each other. This implies that they must have undergone a "rearrangement" from the initial *trans*- Ni(gly)$_2$·2H$_2$O. The copper(II) directed condensation of aminoacids with formaldehyde and nitroethane in basic methanol produces open-chain tetradentate ligands with a pendant nitro substituent[230] after an isomerisation process in which the two amino groups are brought in adjacent position (Scheme 4.4). The capping reaction leads to a chiral tetradentate ligand, which coordinates to the four equatorial sites of a copper(II) atom producing in each case a planar geometry slightly distorted in a tetrahedral fashion and a new six-membered chelate ring in a chair conformation is formed. The condensation is very selective.

Although the template reaction with D/L-alanine might lead to three possible isomers only one is formed, [Cu(S,S-mnpala)]$^{2+}$, where mnpala = 2,5,8-trimethyl-5-nitro-3,7-diazanonanedioate, **4.14**.

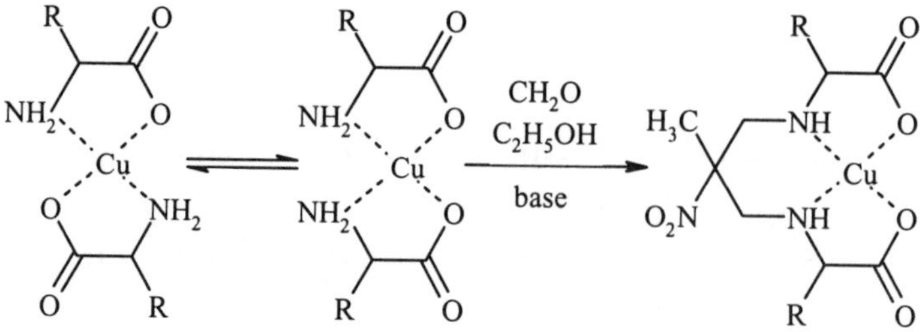

Scheme 4.4

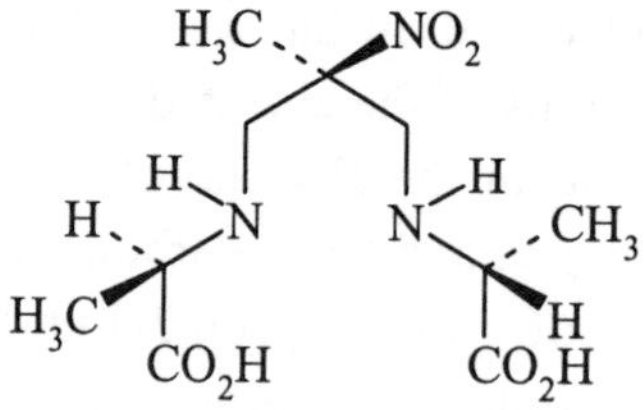

4.14

The obtained product proves that the attack of the nitroethane anion clearly is at the coordinated imines and it is side-selective since the products are different from those of a similar reaction in the absence of copper(II). The explanation for the side-selectivity is coordination of nitroethane to the copper(II) ion at least in an intermediate or transition state, although nitroethane is often used as an "innocent" solvent. The copper(II) ion can be removed and free ligands can be obtained. The free ligands are stable enough as solids in the absence of moisture.

4.2.3. SN donor ligands

New tripodal N_3S_3 ligands **4.15** and **4.16** with terminal thiolates have been obtained using $[Co(H_2N-H_2C-H_2C-S-)_3]$ as precursor and nitromethane or ammonia as padlocks.[17] In order to protect the six sulfur donors, allowing the amine groups to react with the capping reactants, the precursor was coordinated to a Co^{3+} ion resulting in a symmetrical trinuclear cobalt(III) complex $\{[Co(H_2N-H_2C-H_2C-S-)_3]_2Co\}^{3+}$. Two conformers, *meso* and *rac* of the trinuclear precursor have been used and the conformation has been retained after capping.

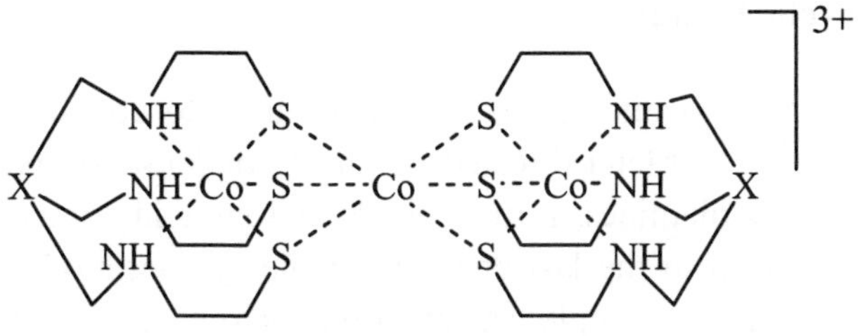

	X
4.15	CNO_2
4.16	N

The ligand field created around the cobalt ions does not seem to be affected by the capping groups. The electronic spectra of the **4.15** and **4.16** as well as those of the trinuclear precursor show two bands at 490 and 550 nm assigned to the two terminal chromophors $Co^{III}N_3S_3$ and central $Co^{III}S_6$. However, the electrochemical behaviour of the capped complexes depends on the nature of the caps. Thus, although the $E_{1/2}$ values due to the central $Co^{III/II}$ couple of the corresponding *meso* and *rac* pair are very similar, these are almost 70 mV more positive than for the uncapped analogue, indicating stabilization of the Co^{II} oxidation state relative to Co^{III} upon capping. There is a small difference between the corresponding values of **4.15** and **4.16**, which may be explained by the electronic-withdrawing nature of the nitromethyl group.

The nickel(II) complex containing unsaturated ligands **4.17** and **4.18** has been obtained only with R = CH_3 or H, respectively. Structural analysis shows that the side chain in **4.18** is not involved in the coordination.[231]

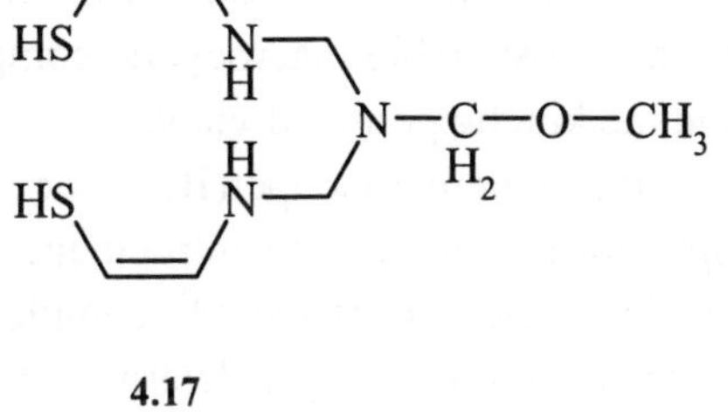

4.17 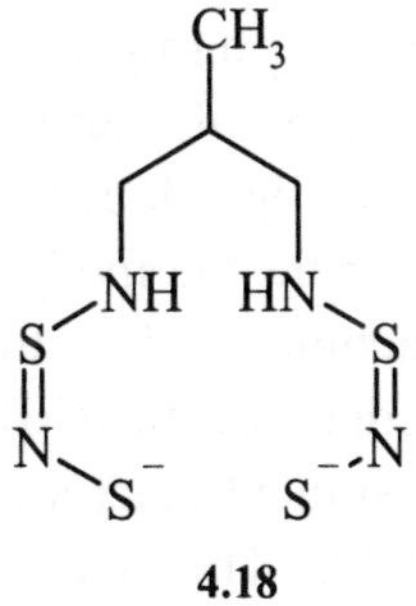   **4.18**

4.3. Monocyclic Ligands

Metal template syntheses has been often used to obtain macrocyclic ligands.[232, 233, 234, 235] Metal directed Mannich condensation between *cis*-disposed primary diamines, formaldehyde and an acid component, AH_2, has been employed in the last decades for the synthesis of macrocyclic polyamine ligands in a simple way avoiding the organic routes which are - as a rule - a multistep task. The macrocycles thus obtained are 13- to 16-membered ring ligands with four nitrogen donors situated in a plane although they can contain five or six nitrogen atoms in the ring. The AH_2 fragment should be a diprotic acid, the acidic hydrogen atoms being bound to a *carbon* atom, e.g. nitroethane, or diethylmalonate, or to a *nitrogen atom* , e.g., a primary amine or amide and, it acts as molecular locking fragment, closing a macrocycle around a labile metal centre with the formation of new six-membered chelate rings. General factors influencing a metal-ion-controlled synthesis[236] when the macrocyclic ligand is produced are also relevant in a Mannich template reaction. Especially the relationship between the size of the metal ion and the size of the cavity and the stability of the precursors should be considered.

Mannich-type reactions, which lead to macrocyclic ligands are generally directed by copper(II) and nickel(II) ions. Both metal ions can form relatively thermodynamically stable square - planar complexes and their metal-nitrogen bond length correspond with those predicted by the sterical approach.

4.3.1. Tetraaza macrocycles

Using metal directed Mannich reactions, macrocycles with various number of nitrogen atoms have been obtained. Among polyaza macrocycles of varying denticity and ring size, the 14-membererd tetraamine ligand, cyclam, forms the most stable macrocyclic complexes from both the thermodynamic and the kinetic points of view.

Reaction of the bis-(ethanediamine)copper(II) or bis-(ethanediamine)nickel(II) cations, formaldehyde and nitroethane yields the substituted macrocycle **4.19** (dinitrociclam) as the M^{II} complexes in high yield.[18] The succes of the condensation is strongly determined by the stability of the amine-metal complex precursor. In this respect, this type of reaction was extended to palladium(II),[237] gold(III)[238] and platinum(II)[239] which form relatively thermodinamically stable square-planar complexes.

Attempts to replace the diamine have been made. Thus, replacing ethane-1,2-diamine for propane-1,3-diamine or butane-1,4-diamine causes the failure of the condensation reaction in the case of copper(II) and nickel(II)[240] whereas with palladium(II) precursors for 16- and 18-membered macrocycle have been obtained. The greather kinetic and thermodynamic stability of the palladium(II) precursors has been considered the principal explanation for the success of condensation reaction leading to **4.25**.

Macrocycles have been obtained which may occupy the four equatorial sites of an octahedral metal ion and, via pendant coordinating functional groups, bind one or both of the axial positions. This subject has been widely reviewed.[241]

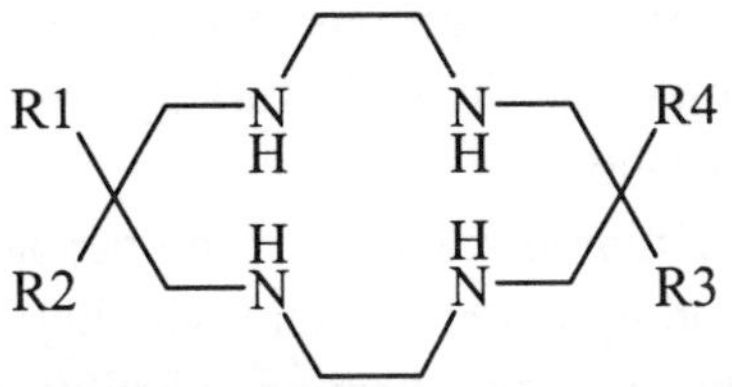

4.19	$R_1, R_3 = NO_2$	$R_2, R_4 = CH_3$
4.20	$R_1, R_3 = NH_2$	$R_2, R_4 = CH_3$
4.21	$R_1, R_2 = N(CH_3)_2$	$R_3, R_4 = CH_3$
4.22	$R_1, R_2 = -CO_2Et$	$R_3, R_4 = H$
4.23	$R_1 = NH_2 \ R_4 = CH_3$	$R_2 = H \ R_3 = CO_2H$
4.24	$R_1, R_3 = NH_3$	$R_2, R_4 = CH_3$

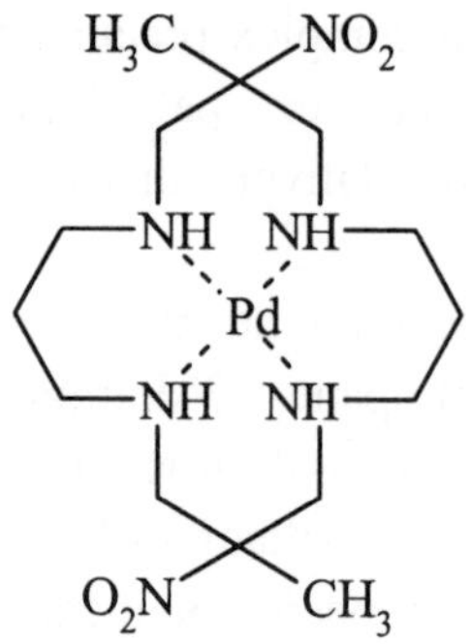

$$H_3C \diagdown \diagup NO_2$$

4.25

Metal directed Mannich condensation is a simple way to obtain ligands with C- or N-pendant arm groups when suitable caping molecules are utilized.

Nitroalkane carbon acids are the best investigated caping agents which lead to $-NH-CH_2C(CH_3)(NO_2)CH_2-NH-$ links between *cis*-disposed primary amines.[242] The formed nitro-substituted macrocycle is a suitable starting material for various derivatizations. The redox and spectral properties may be "tuned" varying substituents by various reactions of the nitro group.

One of the most common reaction of the pendant nitro group is the reduction with zinc and hydrochloric acid, yielding macrocycles with C-pendant amines. In this way, the diamino substituted cyclam macrocycle 6,13-diamino-6,13-dimethyl-1,4,8,11-tetra-azacyclotetra-decane, **4.20** (diam) was obtained as Cu^{II} or Ni^{II} complexes. The ligand **4.20** can be released from the complex ions and complexed with a variety of metal ions, including copper(II), nickel(II),[46] cobalt(III),[243] iron(II)[244], iron(III)[245, 246] rhodium(III),[247] chrom(III),[248] platinum(II), palladium(II),[249] oxovanadium(IV),[250] zinc(II) and cadmium(II).[251, 252]

It was anticipated that two isomeric forms of diam should exist: namely a *trans* and a *cis* isomer where the pendant amines lie on the opposite side or the same side of the macrocyclic ring, respectively. However, the *trans* isomer was almost always reported. The existence of the *cis* isomer was only reported for a copper(II) complex on the basis of subtle differences in the infrared spectra. Bernhardt *et al.*[57] reported the chromatographic separation of *cis*-6,13-dimethyl-1,4,8,11-tetraazacyclotetradecane-6,13-diamine and used it as ligand coordinating with cadmium(II). Since no significant difference between the sterically bulky nitro group and a methyl group exist, the probability for the formation of a *cis* or *trans* isomer is equal. Then, the origin of the stereoselectivity in formation of the *trans* isomer was found in the original template synthesis of **4.19**, namely in the kinetic control of reaction: there is an intermediate step that favours the formation of the *trans* isomer of **4.19**. In a proposed mechanism, the critical step is the introduction of the second nitroethane residue into the mononitro acyclic complex. The orientation of the nitro group at the time that cyclization occurs with the formation of the second six-membered chelated ring determines whether the *trans* or *cis* isomer is formed. While the *trans* isomer is formed preferentially, it is obvious that the coordination of both nitro groups (necessarily in *trans* sites) takes place. Following a similar argument, the *cis* isomer is formed in the absence of axial coordination by at least one of the two nitro groups.

Depending on the degree of protonation of the pendant amino groups and the preferred coordination geometry of the metal ion, the ligand **4.20** may occupy four, five or six coordination sites. The pendant amine groups are uncoordinated for the copper(II)-diam complex. The Jahn-Teller tetragonal elongation commonly observed for this ion, leads to solids where coordination of pendant primary amine groups are energetically disfavoured compared with arrangements in which these groups are engaged in other types of lattice interactions. Octahedral complexes of nickel(II), cobalt(III) and rhodium(III) with diam as hexadentate has been reported. The coordination of pendant amino groups is possible only when the six membered chelate rings carrying the amino groups are folded back and adopt a boat conformation. The coordination of the pendant primary amines in the axial sites of the coordination sphere enforces an equatorial metal-nitrogen compression which is reflected in electronic maxima shifted to higher energy and in the $M^{III/II}$ redox couples.[253] Cobalt(III) complexes with pentacoordinate diam (diam-N_5 or diamH-N_5) or with planar tetracoordinate (diamH$_2$-N_4) have been reported.[254, 255]

Intramolecular condensation[256] between the pendant amine and the macrocyclic N-atoms occurs with the formation of new ligands, **4.26** and **4.27**, for which the general cyclam framework is maintained but new metylene bridges appear. This condensation requires the involvement of the parent α-isomer of [Cu(**4.20**)]$^{2+}$, where the pendant amine and the two adjacent secondary amine hydrogen atoms are all *cis* with respect to each other as part of the same six-membered chelate ring.

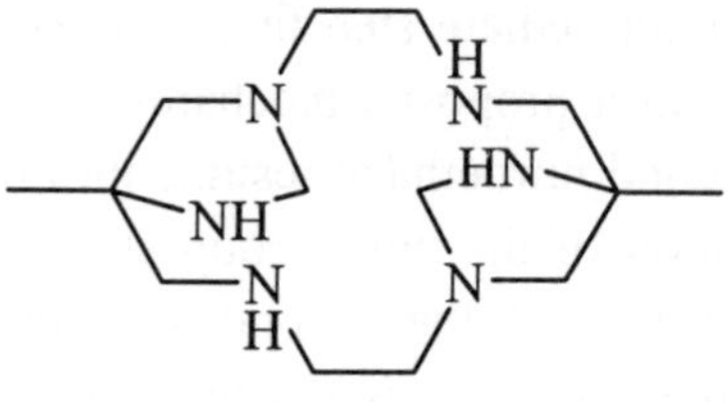

4.26

The structural distortions generated by the formation of the new bridges are partly reflected in the ligand field strength which decreases in the order α-[Cu(**4.21**)]$^{2+}$ > [Cu(**4.27**)]$^{2+}$ > [Cu(**4.26**)]$^{2+}$ > [Cu(**4.20**)]$^{2+}$. Different electrochemical behaviour was observed for [Cu(**4.26**)]$^{2+}$ compared with [Cu(**4.27**)]$^{2+}$ and with the parent α- and β-[Cu(**4.20**)]$^{2+}$ complexes. Thus, [Cu(**4.26**)]$^{2+}$ displays a reversible Cu$^{II/I}$ couple.

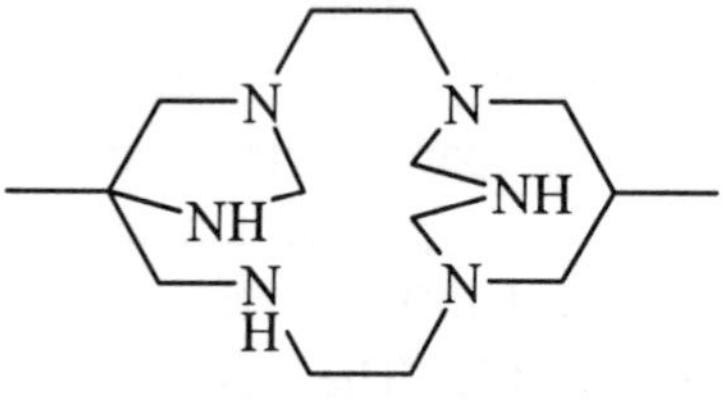

4.27

Partially these differences have been explained in structural terms. Due to the in-plane coordination of the Cu atom with respect to its four N-donors, the parent complexes lead to rapid dissociation upon reduction to monovalent state. In contrast, the copper(II) complexes with methylene bridged ligands where the metal is displaced from the N$_4$ plane (this is typical) lead to reversible Cu$^{II/I}$ couple, regardless of the type of N-donor, being secondary or tertiary. Demetallation of [Cu(**4.26**)]$^{2+}$ and [Cu(**4.27**)]$^{2+}$ leads to intermediate decomposition of the aminal groups - a fact well known in organic chemistry. Thus the aminal groups are unstable unless they are metal-coordinated.

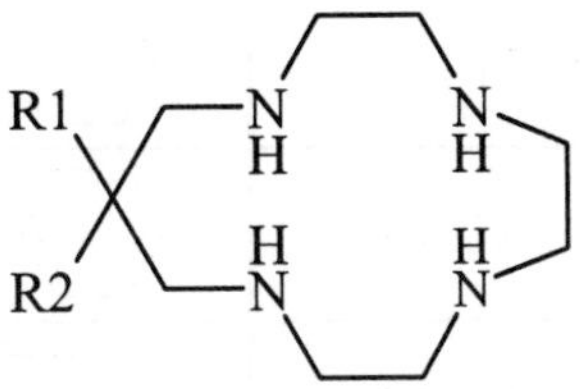

	R_1	R_2
4.28	NO_2	CH_3
4.29	CO_2Et	CO_2Et
4.30	CO_2H	H

Condensation of Ni(II)- or Cu(II)- complexes of linear aliphatic tetraamines with formaldehyde and a nitroalkane yields saturated tetraaza- macromonocycles of 13- to 16-membered monocapped rings, **4.28 – 4.47**, with one pendant arm.[61, 257, 258] The complexes are far less susceptible to dissociation in aqueous acid than the non - cyclic precursors. The carbon side-chain may be varied by alternating the nitroalkane.

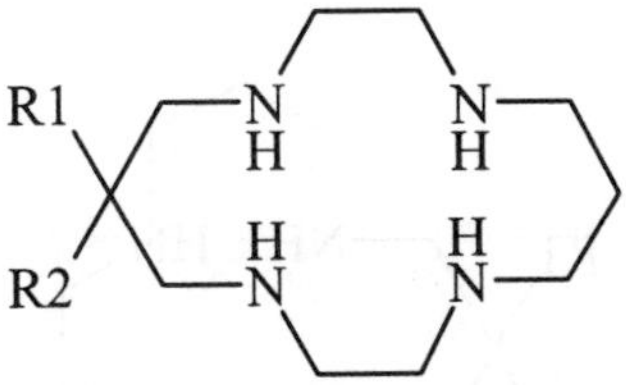

	R_1	R_2
4.31	NO_2	CH_3
4.32	NH_2	CH_3
4.33	CO_2Et	CO_2Et
4.34	CO_2H	H
4.35	CO_2Et	CO_2Et
4.36	NO_2	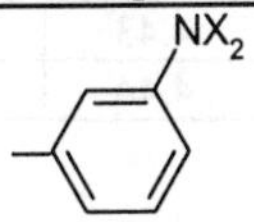
4.37	NH_2	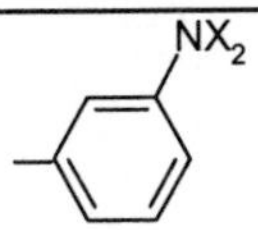

	R_1	R_2
4.38	$R_1 = NO_2$	⬡—NO_2
4.39	$R_1 = NH_2$	⬡—NO_2
4.40	$R_1 = NO_2$	⬡
4.41	$R_1 = NH_2$	⬡

It has been found that the nature of the R group in RCH_2NO_2 has no major influence on the success of the condensation reaction. Lawrance *et al.*[259] performed the condensation with a nitroaromatic padlock resulting in a macrocycle which bears both amino and aminobenzyl or benzyl group attached to the carbon of the macrocycle, **4.36** and **4.37**. Such carbon-substituted macrocycles may be fixed to biological and/or medical interesting polymers (proteins).

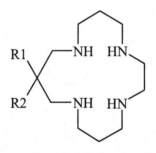

	R_1	R_2
4.42	NO_2	CH_3
4.43	CO_2Et	CO_2Et
4.44	CO_2H	H

R1, R2 substituted tetraazamacrocycle structure

	R_1	R_2
4.45	NO_2	CH_3
4.46	CO_2Et	CO_2Et

Condensation reactions of the copper(II)- or nickel(II)- diamine complexes with diethyl malonate and formaldehyde[260, 261, 262, 263, 264] resulted in the tetraazamacrocycle **4.22**, carrying pendant carboxylate groups attached to a carbon atom. The dicarboxylic acid cation, $[Cu(diacH_2)]^{2+}$, **4.47** can be formed and readily esterified in acidic methanol leading to isomeric 6,13-bis(metoxicarbonyl)substituted complex cations, which can further react with ammonia and amines resulting in 6,13-dicarbamoyl substituted macrocyclic cations. The yields of isolated complexes were lower compared with analogous reactions employing nitroalkanes. The pseudosquare-planar copper complexes exhibit strong ligand field parameters associated with the macrocycle and limited electrochemical reversibility of the $Cu^{II/I}$ couples associated with inadaptability of the rigid macrocycle to the preference of the copper(I) ion to a tetrahedral environment. No interaction of the pendant carboxylate group(s) with the copper(II) was observed although binding of this pendant to metal ions which prefer higher coordination numbers is possible.

The capping process of a diamino complex precursor can be performed in two steps each of them involving different padlocks.[265] Thus, the potentially hexadentate ligand hydrogen13-amino-13-methyl-1,4,8,11-tetraazacyclotetra-decane-6 carboxylate (**4.23**) has been obtained in two steps from $[Cu(en)_2]^{2+}$ as precursor. Diethylmalonate was used as capping agent in the first instance with the formation of the acyclic ligand **4.4**. The macrocycle was then closed with nitroethane. Compared with **4.20** which may exist in both *anti* and *syn* forms, but the *anti* isomer predominates, for **4.23** only the *anti* isomer was put into evidence. All the three ligands are able to act as hexadentate with both pendants coordinated in *trans* axial sites.

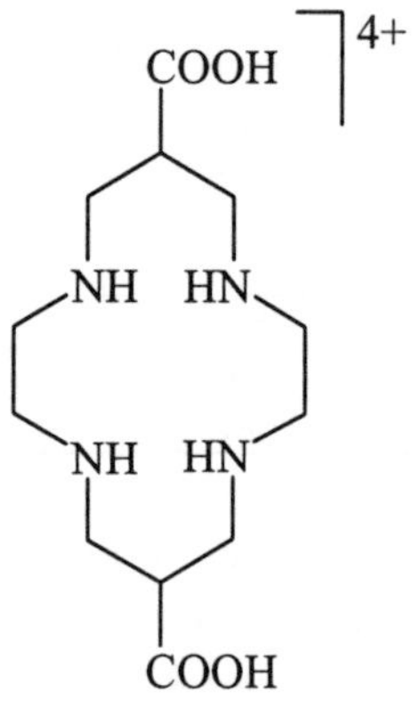

4.47

A series of rings (n = 13, 14 and 15) with either nitro or carboxylic acid groups as pendants were compared by forming the copper complex and observing the variation in $Cu^{II/I}$ redox potential and spectral properties. The obtained values of $Cu^{II/I}$ redox couples indicate that 14-membered rings are the most appropriate to incorporate copper(II) ions. The maxima in the electronic spectra remained unchanged, suggesting that the pendant groups attached to the ring do not have any significant influence.

4.3.2. Pentaaza macrocycles

A wide large of functionalized *azacyclam* complexes have been prepared by a Mannich template reaction using 1,9-diamino-3,7-diazanonane-nickel(II) or -copper(II) as precursors,[266] **4.48** – **4.55**. Primary alkylamine, amides $RCONH_2$, or sulphonamides, RSO_2NH_2, as padlocks have been used to close an open-chain tetramine around the metal ion.

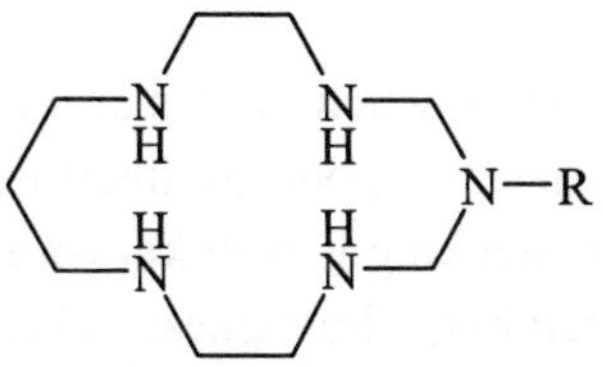

	R
4.48	CH_3
4.49	COH
4.50	$COCH_3$
4.51	SO_2CH_3
4.52	COC_6H_5
4.53	$SO_2C_6H_5$
4.54	$SO_2C_6H_4Me\text{-}p$
4.55	$SO_2C_6H_4CO_2H\text{-}p$

Pyridine,[267] 1,10-phenantroline,[268] ferrocene,[269] benzamide or sulfonyl,[270] subunits have been appended to a metallocyclam when the amide of 4-pyridinecarboxylic(isonicotinamide), 2-(aminocarbonyl)-1,10-phenantroline, ferrocene-sulphon-amide and, sulfonamides were used as the acid. The fifth nitrogen atom introduced by template syntheses exerts a structural function and does not alter the donor properties of the macrocycle cyclam. Also, steric factors prevent the coordination to the template metal ion of the donor atoms belonging to pendant arm. Thus, these atoms can act as donors towards another metal centre, leading to a super-complex like **4.56**.

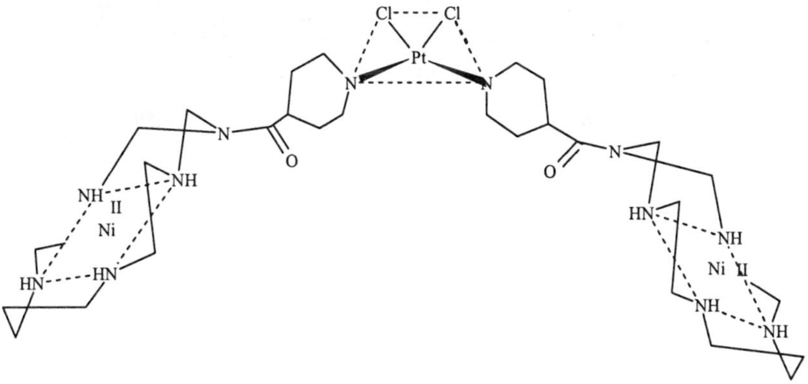

4.56

The synthesis of subunits which posses specific properties and which can be aggregated in a desired mode by intermolecular interactions like hydrogen bonding and π-stacking is well known for organic systems. The principles of self-assembly, host-guest chemistry and molecular recognition have been extended to the study of metal complexes containing ligands with potential hydrogen bonding groups on the periphery. Bernhard *et al*[271] have obtained ligands **4.57** – **4.59** which combine good ligating properties and hydrogen bonding sites utilising aminotriazines like melamine, ammeline and amelide as locking fragments. The charged molecules Cu(**4.57**)]$^{2+}$, [Cu(**4.58**)]$^{2+}$ and [Cu(**4.59**)]$^{2+}$ contain different hydrogen bond donor-acceptor moieties and represent in this way useful components for noncovalent self-assembly.

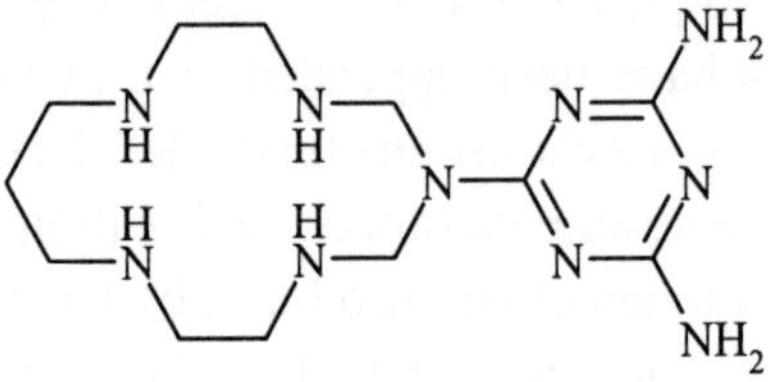

4.57

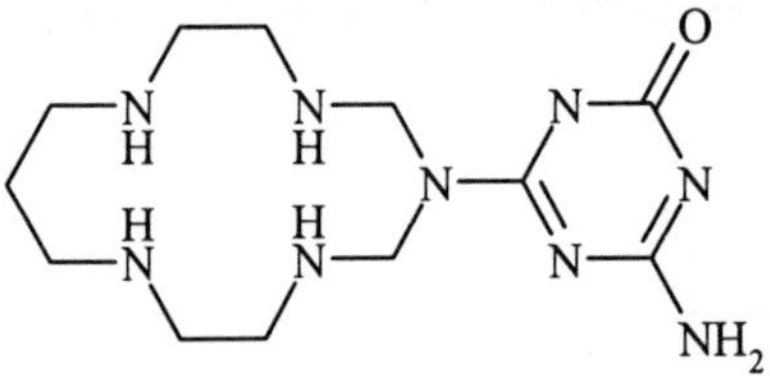

4.58

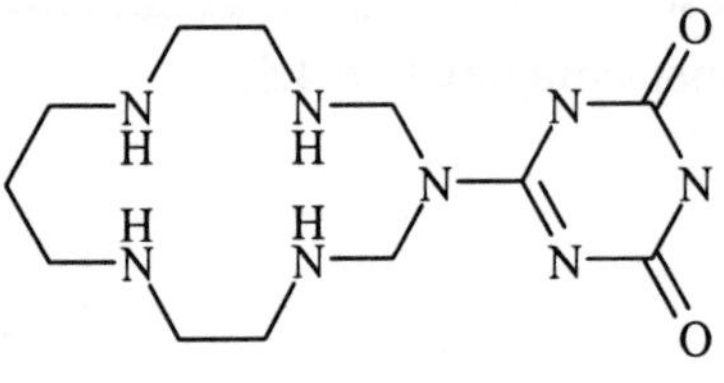

4.59

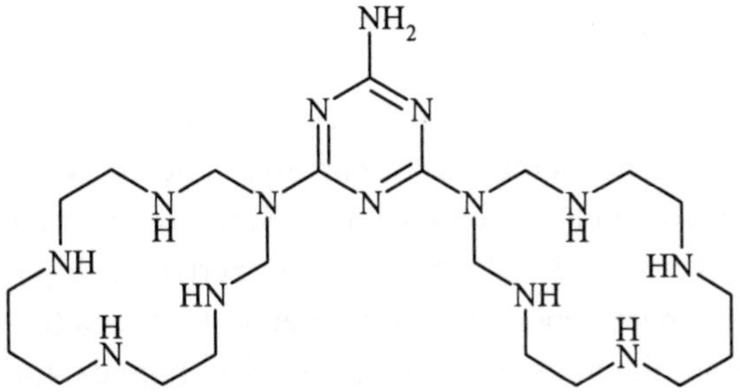

4.60

The presence of additionally three amine groups in the melamine make it appropriate to act as padlock for three macrocycle thus resulting in the bis- and tris-macrocyclic ligans, **4.60** and **4.61**, respectively, along with the monomacrocyclic one **4.57,** as copper(II) and nickel(II) complexes.[272] Both copper(II) and nickel(II) bis-macrocyclic complexes show only the *anti* conformations. For the trinuclear $[Cu_3(\textbf{4.61})]^{6+}$ complex the both possible conformations – *syn, syn* or *syn, anti* – were observed whereas the *anti* conformation was observed for $[Ni_3(\textbf{4.61})]^{6+}$. The structures of these complexes show that melamine fragments form π-stacking pairs.

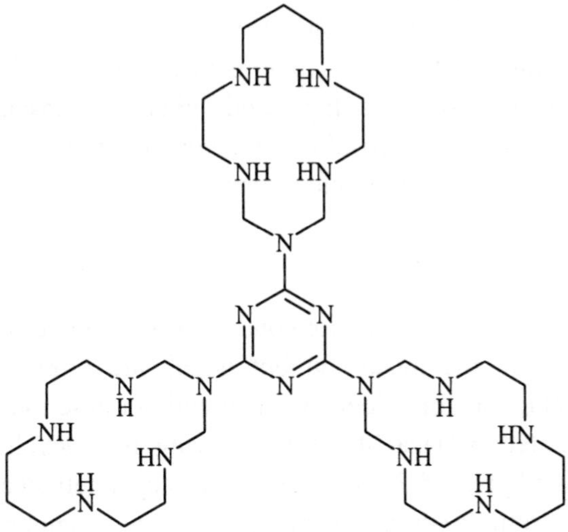

4.61

4.3.3. Hexaaza macrocycles

Goedken and Peng[273] were the first report the synthesis of a 14-membered hexaazamacrocycle, namely, hexaaza-cyclotetradecane coordinated in a nickel(II) complex ion. Since then, the preparation of hexaazamacrocycles **4.63** and **4.64** using copper(II), gold(I) and nickel(II) as template have been reported.[274] The attempt to obtain the complexes of **4.62** failed. Also, the attempt to obtain the free ligands **4.63** and **4.64** by treating the complexes with NaCN, H_2S or strong acid was not successful. Both can be described to the instability of methylenediamine linkages containing uncoordinated secondary nitrogens. These synthesis have been extended [275, 276, 277, 278] to the preparation of 14-membered hexaaza macrocycles which contain pendant functional groups, **4.65**, **4.66**, **4.67** as nickel(II) complexes. The hydroxyl pendant groups do not coordinate intra- or intermolecularly with Ni(II) ions, but, in contrary to the abnormal absence of the reactivity toward electrophiles of uncoordinated OH group in a number of chelate complexes, they react with acetic anhydride. With the appropriate chain length, n=3, the nitrile pendant groups coordinate intermolecular Ni(II) ions in the solid state, forming a coordination polymer. The complexes are oxidized with $(NH_4)_2S_2O_8$, HNO_3, or $FeCl_3 \cdot 6H_2O$ to Ni(III) complexes and isolated as six coordinated species $[Ni(L)X_2]ClO_4$ where X stands for is Br⁻, Cl⁻ or NO_3^- and L = **4.65, 4.66**. This is in agreement with the observation that macrocyclic ligands stabilize high and otherwise unstable oxidation states.[279] The pendant groups allow attachment of the complexes to a solid supports or incorporation into polymer systems.

4.3.4. Octaaza macrocycles

Goedken *et al.*[280,281] utilized for condensation amine groups which are not coordinated to the metal ion but which are more susceptible to nucleophilic attack at formaldehyde. The authors describe the synthesis of the unsaturated octaaza-14-membered ring macrocyclic complexes, 2,3,9,10-tetramethyl-1,4,5,7,8,11,12,14-octaazacyclotetradeca-1,3,8,10-tetraene **4.68** from 2,3-butanedione dihydrazone complexes of Ni(II) Cu(II) or Co(II) precursors. The success in obtaining these complexes is associated with the compatibility of the standard C=N, N-N and C-N distances with the center-to-nitrogen distances of approximately 1.90-1.92 Å. While this distance is compatible with observed metal-nitrogen distances for Cu(II) complexes and for low-spin Fe(II), Co(II) and Ni(II), it is significantly shorter than observed for most Zn(II) and Mn(II) distances.

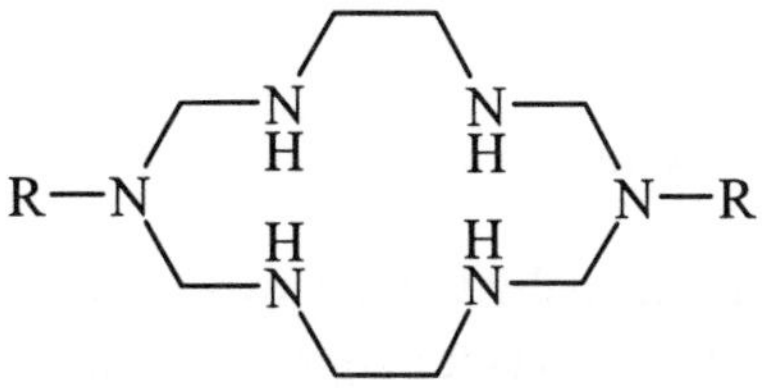

	R
4.62	H
4.63	CH_3
4.64	C_2H_5
4.65	$(CH_2)_n$-OH n=1-3
4.66	$(CH_2)_n$-CN n=1-2
4.67	$(CH_2)_2$-NH$_2$

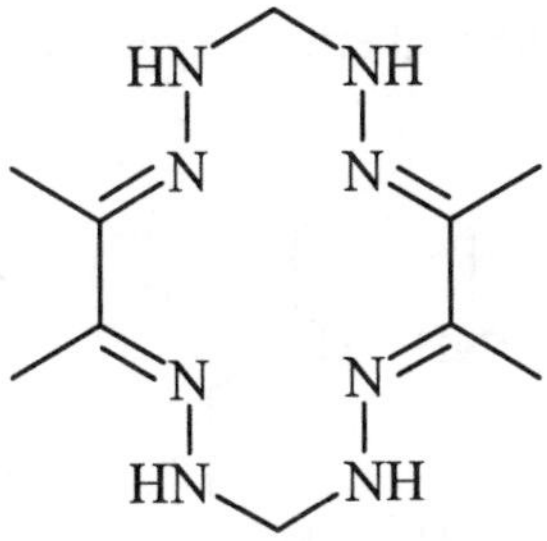

4.68

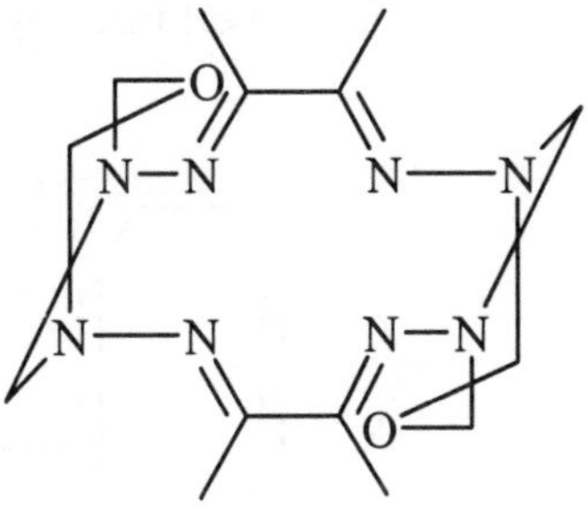

4.69

In the last two cases, the complete cyclization failed due to the large radius associated with d^{10} and, d^5 electron system. The unsaturated macrocycles are strong-field ligands and promote the spin pairing whenever possible. These macrocyclic ligand complexes have lower stability in strong acidic media compared with their tetraaza counterparts and in the presence of amine bases such as pyridine and triethylamine, N-H deprotonation have been observed. The complexes are able to undergo further condensation reaction with formaldehyde resulting in a tricyclic macro-ring ligand, **4.69** with fused rings. The complexes with **4.69** are remarkably inert compared to their 14-membered precursors. Also, the configuration of the nitrogen atoms becomes rigidly fixed upon formation of the tricyclic structures, reinforcing the sp³ character of these atoms and preventing conjugation of the lone pairs with the α-diimine functions.

4.3.5. Azaether macrocycle

Condensation of formaldehyde with two *cis*-disposed primary amine in (1,9,diamino-3,7-diaza-nonane)-copper(II) has been performed[282] under the formation of the saturated oxa macromonocycle, **4.70**. The reaction is controlled by the slight differences in the "bite" of the *cis*-disposed primary amines in the precursor tetraamines and by the preference of the copper ion for a 14-membered macrocycle. Thus the 14-membered macrocycle **4.70** has been obtained in high yield compared with 15- membered macrocycle **4.71**, whereas attempts to synthesize of 13- and 16-membered macrocycles have been not successful. Spectral properties of the oxa macrocyclic complexes show that the donor set is N₄ and that it is not perturbed by the non-coordinated ether oxygen atom.

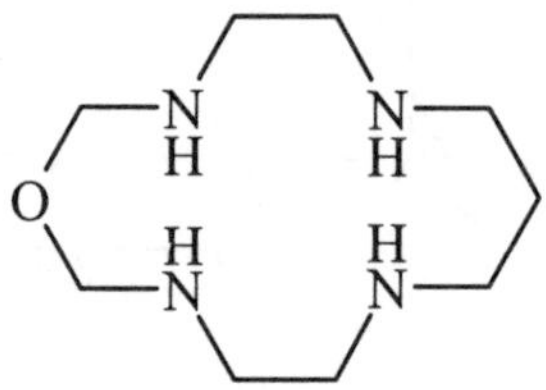

4.70 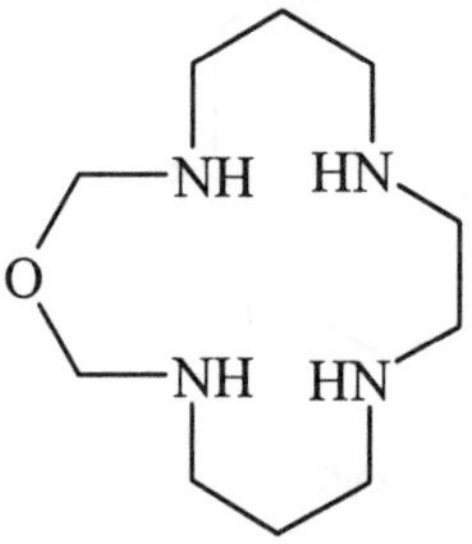　　　　　　　**4.71**

4.3.6. Azathioether macrocycle

The template route to saturated macrocycles utilized for polyamines has been extended for mixed donor azathioether macrocycles. Thus, (3,7-dithianonane-1,9-diamine)copper(II) ion with nitroethane and formaldehyde in basic methanol gives the macrocyclic complex ion (6-methyl-6-nitro-1,11-dithia-4,8-diazacyclotetradecane) copper(II), **4.72**.[283, 284, 285] The ring formation was apparently not inhibited by the presence of thioether groups opposite to the primary amines and, as the polyamine analogous, they show high stability comparing to the non-cyclic precursor.

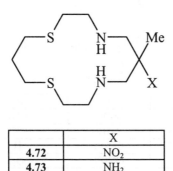

	X
4.72	NO_2
4.73	NH_2

The *cis*-N_2S_2 and N_4 macrocycles are very similar, but some differences arise from the steric hindrance, the sulfur donors being much more pyramidal than the nitrogen donors, so that the chair conformation of the six-membered ring is much more steeply pitched. Also, they are "softer" donors than secondary amine and exhibit a greater capacity to stabilize Cu(I) as a result of their greater π-acceptor ability. These differences are reflected in the electronic spectra and in the reduction potential for Cu(II)/Cu(I) couple which is about 300 mV more positive for the N_2S_2 complex compared to the N_4 complex. The zinc-acid reduction yields the corresponding amino macrocycle, **4.73** as copper(II) complex as well as the free ligand which was used to complex other metal ions, including cobalt(III).[286]

4.3.7. Reinforced macrocycles

The Mannich type condensation has been used as one approach to the structural reinforcement of the macrocycle. This type of macrocycle, first synthesised by Wainwright *et al.*[287,288] , has double bridges between at least one pair of adjacent nitrogen donors, giving piperazine (**4.74** –

4.78), homopiperazine (**4.79**) or 1,5-diazacyclooctane type bridges.[97] As a result, the flexibility of the macrocyclic ring is reduced and therefore, the number of geometric isomers is decreased.

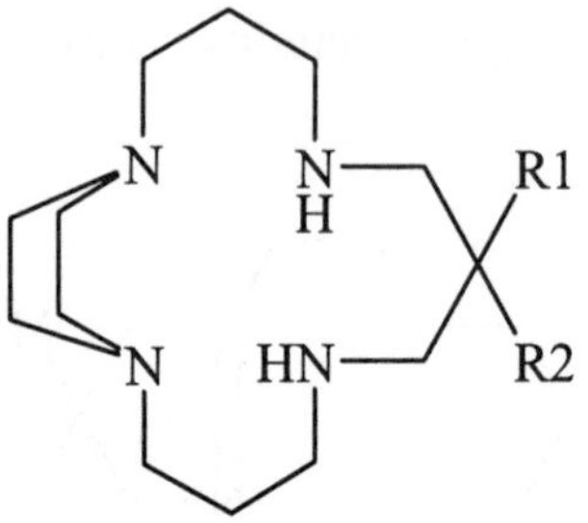

	R_1	R_2
4.74	NO_2	CH_3
4.75	NH_2	CH_3
4.76	CO_2Et	H
4.77	CO_2Et	CO_2H
4.78	CO_2H	H

The reinforced bridging groups have different requirements in terms of the Me-N bond length and N-Me-N bond angle which allow coodination with a minimum steric strain.

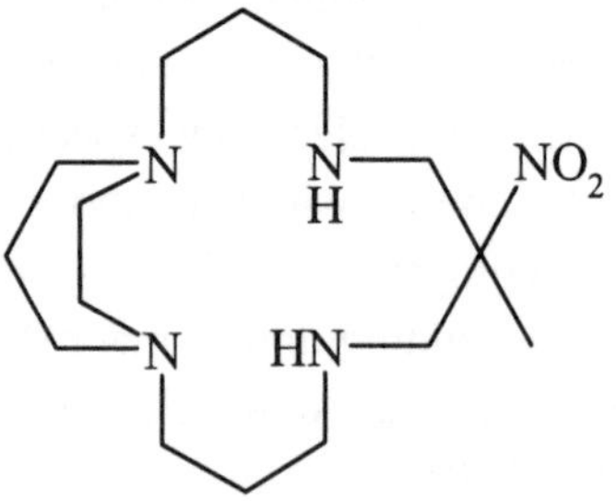

4.79

In this respect the piperazine type of bridge requires Me-N bonds longer, and an N-Me-N angle smaller than in any other metal ion complex which accounts partially for its poor coordinating abilities. The copper(II) complexes of **4.74** have been obtained starting from the Cu(II) complex of N,N'-bis(3-aminopropyl)-1,4-diazacyclohexane as the precursor with formaldehyde and nitroethane.

The X-ray crystal structure analysis of [Cu(**4.74**)(OH)$_2$] indicates square - based pyramidal cation geometry with the copper(II) situated 0.25 Å above the plane of the four nitrogens.[97] The Cu-N distances are slightly longer for the tertiary nitrogens and all the distances are greater than the average Cu-N distance of 2.037 Å found for [Cu(**4.45**)]$^{2+}$. This suggests that the extra strap in **4.74** has produced a slightly larger hole. The electronic spectra and Cu$^{II/I}$ couple for the complexes with **4.74** and **4.45** are very similar indicating that the increase of the ligand strength induced by the tertiary nitrogens is overcome by the increase in the hole size that would act to reduce the ligand field strength.

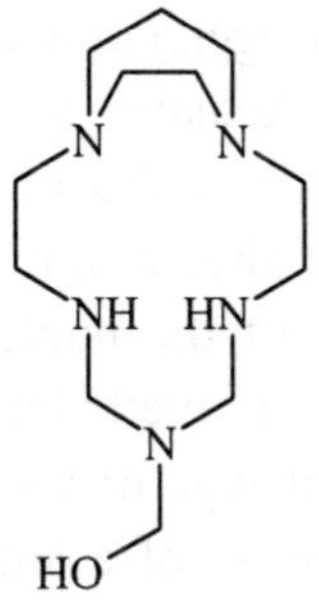

4.80

Hancock *et al.*[289] explained the long Cu-N bond length along with the metal position above the N$_4$ plane observed for Cu(II)- **4.79** complex by the steric requirements of coordination to the homopiperazine part of the macrocycle.

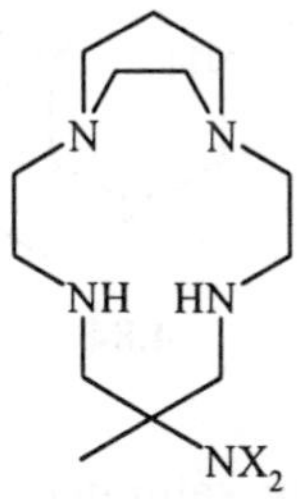

	X
4.81	O
4.82	H

Concerning the ligand field strength, two factors should be taken into account that affects it changing from unreinforced to reinforced macrocyclic ligands: the change of the nature of donor nitrogens from secondary to tertiary and, the decreasing flexibility of the macrocycle. With these factors in mind the very close values of the electronic maxima and $Cu^{II/I}$ couple for the complexes containing related unreinforced and reinforced ligand **4.79** can be explained.

Nickel(II) complexes with the reinforced macrocycles **4.80**, **4.81**, and **4.82** have been obtained by the same Mannich condensation strategy. The mean Ni-N bond lengths are very close to the strain-free Ni-N bond length of 1.91 Å and steric factors permit the accomodation of the Ni(II) ion in the plane of the four nitrogen donors. The result of this planarity is a high ligand field strenght.[290] Actually, for [Ni(**4.82**)]$^{2+}$ the ligand field is one of the highest known for a complex of low-spin Ni(II) with saturated nitrogen donor groups. It was already noticed that in the absence of the appropriate reactant, the role of the nucleophile may be played by an adjacent bound amine. A new methylenediamine linkage is formed together with a concomitant change of some of the secondary nitrogens into tertiary ones and the closing of a small ring, fused to the cyclam framework.

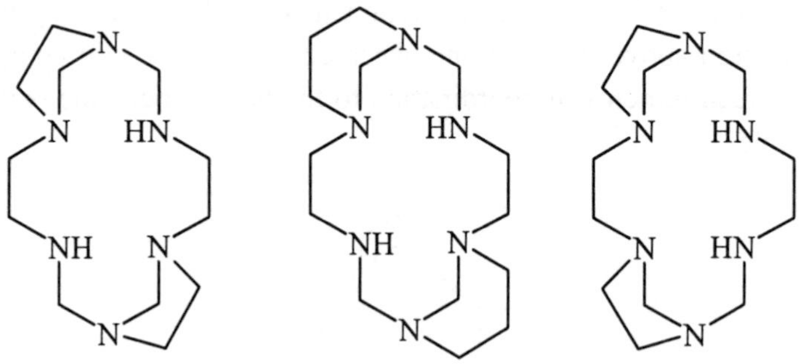

4.83   4.84   4.85

Complexes containing fully saturated hexaazamacrocyclic ligands - **4.83** - **4.86** - with small-ring moieties fused to the macrocycle have been obtained following   nickel(II) or copper(II) directed Mannich condensation of triamines or tetramines with formaldehyde and ethylenediamine.[291,292,293] Although there is no bridge between two donor nitrogens, the fused small-rings affect the metal-donor interactions and consequently, the spectroscopic and electrochemical properties of the complexes. Thus, the ligand field is much lower and electrochemical reductions of $Ni^{III/II}$ and $Ni^{II/I}$ are easier for the complexes involving six member fused ring in the macrocyclic ligand. Thus the Ni(II) complexes of **4.83** - **4.86** have been reduced and their structure has been resolved.[294,295]

The larger hexaazamacrocyclic ligand **4.87** has been synthesized[296] by the Ni(II) directed condensation of triethylenetetraamine with formaldehyde and ethylenediamine. The ligand contains a 1,3-diazacyclopentane ring fused to the macrocycle and acts as a pentadentate ligand closing a rarely observed four-membered chelate ring. These two particularities of the ligand induce sever distortion from the regular octahedral geometry of the metal ion environment.

Tertiary nitrogen donors have been obtained by capping the macrocycle with two additional triaza six-membered chelate rings (structures **4.88** and **4.89**). The nickel(II) complex of the macrohexacyclic ligand 1,3,6,8,10,12,15,17octaazahexacyclo-[15.6.1.1.^{3,17}1.^{6,10}1.$^{8.12}$.0^{18,23}] heptacosane (**4.88**) has been obtained.[297]

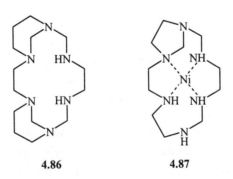

4.86   **4.87**

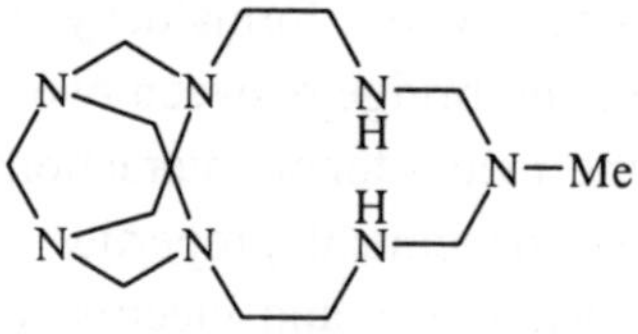

4.88 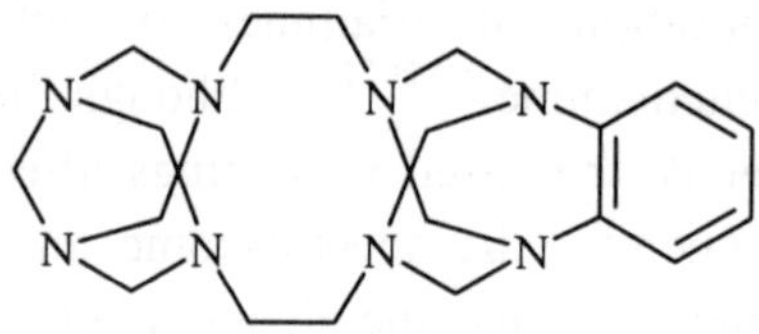   **4.89**

It consist of a 14-membered hexaaza macrocyclic framework to which two additional triaza six-membered chelate rings are fused to each nitrogen atom of the ethylenediamine moieties. These ligands exert a strong ligand field attributed to a constriction effect of the ligands towards the nickel(II) ion as a result of the abnormally short N-N bite distances of the cage ring moieties. The inflexibility does not allow the nickel(III) ion to coordinate in octahedral geometry and to accommodate in the hole. This explains the great difficulty in oxidizing complexes containing these ligands, stabilizing for example the nickel(I) oxidation state.

The reaction of copper(II) complex of 4,7-diazadecane-1,10-diamine with formaldehyde and 2-aminoethanol produced $[Cu(\textbf{4.90})]^{2+}$, in which the ether-oxygen is coordinated (this is a rare event) leading to the N_3O chromophore.[298]

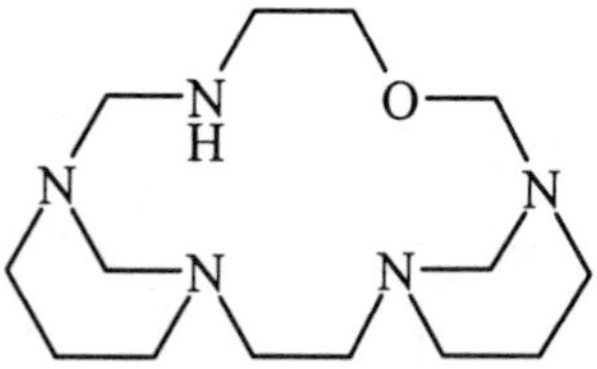

4.90

Two molecules of formaldehyde have been inserted between two nitrogen atoms situated at the end of a three carbon atoms chain closing a six memberd heterocyclic fused rings and producing a more rigid geometry. Structural analysis show that the copper ion lies in a plane of three nitrogens and the ether oxygen with the metal-oxygen distance of 2.008 Å very close to the Cu-N one.

4.3.8. Macromonocyclic dicompartmental ligands

Metal directed Mannich condensation is a simple route to obtain monomacrocyclic ligands which can accommodate two metal ions simultaneously in close proximity. The precursors are binuclear complexes of nickel(II), palladium(II) or copper(II) with 1,5-diaminopentan-3-ol or 1,5-diaminopentan-3-thiol. The copper(II) complex **4.91** has been obtained as the dicopper(II) complex by condensation of the bis(1,5-diaminopentan-3-ol)dicopper(II) ion with formaldehyde and nitroethane.[299]

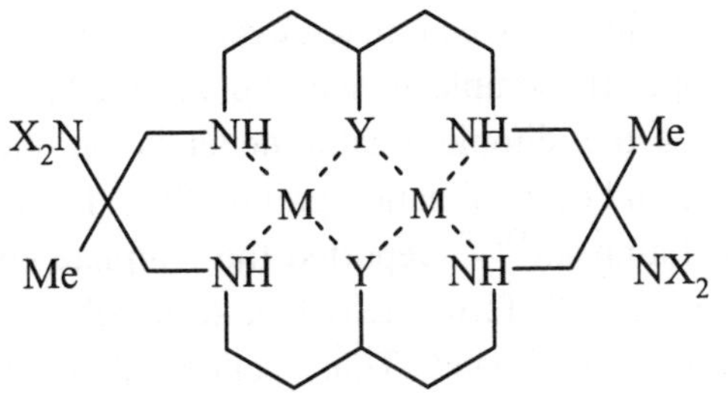

	X	Y	Me
4.91	O	O	Cu
4.92	H	O	Cu
4.93	O	S	Ni

The crystallographic analysis shows that the assembly $Cu_2N_4O_2$ is planar with the substituents nitro and methyl *anti* relative to the donor plane. The perchlorate salt of the macrocyclic complex shows an antiferromagnetic behaviour with a single triplet energy gap of -860 cm^{-1}. Reduction of the dinitro compound leads to the corresponding amino complex, **4.92**.

The condensation of bis(1,5-diaminopentan-3-thiolato)dinickel(II) with formaldehyde and nitroethane leads to **4.93** for which the *syn* isomer was formulated by crystalographic analysis.[300] A consequence, the two N_2CuS_2, planes are folded with an interplanar angle of 162.8°.

4.4. Isolated Dimacrocycles

Special attention should be paid to the molecules containing several spatially separated reactive sites. Such complexes mimic well multimetallic sites of metalloproteins, are models for polynuclear metalloenzymes and are of interest as multielectron reagents and catalysts. The simplest case that of ditopic compounds with two reactive centers is usually realized by binding two polyazamacrocycle by a multistep laborious process. Template condensation of coordinated polyamines with formaldehyde and an acid padlock has been extended to the elaboration of a simple and convenient method for preparation of bis(macrocyclic) structures. In this reaction the padlock should exhibit a double function, simultaneously participating both in cyclization and formation of a bridge between macrocyclic subunits. As precursors, nickel(II) and copper(II)-complexes with polyamine have been employed and the nature of the ditopic ligand depends on the nature of the coordinated amine and of the function group of the diacid.

Rosokha and Lampeka[301,302] reported the template condensation of a *cis*-disposed diamine with formaldehyde and an alkylamine to obtain a bis(pentaazamacrocyclic) ligand. Thus, condensation of $[ML]^{2+}$, M = Cu^{2+} or Ni^{2+}, L = 3,7-diazanonane 1,9-diamine, (danda), with formaldehyde and aliphatic diamines $H_2N(CH_2)_nNH_2$, n = 2-5, results in **4.94** as metal complexes. The success of the reaction was explained by the coordination mode of the tetraamines precursors. Thus, *danda* occupies the equatorial plane in the coordination sphere of metal ions, two *trans* - axial positions being vacant. Such coordination prevents bidentate binding of the diamine added and leads to a process in which each amino group of the diamine undergoes condensation in the coordination sphere of different metal ions. The ditopic complexes are very similar to each other and to the monocyclic analogues. No intramolecular interaction between the metal ions for all the compounds was observed except that of copper(II) with **4.94** for which n = 2. In this last case, the hyperfine coupling constant A_{II}, as well as the electrochemical behaviour indicate Cu(II)-Cu(II) interaction. Oxidation of the dinickel(II) complexes leads to six-coordinated low spin nickel(III) compounds. The similarity of the electronic spectra for bis- and mono-macrocyclic compounds demonstrate the close relationships between the structures of the nickel(III) coordination polyhedrons. As the monomacrocyclic ligands, the bis(macrocyclic) ligands stabilize the nickel high oxidation state. The bis(macrocyclic) complexes suffer a slow destruction in strong acids as a result of acid hydrolysis of NCH_2N bonds in the macrocyclic backbone, as they monomacrocyclic analogues do. The binuclear nickel(II) complexes of **4.95** and **4.96** have been obtained.[303]

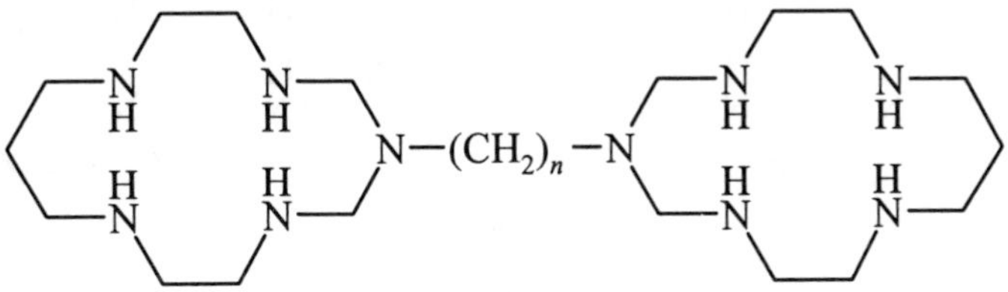

4.94

Likewise other bis(macrocyclic) ligands in which subunits are linked together by a $-(CH_2)_n-$ chain, the two chromophores NiN_4 behave independently. The presence of the fused 1,3-diazacyclohexane ring causes an increase of the Ni-N bond length and, as a consequence, a weaker ligand field strength comparing to that of **4.94**. Also, the alkylation of the coordinated nitrogens of the macrocycle alters its ability to favour the oxidation of nickel(II). Thus, the oxidation potential of $[Ni_2L](ClO_4)_4$ is +0.97 V when L = **4.95** and +1.08 V when L = **4.96**.

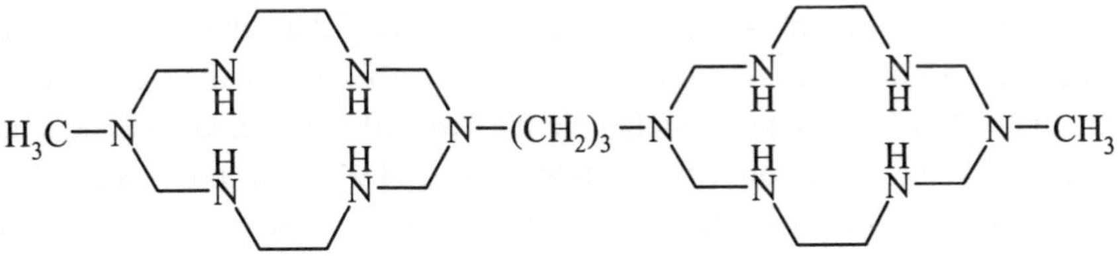

4.95

Utilizing the same reactive species, Suh *et al.*[304] have prepared a mononuclear nickel(II) complex by a simple one-step process of the face-to-face bis(monocyclic) ligand **4.97** in which two saturated 10-

membered tetraaza macrocyclic subunits are linked together by two ethylene chains. The bis(macrocyclic) molecule behaves as a hexadentate ligand through four secondary nitrogen atoms and two of the four tertiary nitrogen atoms; tertiary nitrogen atoms in the N-CH$_2$-N linkages are not involved in the coordination.

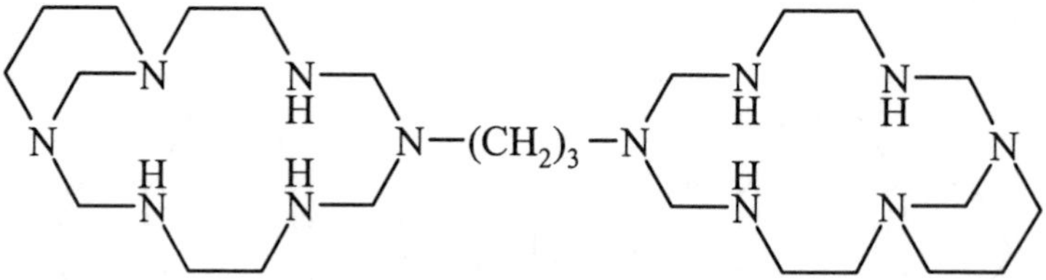

4.96

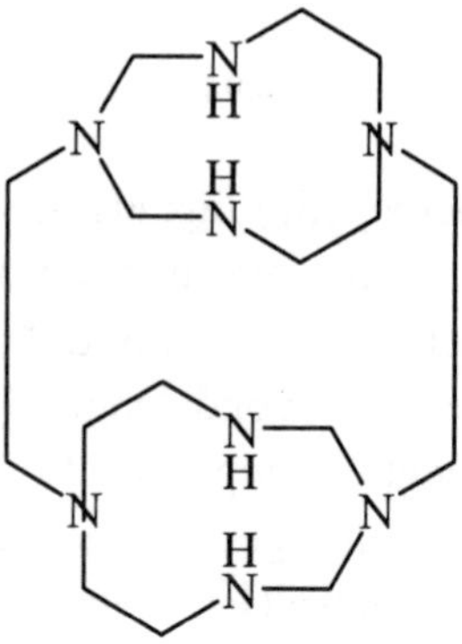

4.97

Structural analysis shows that the nitrogens involved in a N-CH$_2$-N linkage suffer a sp^2 - like hybridization and, that the coordination geometry around the nickel(II) ion is strongly affected by the steric factor of the -CH$_2$CH$_2$- chains which link the two 10-membered macrocyclic subunits. The isolation of the free ligand **4.97** was not successful, indicating the lability of the diaminomethane groups (R$_2$N-CH$_2$-NR$_2$) that are known to be unstable unless both of the nitrogen atoms are tertiary.

The nickel(II) complex of **4.8** - has been used[305] as precursor for the bis(macrocyclic) complexes of **4.98** - **4.100** in form of their perchlorate salts. The spectral studies supported by cyclic voltametry indicate the presence of square-planar Ni(II) and that the length of the bridging chain between the two macrocycles does not significantly affect the ligand field strength and that no metal-metal interactions exists. The oxidation potentials of Ni(II)/Ni(III) for **4.99** and **4.100** differ substantially from those of **4.98** and from a monomacrocyclic analog. The differences has been associated with the free rotation of the long flexible bridge chain of **4.99** and **4.100** which causes difficulty in formation of a six-coordinate Ni(III) species, a preferred coordination number for Ni(III). As in the case of monocyclic compounds, obtained using ammonia or a primary amine as capping agent, the presence of tertiary bridgehead nitrogen atoms in the macrocycle leads to destabilization of high oxidation states of metal ions.[306] The very oxygen and moisture sensitive Ni(I) complexes of **4.98** and **4.99** were prepared by reduction of corresponding Ni(II) complexes, showing that these ligands stabilize two Ni(I) ions.

	R
4.98	$-(CH_2)_2-$
4.99	$-(CH_2)_4-$
4.100	$-(CH_2)-p-(C_6H_4)-(CH_2)-$

The template synthesis of bis(pentaazamacrocyclic) ligands as described above is only feasible for 14-membered macrocyclic products. The formaldehyde - dinitroalkane template reaction is an efficient route for the preparation of bis(macrocyclic) octaamines with other ring size (between 13 to 16) and chain lengths of the alkyl bridge. The ligands thus obtained contain two pendant nitro groups. The reduction of pendant nitro groups to bis(macrocyclic)compounds with pendant amines is possible, but the process is accompanied by a demetallation so that the free ligand is obtained.

Comba *et al.*[307] synthesised the copper(II) complex **4.101**, of with n=2,3 or 4 and the corresponding reduced form **4.102**. Visible spectroscopic studies showed that the geometry of the two copper(II) ions sites are each tetragonal distorted octahedral with an almost planar CuN$_4$

arrangement. No coupling is observed in the EPR spectra denoting that the Cu...Cu separation is large for the solid complexes.

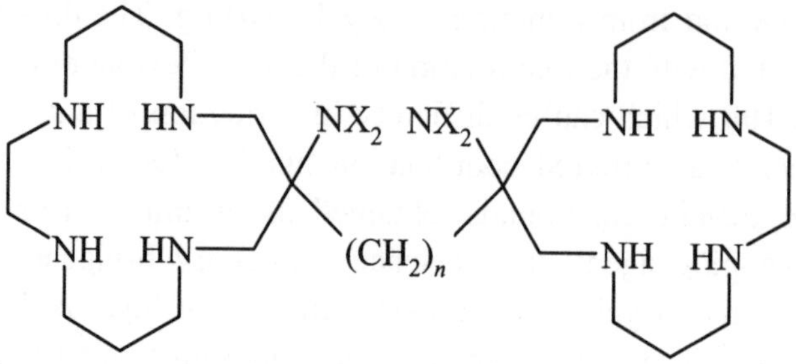

	X
4.101	O
4.102	H

The similar template reaction of (1,9-diamino-3,7-diazanonane) copper(II) with 1,4-dinitrobutane yielded the corresponding dicopper complex of the bis(14-membered) $(CH_2)_2$-linked bis(pendant nitro) bis-macrocyclic ligand **4.103**.

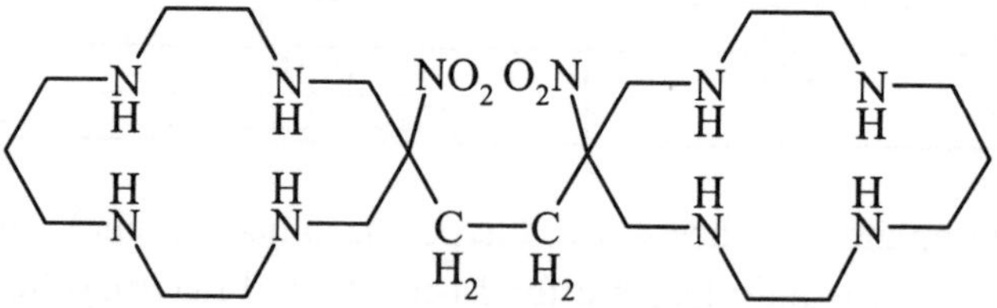

4.103

The dinickel(II) and dicopper(II) complexes, $[M_2(bispen)](ClO_4)_4$ of the bis(pentaazamacrocycle), 2,2-bis(1,3,5,8,12-pentaazacyclotetradec-3-yl)ethyl disulfide (abreviated as bispen) were prepared[308] by template condensation of N,N-bis(2-aminoethyl)-1,3-propanediamine, formaldehyde and 2-aminoethyl disulfide in the presence of the metal ion, according to Scheme 4.5.

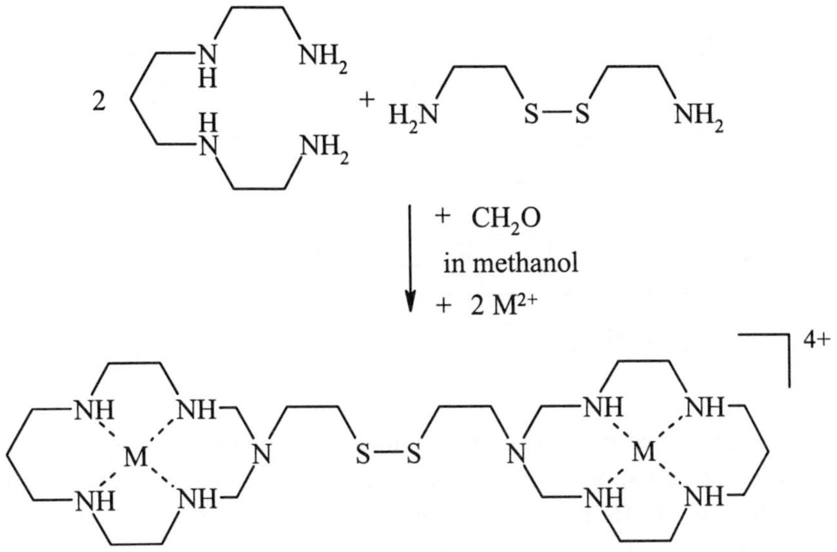

Scheme 4.5

The nickel(II) complexes are soluble in polar solvents in which an equilibrium between the yellow square-planar species and blue-octahedral ones exists. Both the complexes undergo chemical and/or electrochemical reduction in aqueous and acetonitrile solutions. Compared with the corresponding $[M(cyclam)]^{2+}$, the E^{0f} ($M^{3+/2+}$) values in both aqueous and non-aqueous solutions are slightly positive, indicative for a more difficult oxidation to M^{III} and the E^{0f} ($M^{2+/+}$) values are slightly less negative denoting a more easier reduction to M^{I}. This behaviour was attributed to the presence of the uncoordinated tertiary nitrogen atom in the macrocyclic framework.

An important observation is that for complexes containing bis(14-membered) ligands no metal-metal interaction depending on the size of the bridge exist. A combination of molecular mechanics calculations and simulation of the EPR spectra of the dicopper(II) complex of **4.103** in frozen methanol-water solution revealed its stretched conformation comparing to a face-to-face conformation founded for that with **4.94** with n = 2.

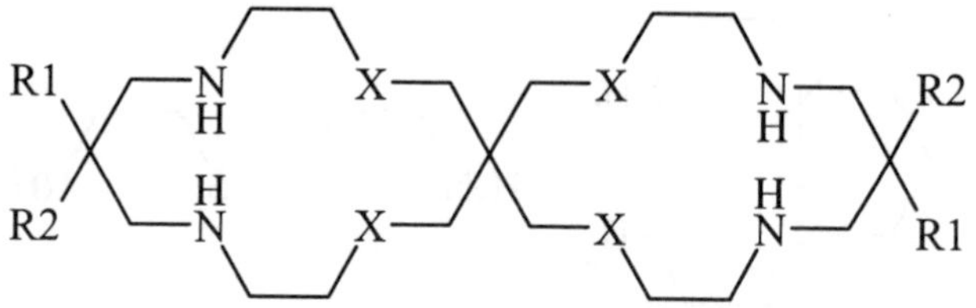

	X	R¹	R²
4.104	NH	NO₂	CH₃
4.105	S	NO₂	CH₃
4.106	NH	H	H

A special type of bismacrocyclic ligands includes two rings share of the same carbon atom. Spiro aza- and spiro thiaaza- macrobicycles **4.104** - **4.106** and **4.116** have been obtained by directed condensation of copper(II) complex of the binucleating ligand 5,5,-bis(4-amino-2-azabuthyl)-3,7-diazanonane-1,9-diamine with nitroethane and formaldehyde.[309] The central spiro carbon atom cancels the free rotation of the two macrocyclic planes with respect to each other and intramolecular coupling of the two metal centres was observed in the low-temperature EPR spectra. Differences in electrochemistry of the dimacrocyclic complexes compared with the analogous mononuclear reflect a significant inductive effect of the neighbouring copper(II) atoms. Thus the cyclic polarogram of [Cu₂(**4.104**)]Cl₄ indicate $E_{1/2}$ of -0.74 whereas for [Cu(metil,nitrocyclam))](ClO₄)₂ the $E_{1/2}$ of -0.56 corresponding to Cu(II)/Cu(I) the redox couples. The electrochemistry of copper(II) complexes changes on going from N_4 to N_2S_2 sets of donor atoms. This explains that the Cu(II)/Cu(I) couple is total irreversible for [Cu₂(**4.104**)]Cl₄ due to rapid dissociation of the copper(I) complex. The Cu(II)/Cu(I) couple is reversible for [Cu₂(**4.105**)](ClO₄)₂, reflecting the great affinity of copper(I) for "soft" donors which confer a greater stability to copper(I) complex.

4.5.Condensed Polymacrocyclic Ligands

4.5.1. Carbon and nitrogen caped amine ligands

Condensed macropolycyclic ligands are strongly selective metal-coordinating agents and among these, the polyamine ligands are suited to bind transition metals. They are nitrogen analogues of the polyether cryptands and attracted interest by their high stability conferred to their metal complexes toward ligand substitution and by offered prospect to study intramolecular rearrangements, electron transfer and spectroscopic properties.

The classical way for preparation large fused-ring molecules is thermodynamically unfavourable. The use of a metal ion as template reduces the problem of the building of large fused cycles to that some smaller chelate rings are entropically favoured. The attention has been focused on multidentate macrobicyclic ligands, which encapsulate octahedral metal ions. The first macrobicyclic ligand **4.107** with six saturated nitrogen donors has been synthesized as its cobalt(III) complex by a metal-assisted Mannich condensation using tris(ethane-1,2-diamine)cobalt(III), formaldehyde and ammonia.[13, 310, 311] The yield is generally very high for a reaction mixture from which multitude of alternating products are possible, suggesting that high symmetry encapsulated complexes must be extraordinarily kinetically favoured.

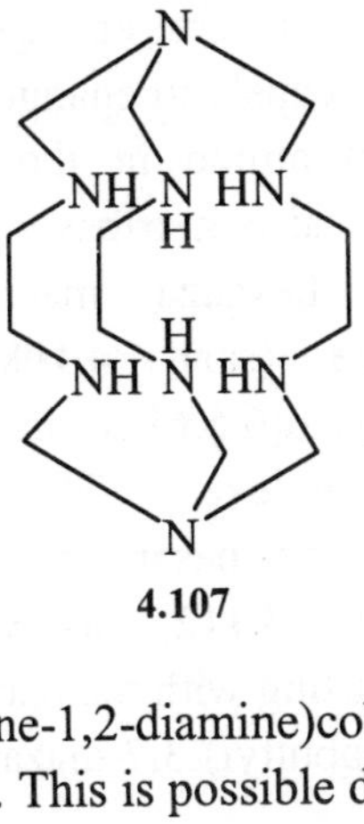

4.107

The capping of tris(ethane-1,2-diamine)cobalt(III) complex ion occurs along the C₃ axis of the ion. This is possible due to the trigonal symmetry about each N atom which is just that required to fit one of the trigonal axis of the tris(ethane-1,2-diamine) metal complex, $[M(en)_3]^{x+}$, (see mechanism). The obtained octaazacryptate ligand, 1,3,6,8,10,13,16,19-octaazabicyclo[6.6.6]-icosane, **4.107**, has been given the trivial name "sepulchrate". Apart from the six nitrogen donor atoms, there are two aza caps, which are not bonded to the metal ion.

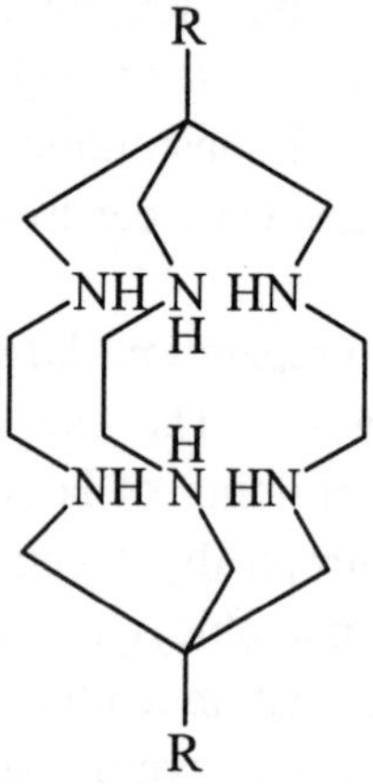

	R
4.108	H
4.109	NO_2
4.110	NH_2
4.111	NH_3^+

This procedure is open to several changes: it is possible to vary the central atoms of the ligand "caps", to change the template metal ion and to change the coordinated amine in the precursor. Substitution of ammonia for nitromethane, also a tribasic acid under the preparative conditions, leads to a hexaaza macrobicycle, 3,6,10,13,16,19-hexaazabicyclo[6.6.6]icosane (sar), **4.108** and to 1,8-dinitro-3,6,10,13,16,19-hexaazabicyclo[6.6.6]icosane (diNOsar), **4.109,** in form of their cobalt(III) complexes. The related cage complex $[Co(azamesar)]Cl_3 \cdot 1.5H_2O$, azamesar = 1-methyl-3,6,8,10,13,16,19-heptaazabicyclo[6.6.6]icosane, **4.112**, was synthesized in essentially the same way as $[Co(sep)]^{3+}$, starting with the partially caged $[Co(sen)]^{3+}$, sen = 5-methyl-5-(4-amino-2-azabuttyl)-3,7-diazanonane-1,9-diamine.

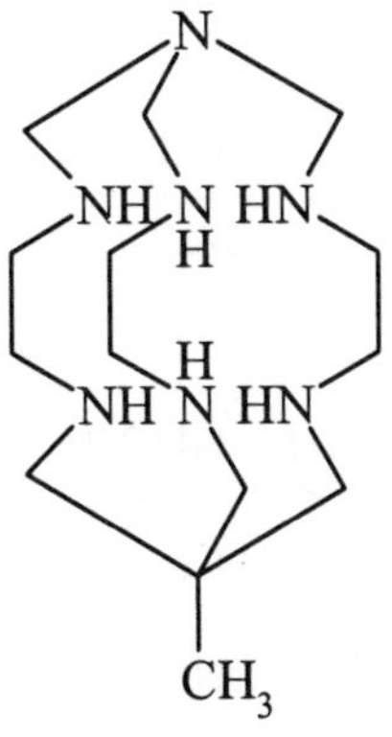

4.112

The stability and the orientation of the first imino group is the most important step that controls the capping process. In this respect, the nature of the template metal ion plays an important role. $[Rh(en)_3]^{3+}$, $[Ir(en)_3]^{3+}$ [312] and $[Pt(en)_3]^{4+}$ [313] have been encapsulated, each of them having an octahedral face containing three primary amine groups properly oriented for cyclization and, an appropriate Me-N bond length. $[Cr(en)_3]^{3+}$ respects the geometrical requirement but the Cr-N bond of 2.09 Å is longer than Co(III)-N (average 1.97 Å) and, as a consequence, as soon as the imine is formed, the bond is broken. The experimental condition should be chosen in a way that the capping process could be made competitive with the Cr-N bond rupture rate. Endicott *et al.*[314] succeeded in finding the experimental condition and reported the synthesis of $[Cr(sep)]^{3+}$ complex. The capping process were aborted in the imine intermediate formation step for $[Ru(en)_3]^{3+}$ and $[Os(en)_3]^{3+}$ when they disproportionate spontaneously the (II) and (IV) oxidation states. With the labile metal ions of Cu(II) and Ni(II) the condensation process is easily diverted to syntheses of square planar macrocyclic complexes. However $[Ni(sep)]^{2+}$ has been obtained in less than 1% yield and its structure established by X-ray crystallographic analysis.[18]

A greater variety of chemistry is possible with the sarcophagine system cages than with the analogous sepulchrate complexes, both because the sar complexes are more stable toward disruption of the cage and because functional groups with a wide range of properties may be attached to the terminal positions of the cage[315].

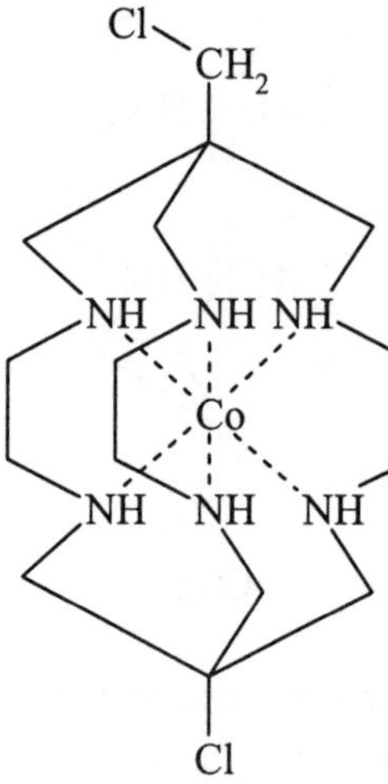

4.113

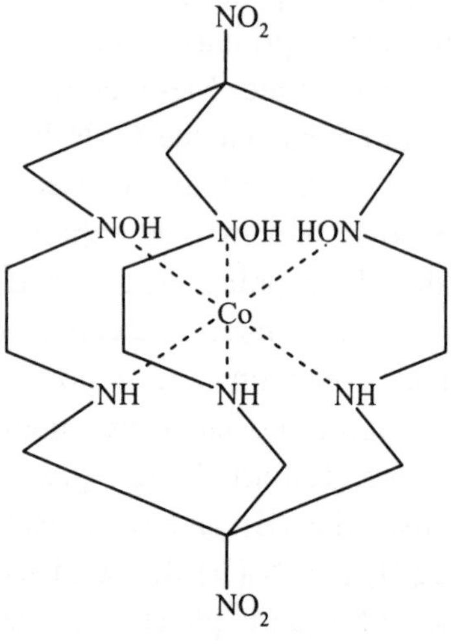

4.114

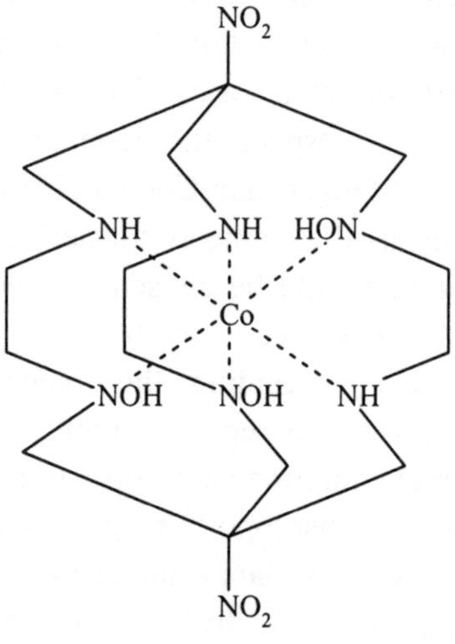

4.115

The derivatization of the nitro group introduce at the 1,8 positions the following substituents: NH_2, NH_3^+,[13,316] $NHOH$,[317] NHR, NR_2, NR_3^+ [318] $NHCHO$, $NHCH(CO_2H)CH_2CO_2H$, CH_3[13], CO_2H, $CH(NH_2)CO_2H$, $CH(OH)CO_2H$,[319] OH,[320] Cl,[13] Br, I, [127,130] H, $ZnCl_3$[321] . Conversion of the di-nitro cage complexes to the other encapsulated complexes illustrate the conventional organic chemistry which does not appear to be largely modified by its occurrence on a metal ion centre. An exception is observed during the nitrosation of Co(diamsar)]$^{3+}$, (diamsar = 1,8-diamino-3,6,10,13,16,19-hexaazabicyclo[6.6.6]icosane, **4.110**, when the formation of a new contracted cage, 3,6,10,13,15,18-hexaazabicyclo[6.6.5]nonadecane, (absar), **4.113**, is formed.

One of the few ways available to derivatize the cage N sites is to oxidize them to hydroxylamine groups. By treatment of the [Co(diNOsar)]$^{3+}$ with H_2O_2 in basic solution, up to three coordinated N sites were oxidized to hydroxylamine groups[138] as in **4.114 – 4.116**.

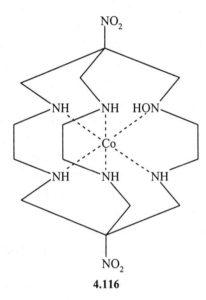

4.116

The cage remains unaffected and the complex shows an increased stability over the unsubstituted hydroxylamine complexes. A somewhat weaker ligand field for the hydroxylamine complexes than for the parent complexes was observed.

Possible conformations arise from stereogenic Co (Λ or Δ) and the six bound N atoms (R or S), the chirality of the conformational labile five-memebered chelate rings (*lel* or *ob*), and the helicity of the trigonal caps (R or S). Despite the number of the chiral centres only one isomer is mostly obtained. The stereochemistry of the precursor and the mechanism of the synthesis accounts for this stereospecificity. Thus, the encapsulation reactions occur with the retention of the [Co(en)]$^{3+}$

precursor chirality. For instance, for $[Co(sep)]^{3+}$, besides the racemic form, both the $\Lambda(S)$ and $\Delta(R)$ forms were isolated when Λ- and Δ-$[Co(en)]^{3+}$ were the precursors, respectively. In most cases, the cage ligands adopt a *lel* conformation of each five-membered chelate ring. Interconversion between the *lel* and *ob* chelate rings is very rare and does not alter any of the stereogenic centres. The *ob₃* conformation has been only observed in the crystal structure of $[Co\{(NH_2OH)_2sar\}]Cl_5 \cdot 4H_2O$, $[Co\{(NO_2)_2sar-H\}]Cl_2 \cdot 4H_2O$, and $[Co\{(NMe_3)_2sar\}](NO_3)_5 \cdot 3H_2O$ complexes although their parent precursors have the *lel₃* conformations.

In order to obtain cages with stable *ob₃* conformations, the *ob₃* and *lel₃* forms of the tris-(R)- or -(S)-propane-1,2-diamine-cobalt(III) were capped with formaldehyde and nitromethane. The steric requirement of the methyl group to remain equatorial in the chelate rings would considered to stabilize the *fac*- and *mer- lel₃*-$\Lambda(S)$- and -$\Delta(R)$-$[Co(pn)_3]^{3+}$. Indeed, with these precursors, the encapsulation gave the analogous $[Co\{(NO)_2, Me_3sar\}]^{3+}$, but the same treatment for Λ- and Δ- *ob₃* isomers did not yield an *ob₃* cage and the *ob₃* - cage has only obtained by an indirect route. This way involves the reduction of *lel₃* dinitro complexes to the cobalt(II) diamine complexes followed by the extrusion of Co(II) with CN⁻ ion and the chromatographic isolation of the *ob₃* conformer. Cobalt(III) was then reinserted in the free ligand *ob₃*-$\{(NH_3)_2, Me_3$-sar$\}$.[322] Although the average Co-N bond length (1.961(6) Å) of the two conformers does not differ greatly, the *ob₃* conformers show a stronger ligand field than the *lel₃* conformers.

The encapsulation reactions effectively trap the metal ion in a tight cage with octahedral donor atoms and the possibility of substitution reactions is very limited. The geometrical constrains act to retain octahedral-type geometry in the course of redox reaction of metal ions or of that which involves the macrobicyclic coordinated ligand. The properties which mirror the structural particularities include unusual stability, chiro-optical phenomena and fast electron transfer reactions.

The cage complexes show high stability constants over a wide pH range, which minimise hydrolysis. In most cases, co-ordinated amines have $pK_a > 14$, denoting a basic character which hinders deprotonation readily in aqueous solution. For the cage complexes with electronegative substituents on the caps the pK_a values are considerable less. The deprotonated species $[Co(diNOsar-H)]^{4+}$ has been isolated and its structure has been resolved.[13] In contrast with the undeprotonated parent ion, a D_3ob_3 geometry has been found. This geometry has been explained in terms of steric causes, as arising from increased valence angle deformation terms in the *lel$_3$* conformations, and these are primarily due to distortions in the C-N-C angles. The N-deprotonation has a significant effect on the bonding of other donor atoms. Thus, it has a considerable shortening effect on the Co-N bond to 1.946(7) Å and concomitant lengthening of the corresponding *trans*-Co-N bond to 2.016(7) Å was observed. These distances are different comparing to the mean of the four other distances, 1.974 Å. The altered bond distances may rationalize the labilizing effect of this deprotonation reaction. For example, the nitro-capped species are susceptible to a retro-Mannich type reaction in base resulting in a contracted cage (absar), **4.117**.

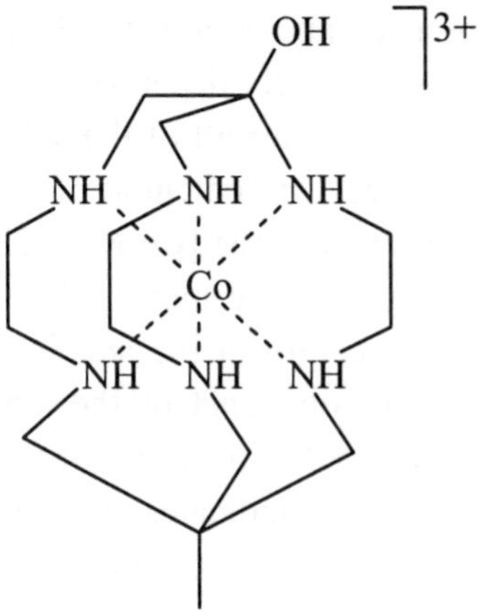

4.117

The acidity of the cage complexes is changed when a cage-derivatization is produced. Complexes with one, two, and three hydroxylamine groups behave as mono-, di-, and tribasic acids (pK_a ~ 3-6), respectively. The introduction of a NOH group at the N -site into the $[Co(diNOsar)]^{3+}$ raises the pK_a of the remaining secondary amines by approximately 1 pK_a unit for a NOH group.

The nature of the cage-substituents affects the acidity of the bonded secondary nitrogens in platinum(IV)-cage complexes too. Thus, the great acidity of the amino protons in $[Pt(dinosar)]^{4+}$ is a consequence of the electronic influence of the nitro groups, since reduction of these groups to amines causes a change in pK_a^1 from < 0 to 2.4.

The cage ligands effect great changes in the redox properties of the metals they encapsulate. All the cage complexes except that of the iridium (III) ion exhibit a reversible one electron couple so that unusual oxidation states can be stabilized. The redox properties of the macrobicyclic hexamine cage complexes of cobalt(III) parallel that of the simple hexamine complexes[323,] since reduction proceeds by an one-electron step to cobalt(II), followed by an irreversible reduction of cobalt(II) to cobalt(I) and Co(0). For cobalt cage complexes a quasi-reversible behaviour is apparent in both aqueous and nonaqueous media[324] and the metal ions are constrained to remain six coordinated upon reduction. The $[Co(sep)]^{3+}$ ion is more easily reduced than the structural similar $[Co(sar)]^{3+}$ ion which was attributed to the more electronegative aza caps as compared to the CH caps. The redox potential of the Co III/II couples is markedly dependent on the cavity size and on nature of the apical substituents for all of the substituted $[Co(sar)]^{3+}$ and $[Co(absar)]^{3+}$ cage complexes. The potential of the $[Co(Me,OH-sar)]^{3+/2+}$ couple is 0.385 V, whereas that of the $[Co(Me,OH-absar)]^{3+/2+}$ couple is -0.551 V, reflecting the significantly increased strain accompanying the reduction of the cobalt(III) centre in the smaller absar cage. When the cavity size of ligand is essentially constant, the substituent effects also are important in influencing the redox potential of the Co III/II couple. The order of the reduction potentials indicates the following trend for carbon substituents in enhancing reduction: $NO_2 \sim NH_3^+ > NH_2 > CH_3 \sim H$ which parallels the electron - withdrawing properties of these groups. The redox properties of the complexes are influenced by the conformational characteristics of the cages. Thus, the ob_3 and lel_3 difference is reflected in the reduction potentials and electron transfer rates of the $[Co\{(NH_3)_2, Me_3sar\}]^{5+}$ isomers. The more negative reduction potential and a 30-fold larger self-exchange rate constants of the ob_3 isomers was observed and explained by the smaller cavity size of ob_3 which fits cobalt(III) better than the lel_3 conformer.

The mechanism for further reduction of the Co(II) species involves initial reduction of the Co(II) sar cage complexes to cobalt(I). The resultant strain leads to a rearrangement to form macromonocyclic square-planar cobalt(I) species. For [Co(sep)]$^{2+}$, ligand rupture and very rapid reduction of the Co(I) transient to Co(0) giving the net two-electron irreversible process has been observed. A different behaviour was assigned to the more robust carbon-capped cage complexes.

The rhodium(III) and platinum(IV) cage complexes do not exhibit one-electron reduction and octahedral monomer rhodium(II), respectively, platinum(III) states are rarely stabilized. A short-lived platinum(III) intermediate was obtained by the pulse radiolysis of [Pt(diamsar)]$^{4+}$ and observed in ESR spectra. By γ-irradiation of [Pt(diamsar)]$^{4+}$ as chloride and trifluoromethanesulfonato salts, the trapping of Pt(III) in crystalline lattice in the solid state was realized. [Pt(sep)]$^{3+}$ ion has been obtained under the same condition; its lifetime in aprotic solvents appears somewhat longer than that of corresponding Pt-sar cages.[123] The reduction potentials of the Rh(III) cage complexes are more negative than their cobalt(III) analogues. It is not unexpected since Rh(III) complexes are normally more difficult to reduce than their Co(III) counterparts.[137] This was associated with the cavity size - ion size relation leading to a strain which destabilize the larger Rh(II) ion. This makes the formation of the Ir(II) analogues complexes impossible.

The rate for the self-exchange reaction between +2 and +3 for cobalt and between +3 and +4 for platinum complexes has been measured and it was found that they are generally much larger than their analogous with NH$_3$[325] or 1,2-ethanediamine[326] ligands. The electron self-exchange rate constant of [Co{(NMe$_3$)$_2$sar}]$^{5+/4+}$ redox couple (+0.05V) was found to be 2 orders of magnitude smaller than that of the [Co{(NH$_3$)$_2$sar}]$^{5+/4+}$ couple under the same conditions. It is assumed that in the second case, the electron transfer process is coupled with a conformational change from *ob$_3$* to *lel$_3$* conformation.

More rigid condensed bismacrocycles, Δ-(4,9,15,20,25,30R)- and Λ-(4,9,15,20,25,30S)-[(1,12-dinitro-3,10,14,21,24,31-hexaazapenta-cyclo-[10.10.-10.0.^{4,9}0.^{15,20}0.^{25,30}]dotriacontane, **4.118** have been obtained as cobalt(III) complexes, Δ-(R,R)$_3$- and Λ-(S,S)$_3$-[Co(diNOchar)]$^{3+}$,[327] when [Co(chxn)$_3$]$^{3+}$ (chxn = trans -1,2-cyclohexanediamine) was the precursor and nitromethane the capping agent. The *lel$_3$*-[Co(diNOchar)]$^{3+}$ ion is formed with complete retention of conformation and configuration imposed by the starting precursor.

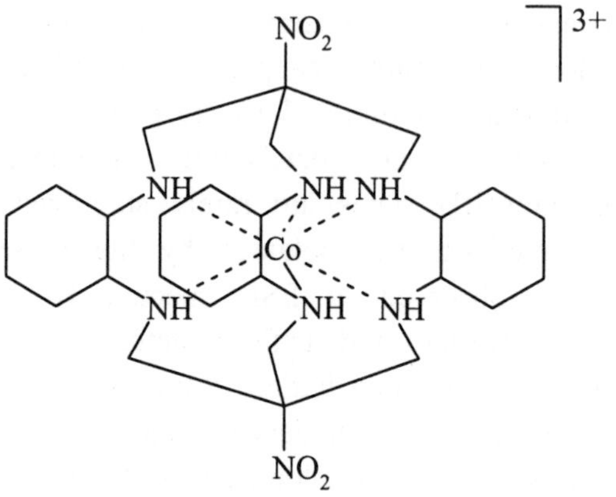

		R
	4.118	NO_2
	4.119	NH_2
	4.120	NH_3^+

The synthesis of the $[Co(diNOchar)]^{3+}$ cage is more difficult than that of the more flexible $[Co(diNOsar)]^{3+}$ analogue. Two factors seem to be responsible for this difference. One of them is the exchange rate of the coordinated N-H protons which is slower for the $[Co(chxn)_3]^{3+}$ than for $[Co(en)_3]^{3+}$. Secondly the conformational flexibility of $[Co(chxn)_3]^{3+}$ and $[Co(en)_3]^{3+}$ is of importance. The changes in the structural and electronic environment of the CoN_6 chromophore on going from the tris diamine parent ion to the caged species are associated with the changes in the visible circular dichroism (CD) and rotatory dispersion spectra, and the trends are very similar with the sar complexes. The reduction of $[Co(diNOchar)]^{3+}$ ion in different ways leads to the corresponding aminoderivatives, $[Co(diNH_2\text{-}char)]^{3+}$, **4.119** to $[Co(diNOchar)]^{2+}$ or, respectively to the $[Co(diAMchar)]^{4+}$ with the retention of configuration and conformation of the parent compound. The protonation of the cap amino groups is accompanied by changes on the CD of the Δ-*lel$_3$*-$[Co(diAMchar)]^{3+}$.

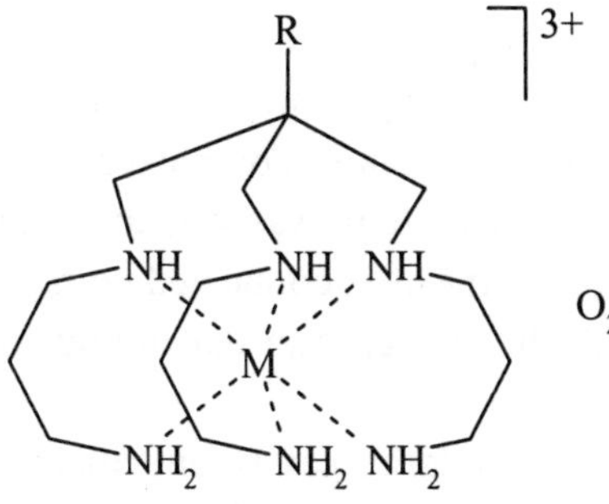

4.121 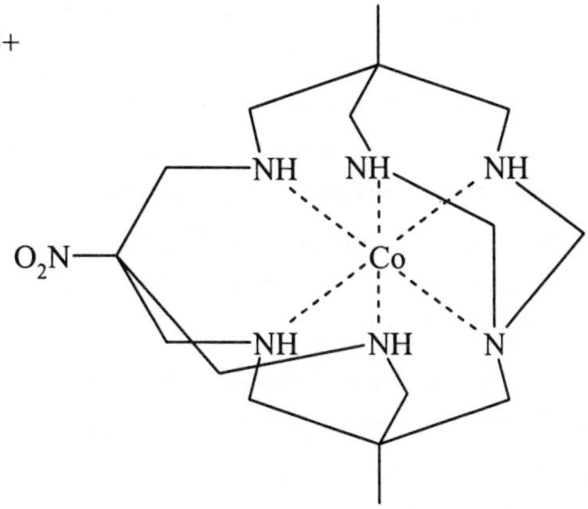   4.122

Since the compound is conformationally locked, the system is very appropriate for studying changes in CD due to both different ionic environment and to different charge distributions within the complex ion. The reduction potentials for cyclohexane systems are similar to those for corresponding $[Co(diNOsar)]^{3+}$ cage species.

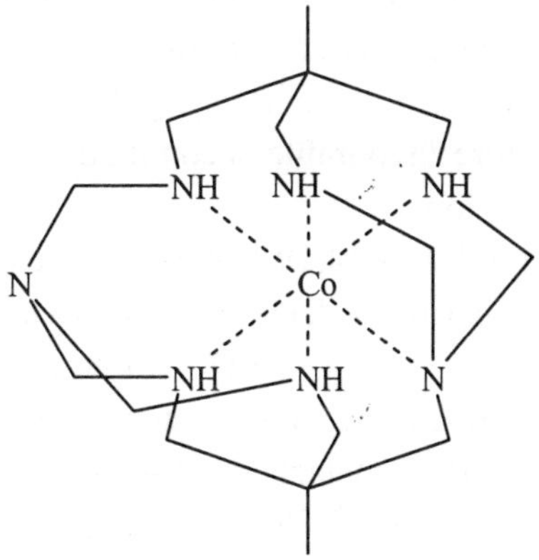

4.123 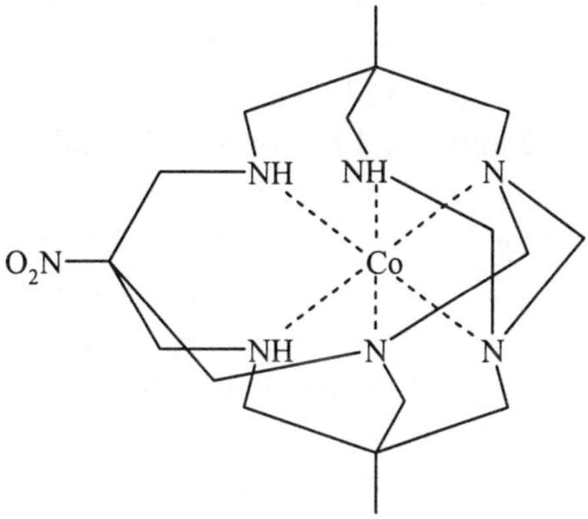   4.124

This was interpreted by the similar cavity sizes of these systems. However, the rigid conformation of cyclohexane cage does not allow the reduction of the Co(II) centre to the larger Co(I) or cobalt(0) states, which would involve a deformation.

In order to accommodate larger low oxidation state ions and to modulate the redox potential of the cage complexes, the increase in the cavity size was considered by extending the capping strategy for tris(1,2-ethane-diamine) complexes to the 1,3-propanediamine complexes or to hexadentatecomplexes of the type **4.121**. This strategy was valuable only in the case of rhodium(III). The success of synthesis is connected with the metal-nitrogen bond length, which should be longer than that found for Co(III)-N of 1.98 Å.

To make a rigid stable cage with a rather large cavity and thereby modulate the redox potential and electron transfer rates of the resulting metal complexes, the encapsulating procedures used for the above presented systems were developed to fuse two chelating groups together. Three major products, **4.122** - **4.124** were obtained in different yields when condensation of $[Co(tame)_2]^{3+}$ with formaldehyde and nitromethane, respectively, ammonia occurs.[142] The molecular structure of these complex ions exhibit gross angular distortion of the $Co^{III} N_6$ chromophore from octahedral geometry due both to the trigonal nitromethane-formaldehyde cap and to the fused four-membered rings on the opposite side. An additional bridge in **4.124** induces a further distortion of the nitrogen donor atoms over the cobalt sphere without significantly increasing of the CoN_6 core size. This effectively opens up one side of **4.124** relative to the equivalent side of the macrotricyclic cage **4.112** and could be conceived to allow more favourable accommodation of the larger but more readily distorted Co^{2+} ion.

Macrotricyclic ligands **4.125** 1,4,7,9,11,14,19-heptaazatricyclo $[7.7.4.2^{4,14}]$docosane, (azasartacn), **4.126** 9-nitro-1,4,7,11,14,19-heptaaza-tricyclo$[7.7.4.2^{4,14}]$docosane, (nosartacn), and **4.127** 9-amino-1,4,7,11,14,19-heptaazatricyclo$[7.7.4.2^{4,14}]$-docosane, (amsartacn), have been obtained starting from $[Co(taetacn)]^{3+}$, taetacn = 1,4,7-tris(2-aminoethyl)-1,4,7-triazacyclo nonane as precursor.[328] The cobalt(III) cage complexes were resolved into their chiral forms and they are optically very stable. As an alternative, $\Lambda(+)_{490}$-$[Co(nosartacn)]^{3+}$ can be obtained from $\Lambda(+)_{487}$-$[Co(taetacn)](ClO_4)_3$. Electrochemical studies show essentially reversible Co(III)/Co(II) couple for all these complexes and the electron self-exchange rate values fall within the range covered by other cobalt-hexaamine cage systems (k = 0.04 – 0.09 $M^{-1}s^{-1}$).

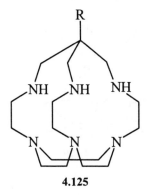

4.125

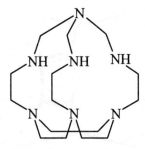

	R
4.126	NO_2
4.127	NH_2
4.128	NH_4^+

The condensation of $[Co(tach)_2]^{3+}$ (tach=1,3,5-triaminocyclohexane) with formaldehyde and nitromethane under basic conditions leads to 11-methylamino-6,16,-dinitro-4,8,14,18,21-pentaazatetracyclo[14.4.2.1.^{3,19} 1^{9,13}]-tetracos-4-enato(6)cobalt(III), **4.128**, tetrachlorozincate hydrate in which a carbanion chelate is stabilized.[329] The rigorous steric requirement of the ligand and the strain it generates for the complex is reflected in the Co-N bond lengths. Five of the six Co-N bond lengths are longer than 2.00 Å which differ from a Co(III)-hexa-amine cage complex (1.96 - 1.98). The sixth Co-N bond of 1.963(5) Å is connected with the deprotonated and delocalized ring structure.

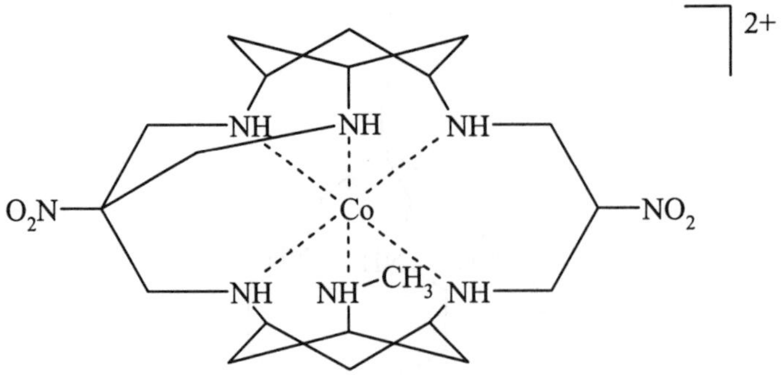

4.129

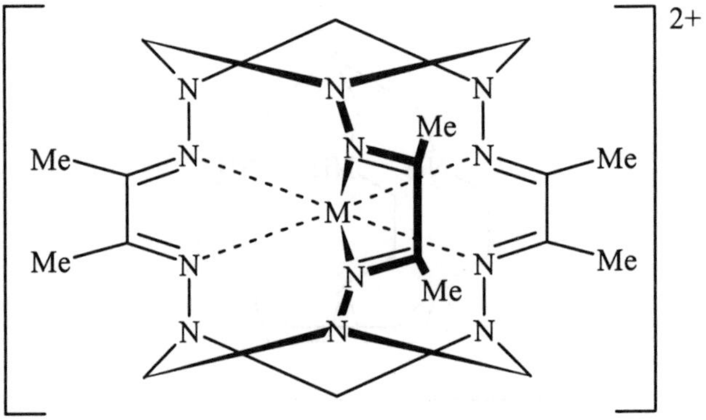

4.130

The ability of secondary amines of complexed ligands to react with formaldehyde to form macrocyclic ligands was employed to yield polycylic ligands with complicated fused rings, **4.130**. The tricyclic and quadricyclic ligand complexes involving nickel(II), cobalt(II) or iron(II) directed condensation of butane-2,3-dione dihydrazone with formaldehyde have been obtained.

Grzybowski *et al.*[330] obtained in a two step process mononuclear and binuclear iron(II) clatrochelates, **4.131**, **4.132**, and **4.133**, utilizing as precursor the labile complex of Fe(II) with 2,3-butanedione-oxime-hydrazone. The first step involves phenyl-boric acid as capping while in the second step the formaldehyde/hidrazone capping reaction occurs.

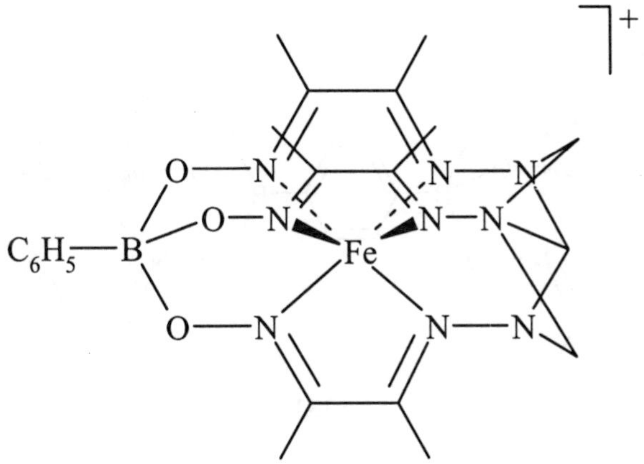

4.131

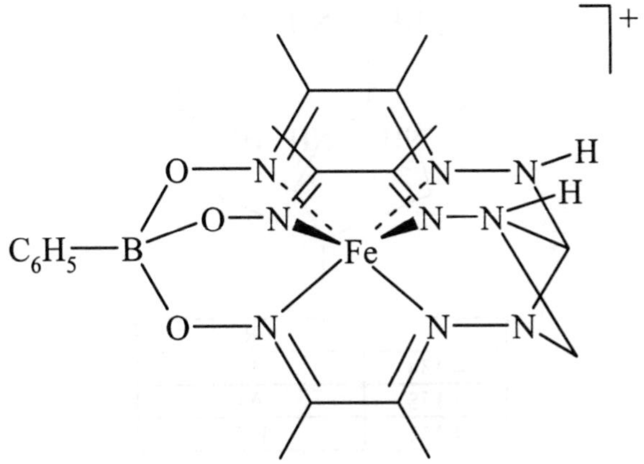

4.132

Electrochemical studies reveal that only the Fe(II) cage complexes undergo quasi-reversible oxidation, which is assigned to the metal centred Fe(III)/Fe(II) couple whereas for the partially capped species the oxidation to Fe(III) is completely. No interactions can be observed between the two Fe centres of the binuclear species **4.133**.

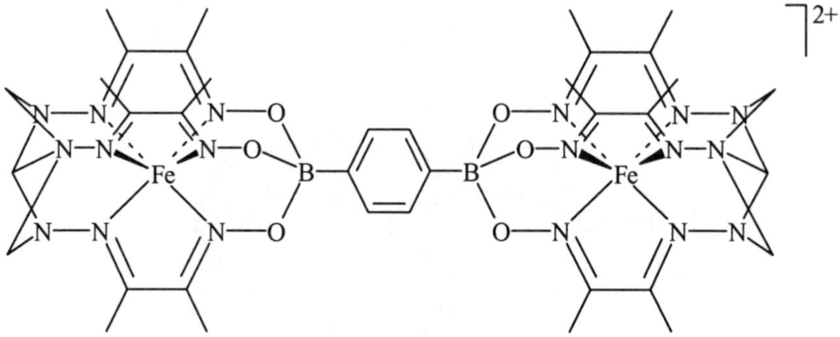

4.133

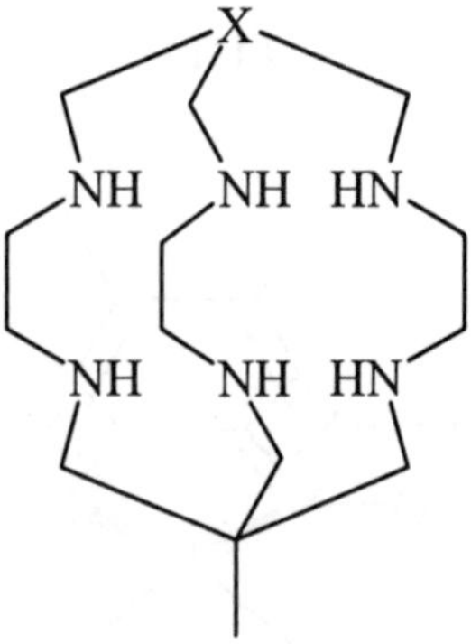

	R
4.134	P
4.135	As
4.136	PO

4.5.2. P-, As amine cage ligands

With the wish to obtain a stabile *ob₃* conformer, the PH_3 and AsH_3 has been used as capping agent for $[Co(sen)]^{3+}$. The Δ-ob_3(SS) $[Co(Mephosphasar)]^{3+}$, Mephosphasar = 8-metyl-1,3,6,10,13,16,19-hexa-azaphospha bicyclo[6.6.6]-icosane), **4.134**, and $[Co(Mearsasar)]^{3+}$, Mearsasar = 8-metyl-1,3,6,10,13,16,19-arsa-hexa-azabicyclo[6.6.6]-icosane), **4.135**, complexes have been prepared in this way.[331, 332] $[Co(Mephosphasar)]^{3+}$ is slowly oxidized in aqueous solution in the presence of air to give the phosphine oxide derivative, 8-metyl-1oxo-3,6,10,13,16,19,1-hexa-azaphospha-bicyclo[6.6.6]ico-sane),

(Me,Ophospha-sar) **4.136**, as Δ- *lel₃*(RR) conformer. Structural data show that in solid state [Co(Mearsasar)]³⁺ may be designated Δ-*ob₃*(SS) and Δ-*ob₃*(RR). Electochemical features of the aqueous solution of the complex ions closely resemble those found for carbon-capped cage analogues which have constrained Δ-*ob₃*(SS) and Δ- *lel₃*(RR) conformers. Structure of the [Co(Me,Ophosphasar)]³⁺ ion also shows that the larger phosphorus cap causes an expansion of the adjacent Co-N bonds [Co-N(Pcap) = 2.003(1) Å] compared with those of the aza and carbon capped analogues. This is reflected in more positive Co$^{III/II}$ redox potentials which denote the stabilisation of low oxidation state.

4.5.3. Thioamine cage ligands

Hexadentate ligand (8-methyl-6,10,19-trithia-1,3,14,16-tetraazabi-cyclo[6.6.6]icosane (azacapten) **4.137** have been obtained by condensing the [Co(ten)]³⁺ [ten = 4,4',4"-ethylidenetris(3-thiabutan-1-amine)] complex with formaldehyde and ammonia.[333]

The analogous capping reaction with nitromethane and formaldehyde resulted in the nitromethyl-caped complex [(1-methyl-8-nitro-3,13,16-trithia-6,10,19riazabicyclo[6.6.6]icosane) cobalt(III) (3+), [Co(**4.138**)]³⁺, and further after reduction, [Co(**4.139**)]³⁺. Only one diastereoisomer has been produced in each capping synthesis, namely that where all the nitrogen and sulphur centres have the same configuration.(e.g. Co-Λ,S-RRR,N-SSS) and the cysteinamine rings adopt the *lel* conformation. As expected, replacing the amine donor groups by thioethers has a marked effect on the electronic and electrochemical properties and stability.[334] The origins of these effects are the longer Co-S (2.17 – 2.24Å) and S-C (1.80 Å) bonds compared with the Co-N (1.98 Å) and C-N (1.50 Å) analogues and, as a consequence, larger cavity size than in hexaamine analogues appear. Further, thioether donors allow the delocalization of the electron density from the metal ion (back-donation).

The complexes exhibit a reversible Co$^{III/II}$ couple and at more negative potentials, a Co$^{II/I}$ couple compared with N₆ cages. Reduction of [Co(azacapten)]³⁺ results in the extremely oxygen-sensitive [Co^{II}(azacapten)]²⁺. The triflate or dithionate salts of [Co^{II}(azacapten)]²⁺ was found to exhibit a magnetic moment of 1.89 μ$_B$ indicative for a low-spin ^2E_g, $t_{2g}^6 e_g^1$) configuration.[335] This shows the relative high ligand field created by azacapten compared with $[Co(sar)]^{2+}$ and $[Co^{II}((NH_3)_2sar]^{4+}$.

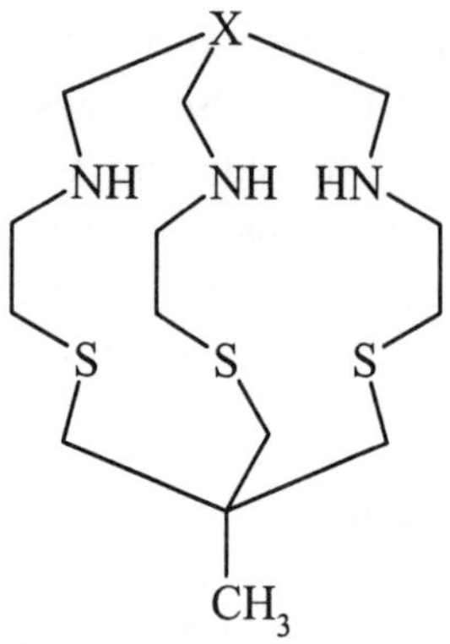

	X
4.137	N
4.138	C-NO$_2$
4.139	C-NH$_4^+$

4.5.4. Cage complexes of other metal ions

The metal-cage complexes are inert to substitution. For example Ni(II) sepulchrate is extraordinarily stable against ligand exchange with donors such as NCS⁻, Me$_2$SO, H$_2$O, or CH$_3$CN whereas $[Co\{(NO_2)_2sar\}]^{3+}$ and its Co(II) analog are completely inert to ligand substitution and sometimes even ligand decomposition occurs before exchange is observed. Free sarcophagine and its amine-derivatives have been obtained by reduction of their cobalt(III) complexes to the corresponding cobalt(II) from which the ligands were removed in concentrated acids or in hot aqueous solutions containing an excess of cyanide ions. The free sarcophagine and (NH$_2$)$_2$-sar ligands are strong bases, accepting up to four and five protons , respectively, in aqueous solution.[127] The pK values of pK$_1$ = 11.95, pK$_2$ = 10.33, pK$_3$ = 7.17 and pK$_4$ ≅ 0 for sarcophagine, and pK$_1$ = 11.44, pK$_2$ = 9.64, pK$_3$ = 6.49 and pK$_4$ = 5.48 and pK$_5$ ≅ 0 for diamino-sarcophagine show that they are more basic than comparable macromonocyclic tetramines being in agreement with the idea that protonation may in some way be facilitated by the bicyclic structure. The simple solution ^{1}H NMR spectra of the free ligands show that they are conformational labile. Consistent with this is that the rate of formation of their metal complexes is generally high. However, complexes formed by a direct reaction show remarkable thermodynamic stability. For instance, the formation constant for [Hg(sar)]$^{2+}$ has been evaluated as $\sim 10^{29}$ at 298 K. The ligand exchanges very slowly with water, $t_{1/2}$ being much greater than 24 hrs for copper(II)[336] manganese(II) and cobalt(II).[119]

Once the ligands are free, other metal ions can be encapsulated. Direct metal ligand reaction can be used to encapsulate almost all of the first row bivalent and trivalent metal ions revealing a strong tendency toward hexadentate coordination of these ligands to a wide variety of metal ions.[155, 337, 338] For most of the sar and (NH$_3$)sar^{2+} complexes as well as for free sar, the X-ray crystal structure has been resolved and the same overall lattice has been identified. The hexacoordination of the N$_6$ cage ligand and the *lel* conformation of each five-membered chelate ring was established except the [Ni(NH$_3$)$_2$-sar]$^{4+}$. In this case, the chelate rings are (*lel*)$_2$(*ob*) for chloride salt and (*lel*)$_3$ for nitrate salt. Only small variations in the cavity size defined by the bicyclic ligands were observed and similar overall lattice has been identified. However, a trigonal twist distortion at the cap fragment has been noticed.[339] The degree of trigonal twist has been shown to depend on the metal ion and the specific requirement of the ligand. In terms of their trigonal twist angles, ϕ, about the C$_3$ axis, the cage complexes fall into two groups. The first group has ϕ $\sim 28°$ and includes the cage complexes of bivalent metal ions except that of Ni(II). The second group includes the complexes of Cr(III), Fe(III), Co(III) and Ni(II) with a ϕ lying between 46 and 60°.

An additional tetragonal distortion is superimposed in the case of copper(II) complex [Cu(NH$_3$)$_2$-sar](NO$_3$)$_4$·H$_2$O due to the Jahn-Teller distortion well known for the d^9 configuration in octahedral ligand field. According to this [Cu(NH$_3$)$_2$-sar](NO$_3$)$_4$·H$_2$O, the six Cu-N bonds are not equivalent and an unusual trigonal twisting angle of 18^0 appear in [V^{IV}(NH$_3$)$_2$-sar-H$_2$](S$_2$O$_6$)$_2$·2H$_2$O is observed. The cage ligand is deprotonated and therefore the comparison with the other systems should be made with precaution. Also steric factors and metal-ligand π bonding may be responsible for the preference of trigonal prismatic geometry. The reduced symmetry observed crystallographically for the individual metal ions is reflected by the electronic and magnetic properties.[340]

A special attention has been paid to iron(II) cage complexes. Magnetic moment of colourless solids $[Fe(sar)](CF_3SO_3)_2$ and $[Fe((NH_3)_2sar)]Cl_2Br_2.4H_2O$ established they are both high-spin with $^5T_{2g}$ ground state.[341, 342, 343] In contrast the complex $[Fe((NH_2)_2sar)](CF_3SO_3)_2$ is a deep-blue diamagnetic crystalline solid which confirms the low-spin $^1A_{1g}$ state. These major differences of colour and magnetism indicate that the symmetry and strength of sar-type ligand field are placing divalent iron in the saturated N_6-cage environment close to its ligand field spin crossover region.

Chapter 5

Self Condensation of Nitriles

5.1. Phthalocyanines

The metal phthalocyanines are one of the most known and studied compounds. There are many articles and books that refer to this field/topic.

The classical methods to obtain phthalocyanines and metallophtalocyanines give low yields, a mixture of isomers, and require poorly accessible starting materials and high temperatures. Several general methods to obtain phtalocyanines using metal ions as template have been successful developed and nowadays they became the most common methods.[344, 345, 346] Scheme 5.1 presents the precursors employed to obtain metallo- phtalocyanines (MPc) when metal ions are used to direct the course of the reactions. Among these methods, the most utilized one involves the action of a metal salt upon phtalonitrile in dry or in an adequate solvent. For example, copper(II)-, platinum(II)-, palladium(II)- or rhodium(III)-phtalocyanines have been obtained when the corresponding chlorides were heated with phtalonitrile. By heating thorium(IV) tetrachloride in quinoline at 260 °C resulted in the corresponding metallophtalocyanines. Ruthenium(II) phtalocyanine has been obtained by heating RuCl$_3$ with o-cyanobenzamide, according to equation 2 in Scheme 5.1. The product sublimes and forms easily adducts to reach six-coordinated environment. This method has also been employed to obtain europium(III), gadolinium(III) or ytterbium phtalocyanines.

Metal carboxylates were also used. For example, samarium formate, manganese(II) or chromium(VI) acetates were used as templates.

Sometime, the reaction takes place in the presence of air or oxygen; this is the case of titanium-, zirconium or hafnium tetrachloride which leads to the corresponding square pyramidal metallophtalocyanines formulated as oxy derivatives.

The treatment of phthalonitrile, 1 or 4- substituted phthalonitriles, 4-neopentoxyphthalonitrile or 4-nitrophthalonitrile with zinc acetate gave the appropriate phtalocyaninato zinc(II), **5.1-5.3**.[347] Relative high yield of pure single isomers, until 56 %, have been obtained at for 7-10 day. Using zinc acetate as template, it has been possible to work at low temperature of 20 °C when temperature stable substituents and precursors can be used. Also, the possibilities to obtain a pure isomer and to eliminate the formation of impurities, which are difficult to remove, are enhanced. The phtalocyanine macrocycle is found in the very rigid planar conformation. However, the phtalocyanine dianion seems to be quite flexible, especially in complexes containing oxygen donor substituents.

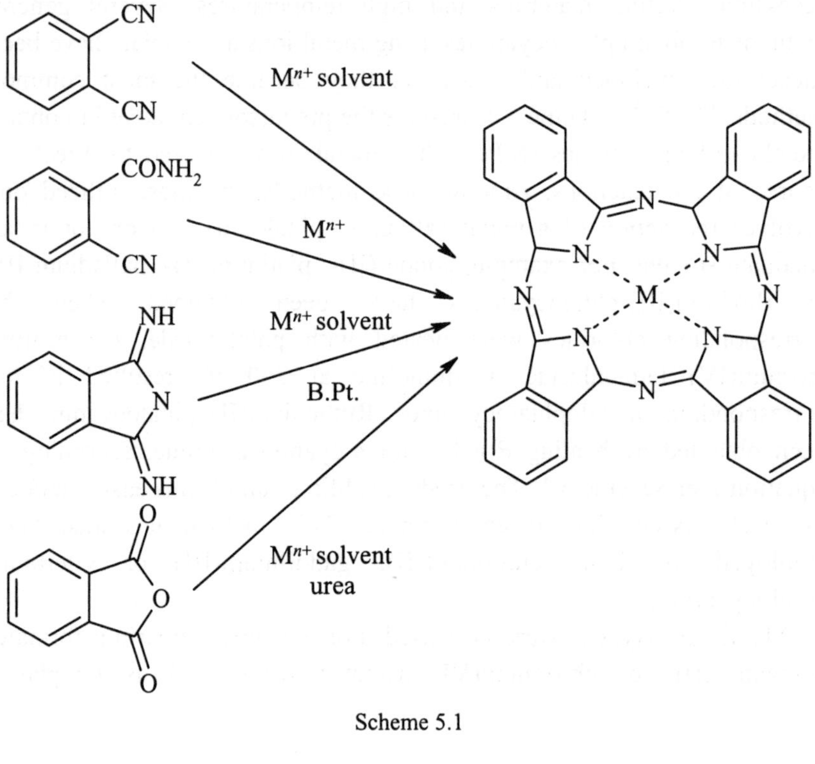

Scheme 5.1

Several metallophthalocyanines (MPc: M=Ni(II), Co(II), Zn(II), Pb(II), Fe(II), Cd(II), and Mn(III)) were obtained by heating phthalonitrile with metal salts in alcohol in the presence of 1,8-diazabicyclo[5.4.0]undec-7-ene. Metal acetylacetonates as well as metal halides were available as metal sources for preparation of metallophthalocyanines by this method.[348]

Substituted phtalocyanines with long hydrocarbon chains possess properties which make them suitable thermotropic liquid crystals. For example, the octasubstituted copper(II)-phtalocyanines **5.5** have been obtained directly from **5.4** by reaction with an excess of CuCN. The solid compound shows discotic mesophases with the distances between neighbouring columns of ~ 34 Å. It presents at 53 °C a viscous birefringent phase which is stable up to about 300°C.[349]

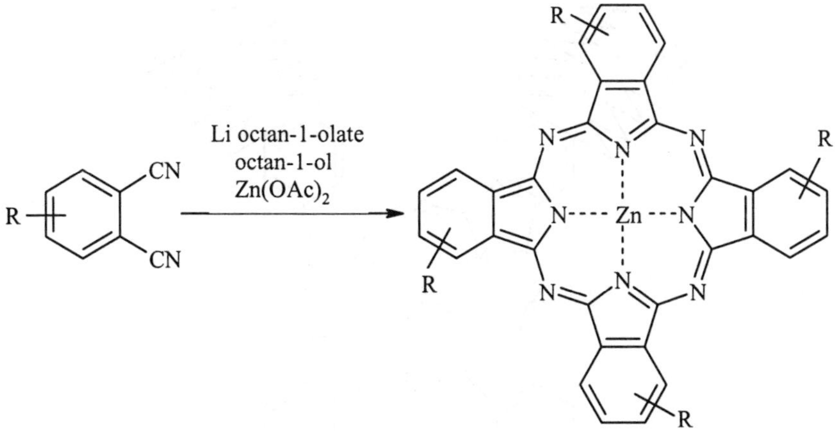

	R
5.1	3-OCH$_2$CMe$_3$
5.2	3-OCH$_2$Ph-*p*-Bu
5.3	3-OMe

Scheme 5.2

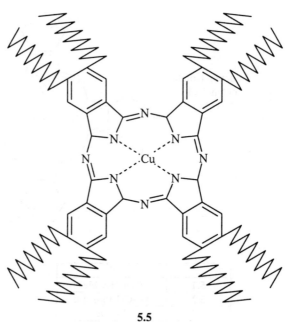

Scheme 5.3

Supramolecular structures of bis(phtalocyanato)lanthanides have been obtained. For example, from a mixture which contains dicyanobenzo-15-crown-5, phtalonitrile, 1,8-diazabicyclo[5.4.0]-7-undecane and lutetium acetate it has been isolated the compound **5.6** which is mono-15-crown-5-substituted bis(phtalocyaninato)lutetium(III), Lu(CR$_1$Pc)(Pc).

5.5

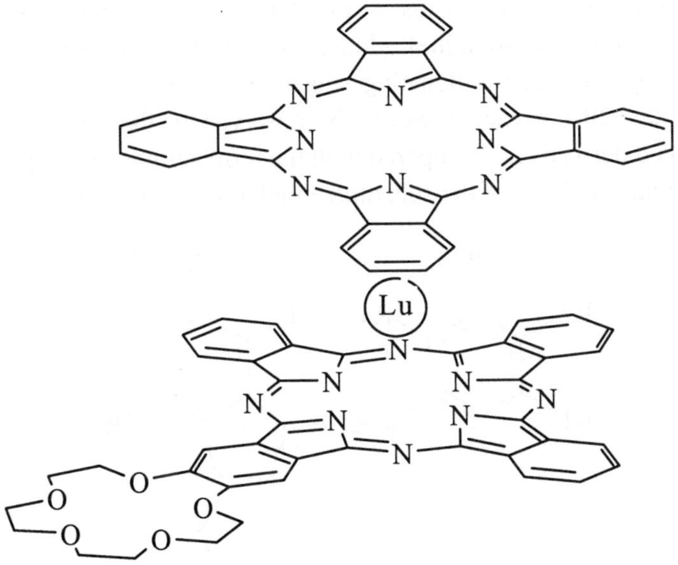

5.6

In a similar way it has been obtained tetra-15-crown-5-substituted bis(phtalocyaninato)-lutetium(III) complex, $Lu(CR_4Pc)(Pc)$.[350] The monocrown dimer gave a K^+-induced supramolecular structure with freedom rotation around potassium cation.

The actinides, with their high ionic radii and/or unusual coordination geometry can produce and stabilize expanded macrocyclic ligands, as in the preparation of superphtalocyanines. Thus, by warming phthalonitrile with uranyl chloride in dmf at 180°C resulted in the uranyl-phthalocyanine.

The preparation of substituted phthalonitriles involves the sequence dicarboxylic acid → anhydride → imide → amide → nitrile. This is the reason for which it is sometimes easier to obtain substituted metallo-phthalocyanine using substituted phthalic anhydride as starting reactant (Scheme 5.1, d). In these cases, high-boiling solvents like nitrobenzene 1-chloronaphthalene, nitrobenzene and 1,2-dichlorobenzene are used. Series of cobalt and iron complexes **5.7 – 5.10** have been obtained using appropriate anhydride, urea and metal salts in the presence of ammonium molibdate as catalyst.[351] The thermal stabilities, electronic spectra and the behaviour against donor solvents are affected by the electronic influence of the peripheral groups. Thus, in pyridine solvent, both iron and cobalt complexes form bis(pyridine) adducts whose electronic spectra are not affected by the electron-donating substituents. Displacement of the Q-band with ~20 nm can, however, be observed in the chloro- or nitro-substituted derivatives. Compared with the unsubstituted analogues, MPc, the thermal stabilities of the substituted phthalocyanines are smaller.

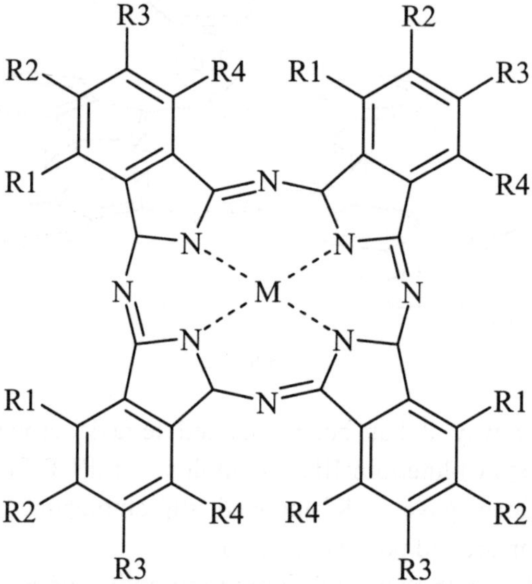

	M	R^1	R^2	R^3	R^4
5.7	Fe	H	$C(CH_3)_3$	H	H
5.8	Fe	Cl	Cl	Cl	Cl
5.9	Co	H	$C(CH_3)_3$	H	H
5.10	Co	H	NO_2	H	H

5.1.1. Mechanism

The role of the metal ions in the cyclization reaction leading to the macrocyclic phthalocyanine has been studied following an electrosynthesis strategy at -1.6 V. Phthalonitrile was placed in the cathode compartment where the formation of the anion radicals in Scheme 5.4 has been observed.

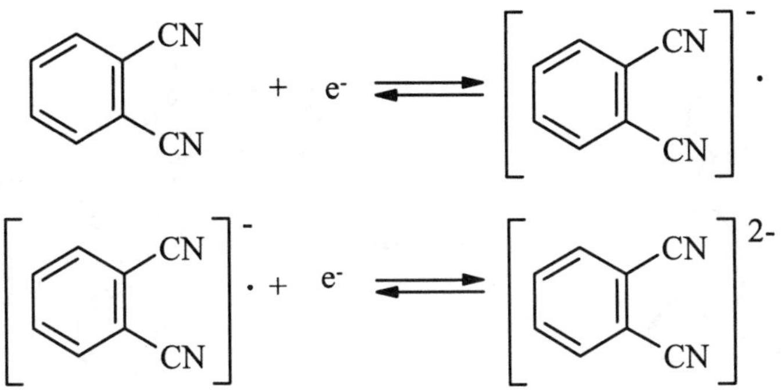

Scheme 5.4

The metal ions placed in the anode compartment migrate in the cathode space where they react with the two anion radicals and two phthalonitrile molecules.[352]

The thermal behaviour of the reaction of phthalonitrile with a range of copper compounds has been investigated using differential scanning calorimetry with product analysis by infrared spectroscopy. It has been concluded that the reaction with copper metal to give copper phthalocyanine is exothermic and requires the presence of air or a small amount of a copper(II) salt. Contrary, reaction with copper(II) chloride takes place smoothly either in an air or a nitrogen atmosphere to give a copper monochlorophthalocyanine. Contrary, the reaction with copper(I) chloride is air-sensitive. A general mechanism for this industrially important series of reactions is proposed to account for the results of the thermoanalytical investigation according to Scheme 5.5.[353] The reaction is initiated by attack of a nucleophile. Two types of nucleophile agents may be involved: *i*) Y⁻ which is the chloride ion or/and another phthalonitrile molecule, at one of the cyano groups activated by coordination of the nitrogen atoms with copper(II) ion. Copper(II) ions also play an important role in retaining the structure of the intermediate **5.11** in a conformation favourable for ring closure by intermolecular nucleophilic attack to form intermediate **5.12**. Further, it is the possibility that two electrons are transferred from the metal ion, thereby allowing the elimination of Y⁻ to form the very stable 18 p-electron copper phthalocyanine Cu(Pc).

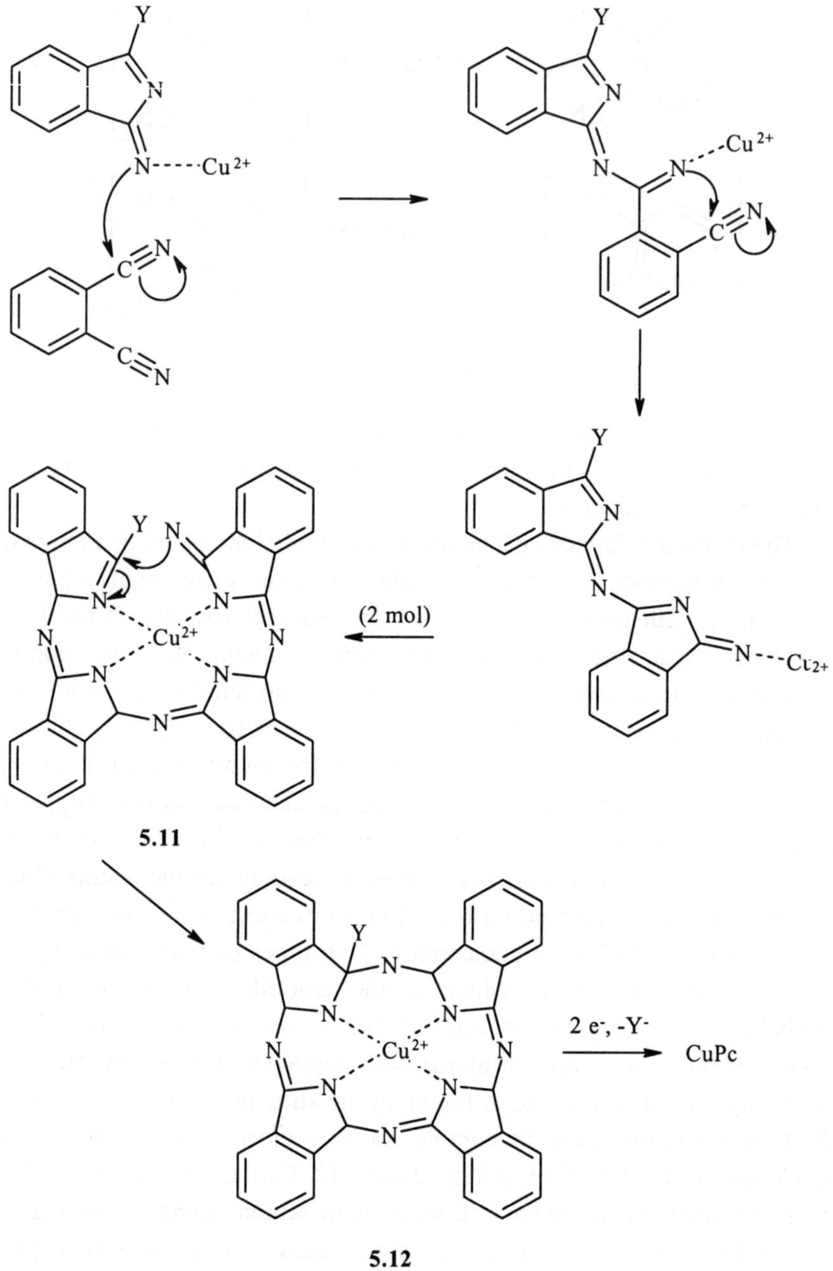

5.11

5.12

Scheme 5.5

5.2. Porphyrins

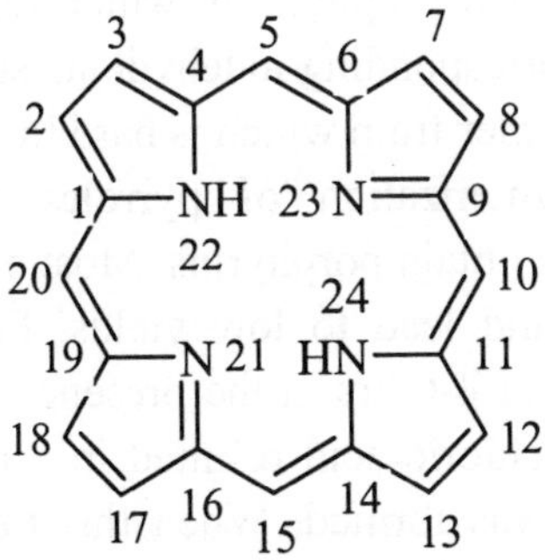

5.13

Template condensation of maleonitrile to form porpyrine is the analogue method to the condensation of phthalonitrile to phthalocyanine.[354] Reaction of maleonitrile with magnesium *n*-propanoxide in *n*-propanol leads to magnesium complex in yields of up 15 %. It has been found that nickel(II) and copper(I) are not so effective as magnesium. When maleic nitrile was heated with urea, anhydrous nickel(II) chloride and ammonim molybdate as catalyst, nickel(II) tetrazaporphyn (Ni(TAP)) of the general type **5.14** was obtained in yield of 2.5%. Traces of copper(II) tetrazaporphyn were obtained in the presence of cuprous chloride instead of nickel salt. Due to their higher stability, the utilization of the substituted maleic nitriles, in the preparation of the peripherally substituted tetraazaporphyrins is much more conveniently than the unsubstituted parent macrocycle.

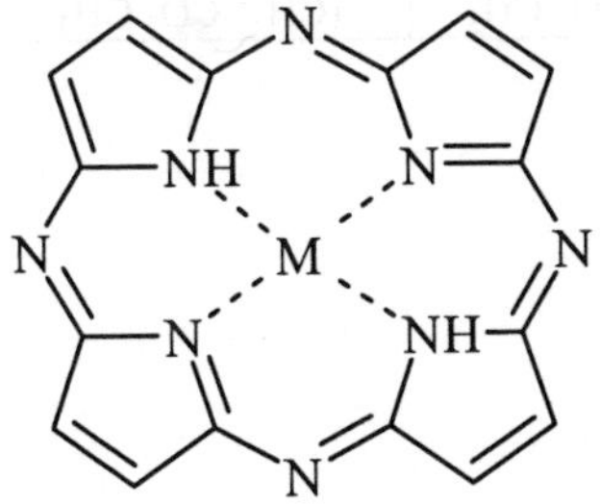

5.14

Methods based on the pyrrole reactions were developed by Rothemund.[355] The author obtained porphine and substituted porphine by condensation of pyrrole with formaldehyde and other aldehyde. In the last cases, it has been obtained porphyne in which the *meso*-positions carry the residue from the corresponding aldehyde as substituent. The method leads to a mixture of isomer from which is hard to separate one of them.

Acid-catalized polymerization of pyrroles has been extensively studied with the hope to obtain porphyrins. Most of these synthetic routes are long and tedious and lead to low yields. For example, when the pyrrole **5.15** is heated for 4-6 hrs in the presence of varying amounts of acetic acid or of hydrochloric acid resulted in a mixture in which trace amounts of porphyrin was formed. When this mixture was neutralized with ammonium hydroxide and treated with zinc acetate, zinc octamethylporphin has been deposited. Experiments have shown that metal ions like nickel(II) or iron(III) are ineffective but salts of zinc and cobalt(II) give small yields of porphyrin. The best results were obtained with copper(I) acetate. Thus, copper(I) octamethyl porphin **5.18** was obtained with 20 % yields.[356]

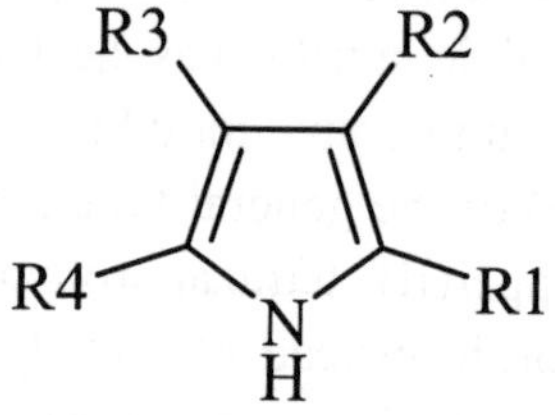

	R1	R2	R3	R4
5.15	CH_2NH_2	CH_3	CH_3	H
5.16	CH_2NH_2	CH_3	CH_3	COOH
5.17	$CH_2CO_2\,CH_3$	$(CH_2)_2CO_2\,CH_3$	CH_3	COOH

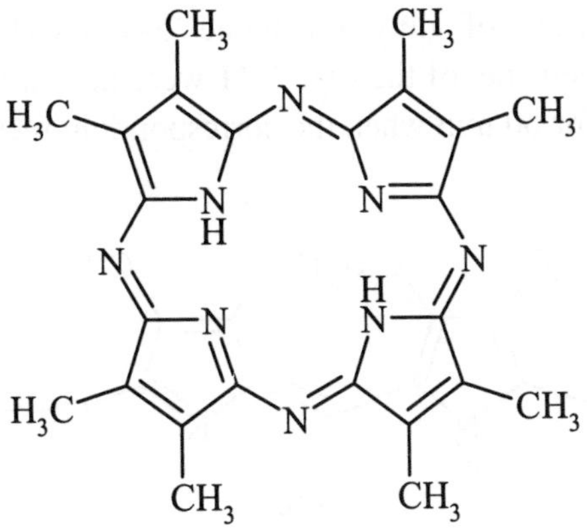

5.18

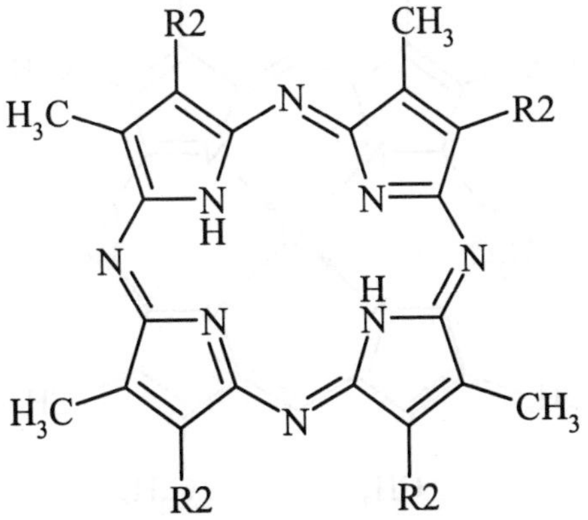

5.19

A coproporphyrin tetramethyl ester has been obtained in 52% yield when the tetramerization of **5.17** is carried out in the presence of cobalt(II) chloride, the cobalt complex of the porphyrin being obtained. In an alternative synthesis coproporphyrin **5.19** tetramethyl ester (R = $CH_2 \cdot CH_2 \cdot CO_2Et$, R' = Bu) was obtained in the presence of zinc acetate, as zinc complex but in lower yield.[357]

Dipyrromethenes have been used as precursors in the preparation of the porphyrins after it had been observed that the action of palladium-strontium carbonate on 5-bromo-5'-bromomethyl-3,4'-diethyl-4,3'-dimethyldipyrro-methene hydrobromide (**5.20**) in ethanolic solution gave the palladium complex of an aetioporphyrin. The cobalt, nickel, copper and zinc complexes of dipyrromethene 5-bromo-3,4'-diethyl-4,5,3'-trimethyldipyrromethene, of the type **5.21** were heated under reflux in *o*-dichlorobenzene for 60 min when the corresponding metal porphyrin was formed.

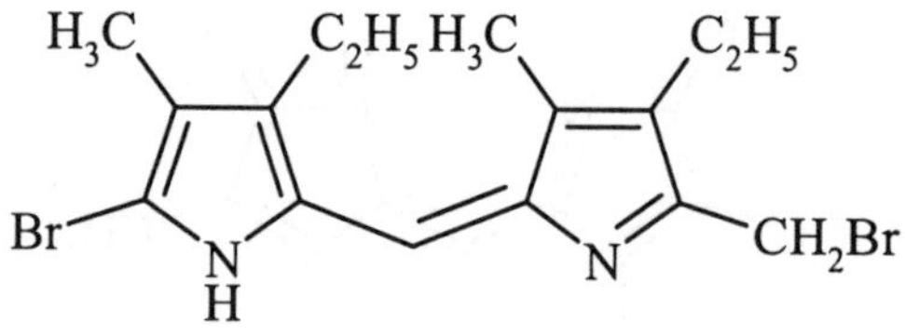

5.20

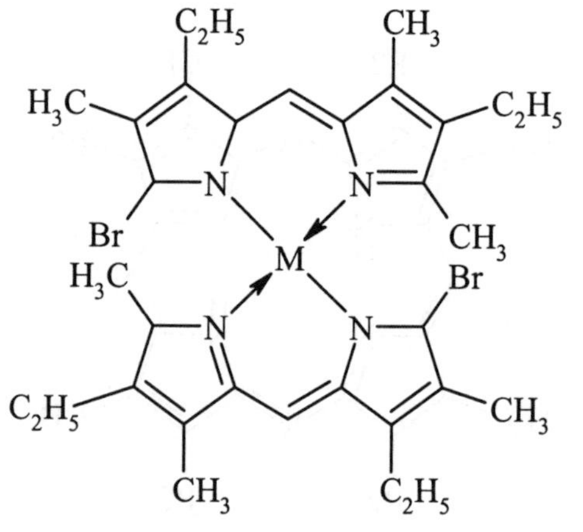

5.21

Cyclization of alkyl-substituted tetrapyrrenes has been developed as method to prepare porphyrins.[358] Thus, α,α'-dimethyltetrapyrrenes were successful closed in the presence of copper acetate in methanol. Following this procedure, unsymmetrical substituted diacety-, diformyl- and dicyanohexamethylporphyin have been obtained in relative lower yields by cyclization of the corresponding tetrapyrroles **5.22-5.24**.[359]

Spectral and chemical evidence proved the surviving cyclisation of the acetyl and formyl groups. Thus, the magnitude of bathochromic shifts of the ultraviolet and visible absorption bands indicate the number of these groups. Also, acetyl or formyl groups can oximated and further, the magnitude of the hipsochromic shifts of the bands indicates the number of these substituents. Two free porphyrins were obtained when the crude product resulted in the cyclization of dicyanotetrapyrrene with copper acetate was treated with concentrated sulphuric acid to remove copper(II). The first porphyrin was a monocyanoporphyrin and the second, a dicyano- one. It result that a cyano group was eliminated during the cyclization. This fact was attributed to a hydrolysis reaction which further leads first to decarboxylation.

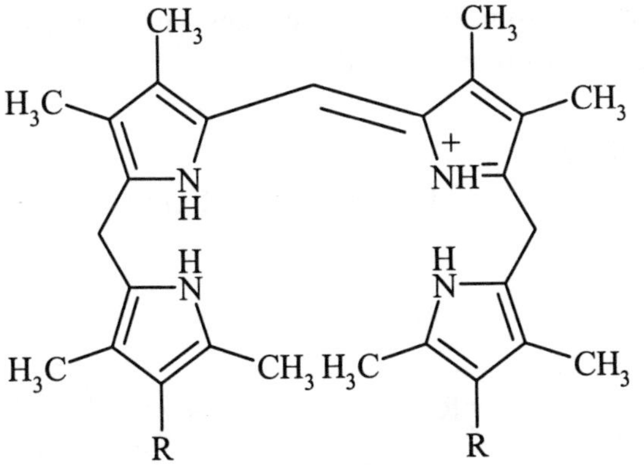

	R
5.22	$COCH_3$
5.23	COH
5.24	CN

Cyclization of a suitable substituted linear tetrapyrrolenes is the classical organic method which leads to mixture of product of various degree of polymerization. Metal-assisted template syntheses of 5,10,15,20-tetraalkylchlorin and tetraalkylporphyrin complexes of transition metals are described. For example, with cobalt(II) as template, only the porphyrins are obtained; with copper(I) only the chlorines whereas, with nickel(II) a mixture of chlorin and porphyrin is obtained depending on alkyl and added anhydride. As opposed to the higher alkyls, (5,10,15,20-tetramethylporphyrnato)nickel(II) dimerizes in solution, a dimerization constant of $3.9+/-1.3$ M^{-1} being derived from 1H NMR data.[360]

The problem which arises in the synthesis of porphyrins in this way is their contamination with undesired products. For example, the

5,10,15,20-tetramethylpophyrinatonickel(II)   (Ni(TMP))   **5.26**   is
impurified with 5,10,15,20-tetramethylchlorinatonickel(II) (Ni(TMC))
5.25. The amount of this contamination is found to depend on the nature
of the metal ion and that of the substituents. The metal directed synthesis
of metal-porphyrins according to Scheme 5.6 shows that when R = CH$_3$,
cobalt(II)   and   copper(II)   lead   to   the   corresponding   *meso*-
tetraalkylmetalloporphyrin whereas for nickel(II), the ratio Ni(TMC) **5.25**
to Ni(TMP) **5.26** is 4:1.[361] However, when R is ethyl, propyl the
(tetralkylporphyrinato)nickel(II) complexes are formed whereas no
copper(II)porphyrinates are formed.

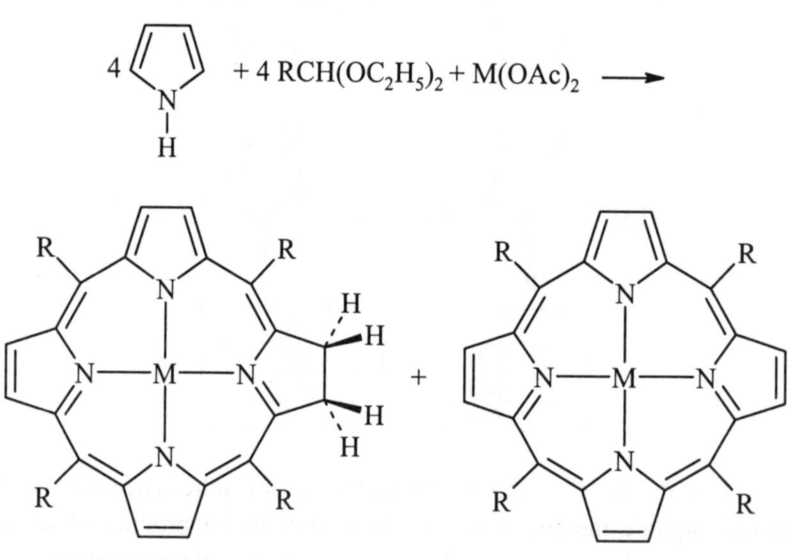

$$4 \; \text{(pyrrole)} \; + 4 \; RCH(OC_2H_5)_2 + M(OAc)_2 \longrightarrow$$

H$_2$TMC   H$_2$TMP
5.25   **5.26**

Scheme 5.6

Several new skeletal relatives of Pc have been obtained in which
nitogen-sulfur heterocycles replace the phenyl rings. Thus, tetra-2,3-
thiophenoporphyrazine has been obtained as an impure copper complex
in   1937.   A   second   complex,   the   copper(II)   complex   of
tetrakis(thiadiazole)porphyrazine **5.27** is the second example obtained
from 3,4-dicyano-1,2,5-thiadiazole as organic precursor.

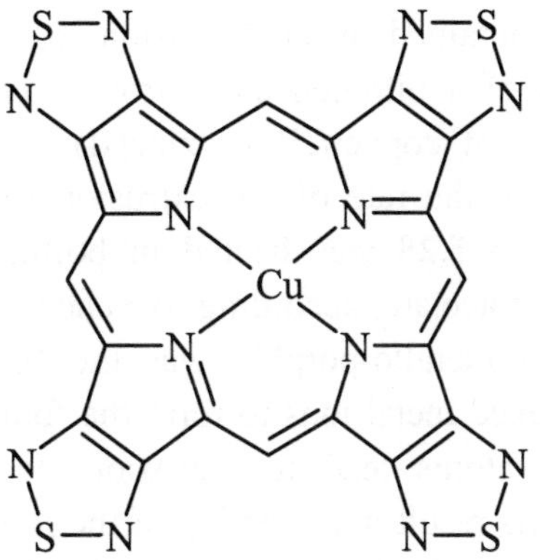

5.27

In the last years, a special attention has been payed to assembling metalloporphyrins into oligometalloporphyrins due to their potential as biological mimetics and, in the same time as advanced materials.[362]

5.3. Corrole

Corrole – the one-carbon short analogue of porphyrin, is a tetradentate macrocyclic ligand where three of the four nitrogen atoms carry replaceable hydrogen atoms as it result from the tautomeric forms **5.28**. As porphyrins, corroles contain an aromatic 18 p-electron chromophore. The ring is numbered in the same manner as porphyrins as it results in formula **5.28**.

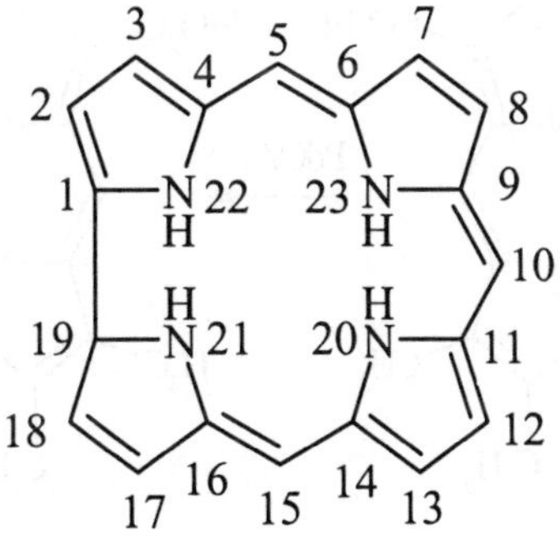

5.28

Corroles can be obtained by an extrusion reaction from a porphin. Thus, the expulsion of a nitrogen atom from a *meso*-homoazaporphin **5.29** in the presence of copper(II) or zinc(II) was observed. Also, a remarkable increase in the rate of the extrusion reaction was observed when *meso*-thiaphlorin **5.28** was heated in boiling acetic acid in the presence of palladium acetate, according to Scheme 5.7.[363] In both cases the formation of the metallo-porphin was the first step. Further, the efforts of the complexed metal ions to form the four equivalent bonds to the ligand nitrogen atoms lead to extrusion. In the case of *meso*-thiaphlorin, this efforts result in a twisting of the *p*-orbitals on the carbon atom flanking the sulphur bridge in a manner which favours a disrotatory cyclisation.

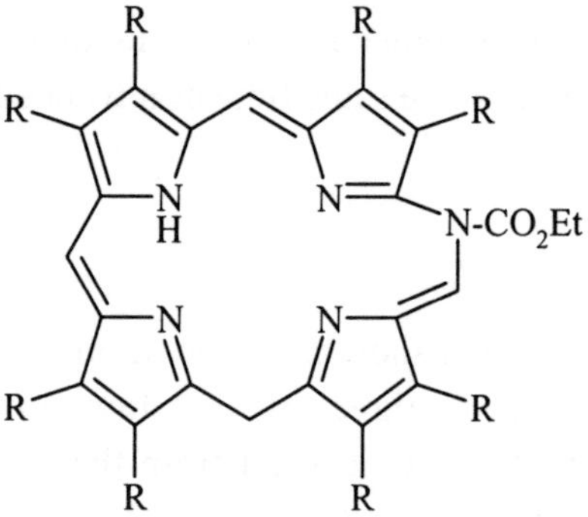

5.29

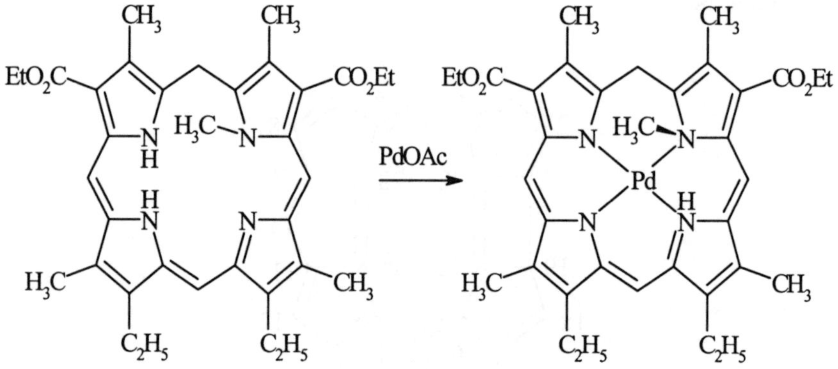

5.30   **5.31**

Scheme 5.7

However, two general reactions have been developed for the cyclisation to corroles. The first one involves the cyclization of 5,5'-bis(5'-bromodipyrromethenes), **5.32** with introduction of an extra *meso*-carbon. Starting from 5,5'-bis(5'-bromodipyrromethenes)palladium(II) derivative, oxa-, imino-, and thia-analogues of corroles have been obtained. For example, the cyclization of palladium derivative of 5,5'-bis(5'-bromodipyrromethenes) with formaldehyde and hydrochloric acid resulted in the cyclic ether as ligand.

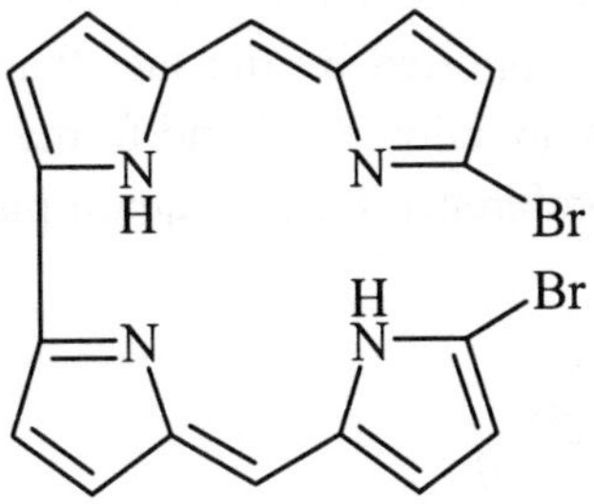

5.32

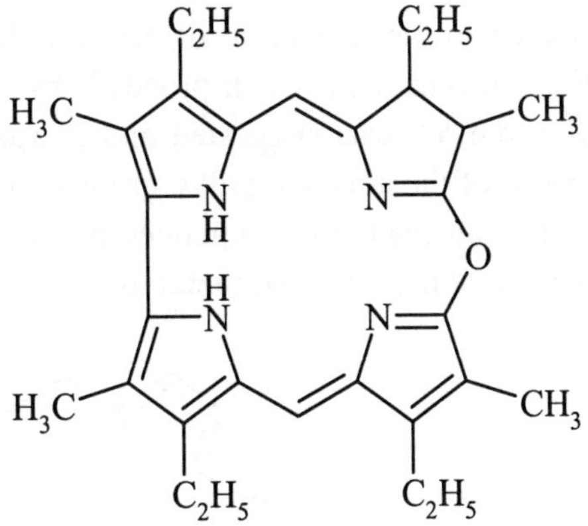

5.33

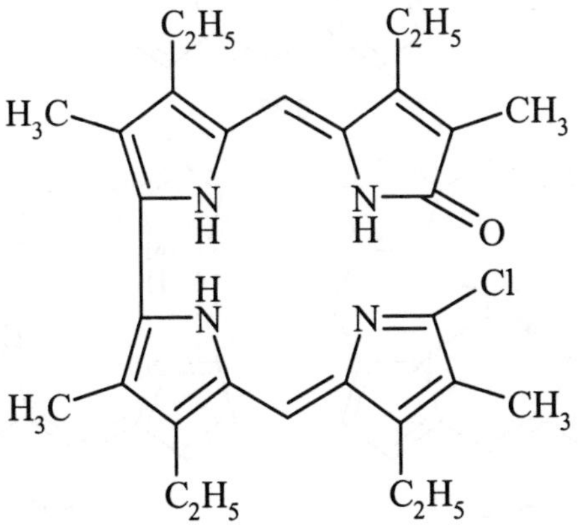

5.34

Copper(I) and nickel(II) salts act as template to form **5.33** only when the sarting precursor is the monoamide **5.34**.[364] It is assumed that an initial step involves the tautomerism of the monoamide to the lactim form which then forms a metallic complex. The mechanism finds analogy in the formation of the vitamin B_{12} lactone.

In the second approach, the cyclization of 1',8'-dideoxybilenes-b, **5.35**, occurs around metallic ions by irradiation. As metal ions, nickel(II), copper(II) or zinc were used. Based on the visible absorption spectra it has been established that these green corroles are derived from the tautomeric non-aromatic base, **5.36**. However, addition of alkali to the nickel complex causes the appearance of a red solution which was attributed to the formation of the complex derived from aromatic anion **5.37**. A mechanism which proceeds by electron shifts like that depicted in Scheme 5.8 was suggested and it was supported by the fact that the presence of the adjacent bulky groups does not allow the bond formation. Also this mechanism is explained by the preference for the square-planar geometry of the involved metal ions.

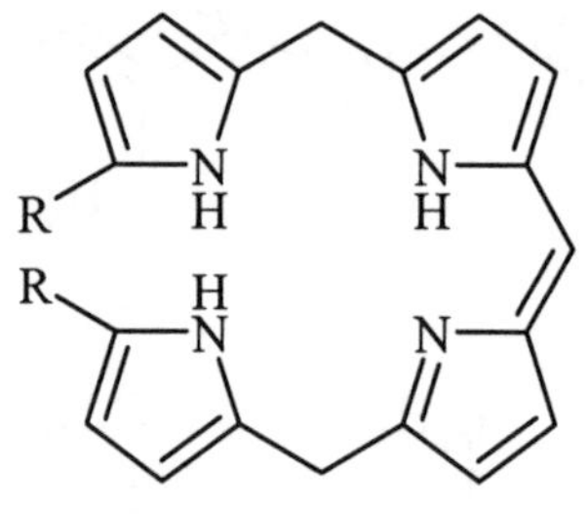

5.35

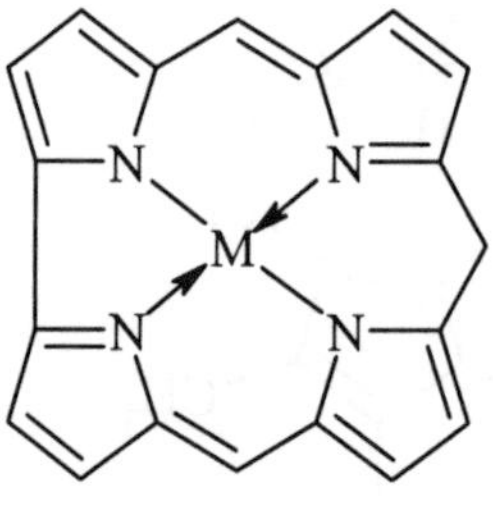

5.36

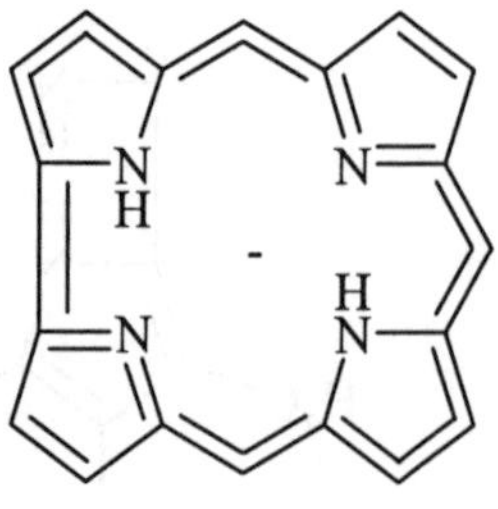

5.37

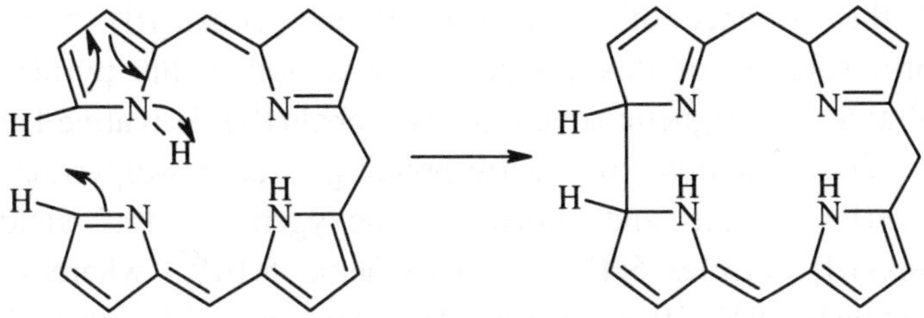

Scheme 5.8

Complexes of zinc are known to exist in a tetrahedral configuration in most cases; so, it has been expected that in the presence of zinc(II) the cyclization is not favoured because in the 1',8'-dideoxybiladiene-ac zinc complex, the terminal groups will not have the adequate positions. The lower yield and the stability of the zinc corroles have been explained in these terms.

This second approach also occurs when A/D-secocorrin ring **5.38** is closed to corrin **5.39** around nickel(II) *via* cycloisomerisation as in Scheme 5.9.[365] The reaction is initiated either by light excitation or by a redox reaction in dark. In both cases the square-planar geometry of the molecule maintained by nickel(II) is important.

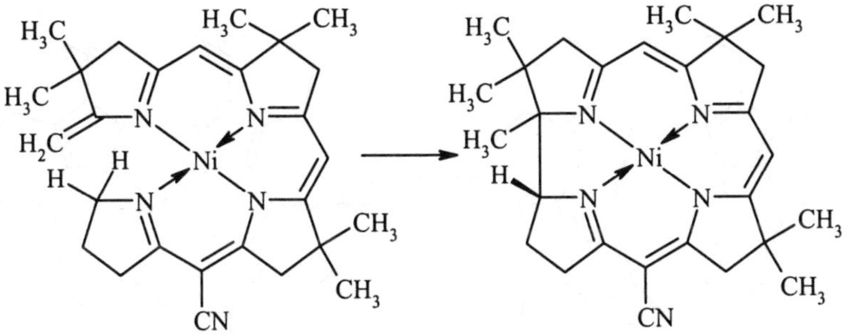

5.38   **5.39**

Scheme 5.9

It is known that corroles stabilise unusually high oxidation states of transition metal centres like iron(IV), cobalt(III) or copper(III). For example, cobalt(III) corroles have been obtained by cyclization of 1,19-dideoxy-biladienes-*ac* in hot methanol containing cobalt(II) acetate. The infrared spectra of the pyridine adduct of the resulted complex **5.39** supports the presence of the pentacoordinate cobalt(III).[366] When a chloroform solution of this compounds was boiled, the pyridine was detached and paramagnetic square planar cobalt(III) derivative has been formed. The reaction of corresponding 2,2'-bis-dipyrrins with manganese(II) acetate and molecular dioxygen in dmf yields the manganese(III) corroles **5.41** – **5.43** (Scheme 5.10[367]) whereas Pd(II) leads to palladium(II)-10-oxacorrole. The manganese(III)corrole **5.41**, is oxidated by chlorinated solvents to manganese(IV)corrole or can be demetallated with HBr in acetic acid.

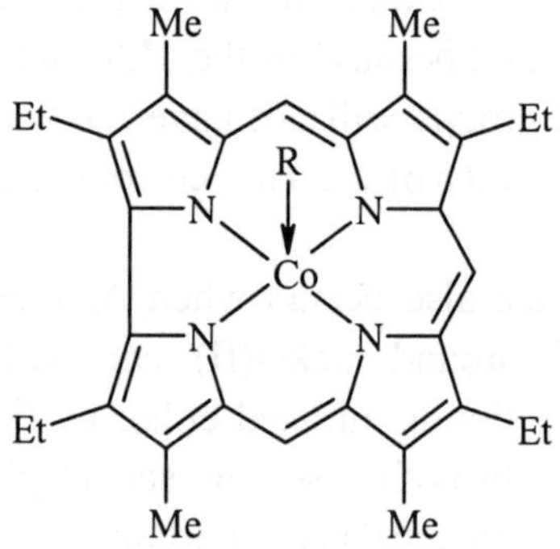

5.40

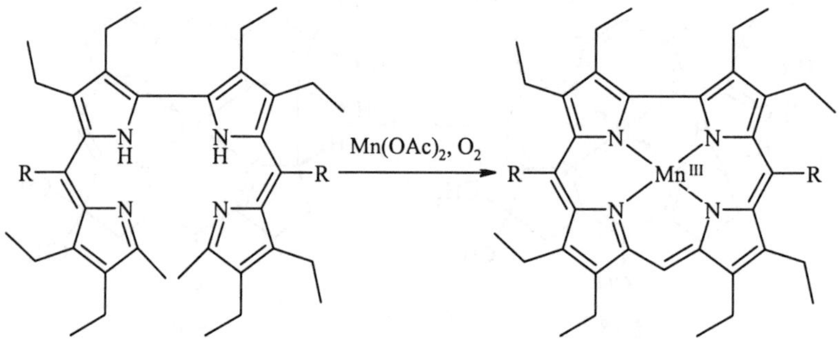

	R
5.41	H
5.42	C_6H_5
5.43	p- $CH_3 C_6H_4$

Scheme 5.10

Macrocyclic corrole rings have been obtained through reactions of the 2+2 type. For example, when a mixture of 3,3'-diethyl-4,4'-dimethyl-2,2'-bipyrrole-5,5'-dicarboxylic acid **5.44** and 3,3'-diethyl-5,5'-diformyl-4,4'-dimethylpyrromethane **5.45** in methanol was acidified at 0 °C with hydrobromic acid and further heated with cobalt(II) acetates and triphenylphosphine, resulted in cobalt(III) 2,8,12,18-tetraethyl-3,7,13,17-tetramethylcorrole containing an axial triphenylphosphine ligand. The attempts to cyclise such compounds in the absence of complexing metal, were unsuccessful thus stressing the importance of the cobalt, both for stabilisation of the intermediate and in its template action in holding the reactive sites in proximity.

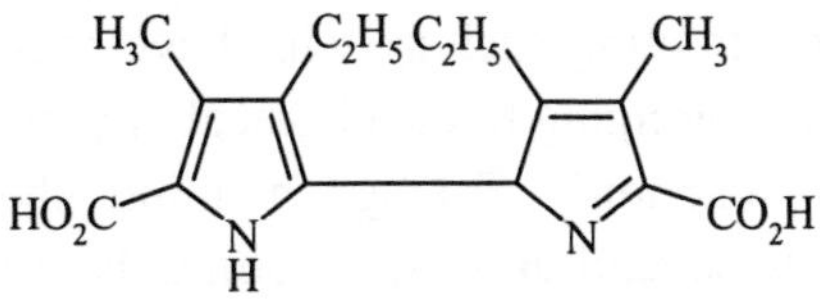

5.44 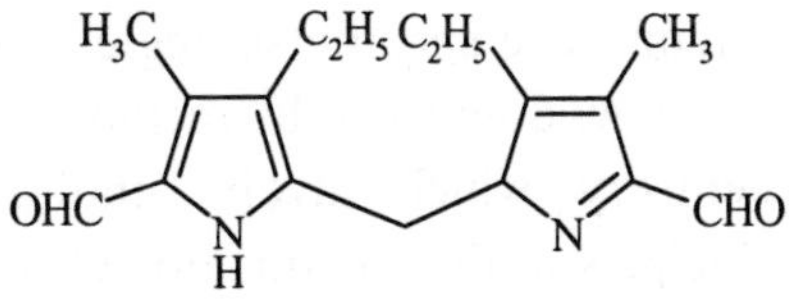   **5.45**

Chapter 6

Self-Assembled Systems

Discrete molecular architectures play an important role in the advanced material sciences, in the field of host-guest chemistry as well as in the understanding of complicated structures of native biological systems. The specific strategies for constructing such systems have been up to now dominated by methodologies belonging to the classical organic chemistry which involve multistep reactions. Taking in mind the principles of the coordination chemistry, several groups, including that of Lehn,[368] Sauvage[369], Constable[370] or Fujita[371] developed an elegant method based on the use of metal templating combined with the appropriate ligand design, and the employment of noncovalent interactions. The method refers to the one of the earliest examples of molecular self-assembly, a basic concept of supramolecular chemistry.[1] Although the metal ions with specific coordination geometries are the key organizing elements, the self-assembly processes are guided by electronic and structural information stored in the molecular component units and display a remarkable dependence on the choice of the experimental conditions. The obtained structures display a range of novel physicochemical properties not expressed by their isolated component part such as redox, magnetic, optical and catalytic functions.

Application of template method is quite attractive since the usual synthetic method is frequently tedious and low-yield process. Recent progress in this area has generated an explosion of reports. For the purpose of this review, we will give a short survey on the progress made in the metal-mediated synthesis of panels, catenanes, rotaxanes, helicates, panels and cages, ladder and grids.

6.1. Catenanes

6.1.1. Two interlocked rings

Catenanes consist of interlocked rings, usually able to glide freely one into the other. They are mechanically interlocked molecules and are topological isomers of their constituents. Following the pioneering work of Schill and Wasserman[372, 373, 374] template-directed strategies that use transition metal complexation developed by Sauvage *et al.* have made the synthesis of interlocking rings more accessible.[375, 376] The three-dimension template synthesis around a transition-metal ion involves two strategies as they are presented in Scheme 6.1. As can be noticed, in both strategies, the intermediate is a metallocatenane (catenane) which further by demetallation leads to the free ligand - a catenand.

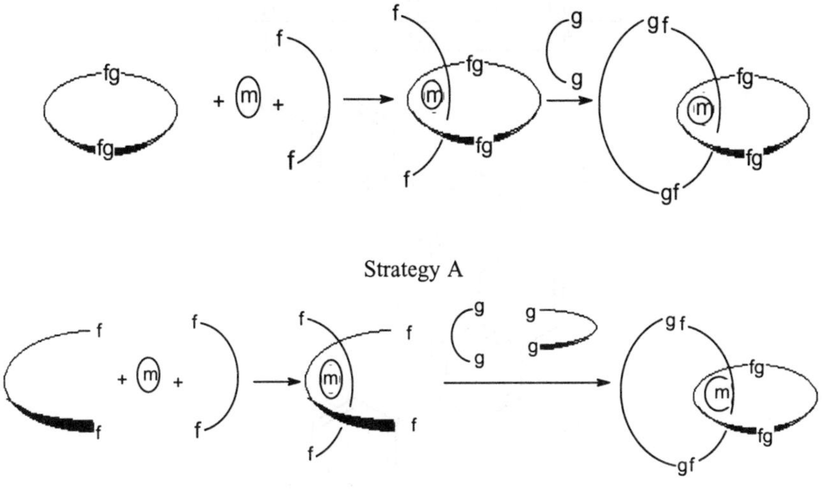

Strategy A

Strategy B

Scheme 6.1.

The proposed strategies are based on the fact that 1,10-phenantroline-based ligands coordinate to copper(I) in a tetrahedral manner so that they are mutually perpendicular and arranged into a suitable mode for a further cyclization.

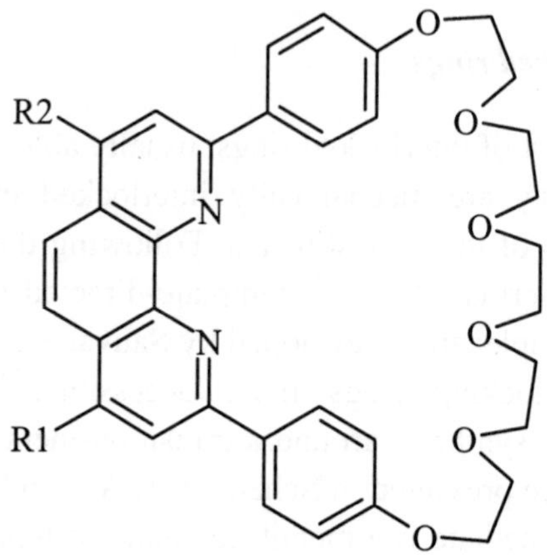

	R¹	R²
6.1	H	H
6.2	C_6H_5	C_6H_5

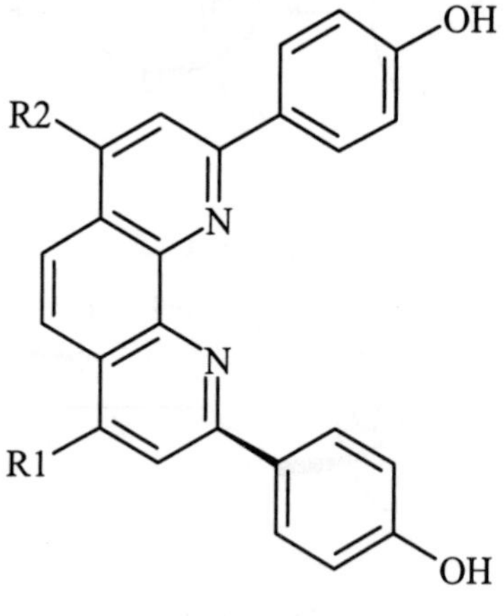

	R¹	R²
6.3	H	H
6.4	H	C_6H_5

The copper(I)-catenane **6.6** was obtained from the macrocyclic complex [Cu(**6.1**)]⁺ which was allowed to react with **6.3**. In a second step, the cyclization of **6.5** with 1,16-diiodo-3,6,9,12-tetraoxatetradecane in the presence of Cs_2CO_3 according to the strategy A in Scheme 6.1 results in the interlocked rings.[377] Following the second strategy, an one pot synthesis, the ablility of Cu(I) to bind two molecules of a 2,9-disubstituted 1,10-phenantroline, **6.3**, in a tetrahedral complex was used and, in the resulted organization, the cyclization of each phenantroline fragment occurs and leads **6.6**.[378] The copper(I)-[2]-catenane **6.6** was demetallated resulting in the corresponding catenane.

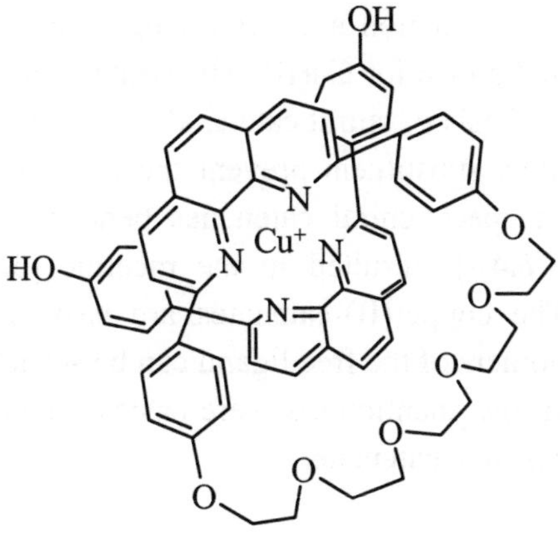

6.5

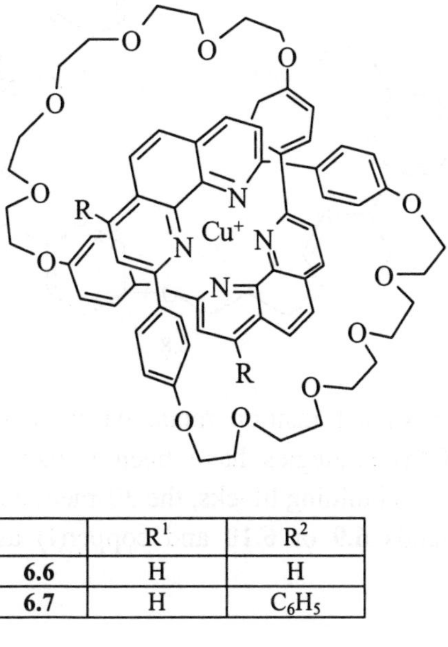

	R^1	R^2
6.6	H	H
6.7	H	C_6H_5

Although the two species differ only by a copper atom, their respective molecular shapes are totally different, as shown the X-ray study.[379] Recomplexation of the catenane by copper(I) is quantitatively carried out, the rearrangement of the molecular geometry being reversible.

Using a ring containing two different subunits, **6.8**, and **6.3**, a copper(I) [2]-catenate which shows a molecular isomerism has been obtained.[380] The isomerism can be induced electrochemically and it is based on preferred coordination number for the two redox states of the metal: 4 for Cu(I) and 5 or 6 for Cu(II). The strategy B was developed to realize templated synthesis of chiral catenands[381] when **6.4** has been used as ligand, as the bulky substituent prevent the rings to glide freely one into other. In this respect, chiral catenanes behave like a rotaxane.[382] Cyclization of [Cu(**6.4**)$_2$]$^+$ resulted in the racemic [**6.7**][BF$_4$] as deep crystalline solid. The copper(II)-catenates are easy demetallated with KCN and the enantiomers of the free ligand can be separated. However, it has been shown that the enantiomers were easiest separated at the stage of copper(I) catenates than catenane.

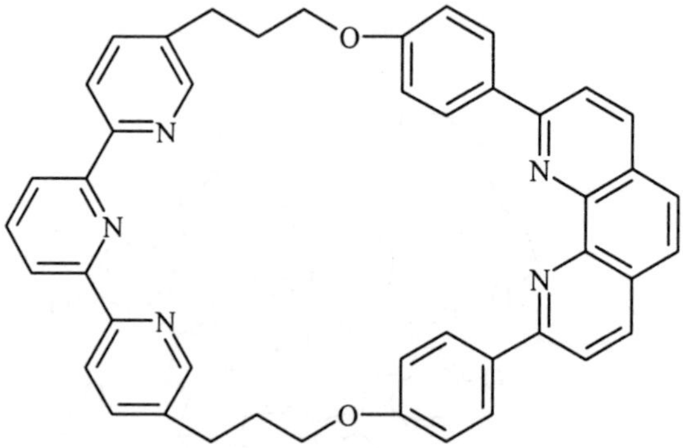

6.8

Combination of a transition metal based template and ring-closure metathesis (RCM) strategies have been utilized to provide access to [2]catenanes.[383] As building blocks, the 30-membered macrocycle **6.1** and the acyclic ligands **6.9** or **6.10** and copper(I) as template were used.

Further, the obtained complex precursors suffer intramolecular RCM to yield the corresponding catenanes **6.11**.

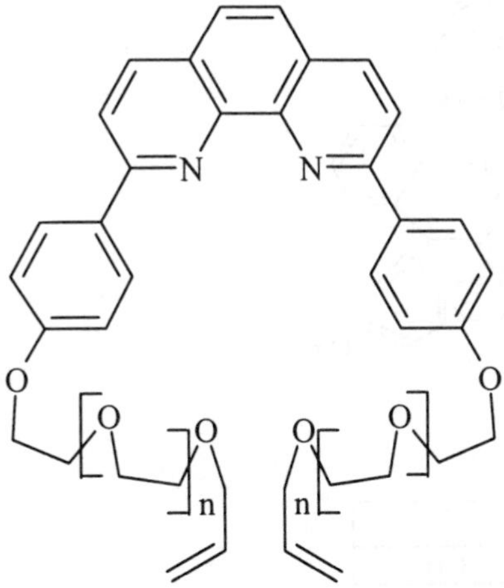

	n
6.9	1
6.10	2

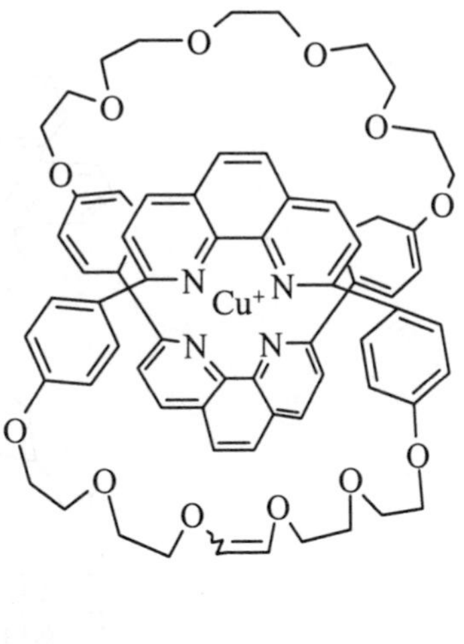

6.11

Reaction of the Cu(I) with the ligands **6.9** or **6.10** in molar ratio 1:2 yielded the corresponding complexes of the type $[Cu(L)_2]^+$ which further lead to the interlocked molecules **6.12** by the same RCM reactions.

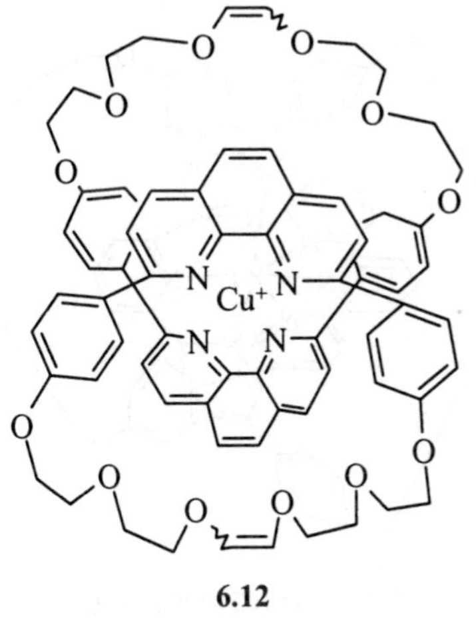

6.12

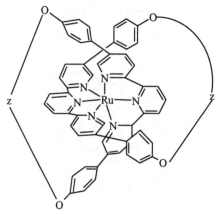

	Y
6.13	CH_3
6.14	H

Taking advantage of the high stability of the complexes of the $[M(terpy)_2]^{2+}$ type, where terpy stands for 2,2':6',2"-terpyridine, ligands containing this coordinating motif have been designed with the aim to obtain catenanes.[384] Three-dimensional template synthesis of two interlocked macrocycles have been obtained around octahedral Ru(II) leading to the precursors **6.13** and **6.14** and further to $[\mathbf{6.15}][PF_6]_2$.[385]

6.15 $Z = -CH_2(CH_2OCH_2)_5-CH_2-$

Ruthenium(II) ion serves as the template in this case in the same way as copper(I) serves as the template in the syntheses of the phenantroline-based catenands. The key step in the synthesis involves reaction of the acyclic precursors with two equivalents of the diiodo derivative of hexaethylene glycol under high dilution conditions in DMF. The position of the terpy substituents is very important in this stage. Indeed, the 6,6'-disubstituted terpy ruthenium(II) complex shows a hindered environment which results in a relatively weakly bound and highly labile complexes. By contrast, in the 5,5'-substituted terpy, the substituents do not interfere at all with metal-ligand binding and favours the steric requirement for the phenol oxygen atoms for cyclization. As in the previous syntheses of catenands, the Cs_2CO_3 acts as a convenient deprotonating agent for the phenol functions, the cesium phenolat so obtained being particularly active nucleophilic species.

The incorporation of a coordination angle into metal-organic network, a strategy developed by Fujita's group, has been used to obtain interlocked rings. Pd-[2]catenane **6.16** was obtained by reaction of Pd(en)(NO$_3$)$_2$ with appropriate ligand.[386] **6.16** is stable at high concentrations of polar media whereas, at lower concentrations, a rapid equilibrium between **6.16** and **6.17** was observed. For the more stable Pt(II) analogue, the catenane structure was confirmed by crystallographic study.[387]

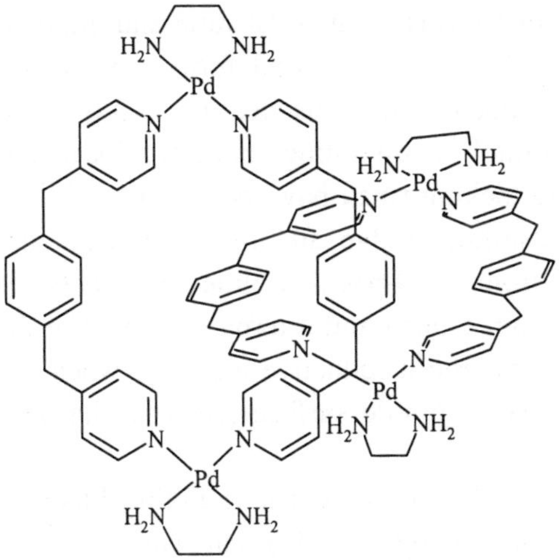

6.16

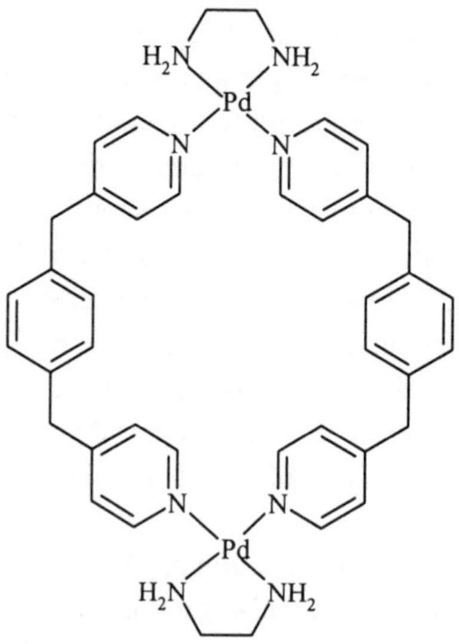

6.17

Polycatenanes containing oxygen donors has been obtained by Williams *et al.*[388] using manganese(II) as template and N,N'-*p*-phenylenedimethylenebis(pyridin-2-one)], (*p*-XBP4), as ligand. The complex {[Mn(*p*-XBP4)$_3$](ClO$_4$)$_2$}$_n$ contains the ligand molecules in two distinct geometries. The two forms are in a 2:1 ratio and the first form creates an open 34-membered ring. The second form acts as bridge which links pairs of the 34-membered ring, thus creating a set of 68-membered ring.

Disulfide and thiol incorporating copper(I)-[2]catenanes consisting of two different interlocking rings have been obtained.[389] One of the rings incorporates a disulfide bridge and the other contains either a single chelate arising from 1,10-phenantroline fragments or two different chelates arising from a phen and a terpy moiety. The disulfide bridge is enough labile to undergo a reductive cleavage when the complexes are deposited on a gold electrode surface whereas, the phen or terpy rings remains unchanged.

Porphirins have also been used as building blocks for constructing novel structures and topologies such as catenanes.

6.1.2. Three interlocked rings

Three interlocked macrocyclic subunits have been obtained following two strategies which use the generalized template effect around a metal ion. The first one involves the eight centres to be linked and is similar to strategy A in Scheme 6.1. In this case, the cyclization occurs by an alkyation with an alkyl-diiodide or –dibromide. The second one is based on acetylenic oxidative coupling and requires four reacting centres. It has been proved that the second strategy leads to the higher yields. Following the first strategy, the dicopper(I)[3]-catenates **6.18** or **6.19**, and the corresponding copper(I)[2]-catenate, as by-product have been synthesized[390] when the starting complex **6.5** reacted with $BrCH_2(CH_2OCH_2)_2CH_2Br$ or $BrCH_2(CH_2OCH_2)_3CH_2$ Br in DMF under high-dilution condition in the presence of cesium carbonate. The free ligands were obtained by demetallation and the structural changes induced by the presence of the third ring were examined by comparing the [1]NMR spectra of both the catenands and crystal structure of the [3]catenand.[391]

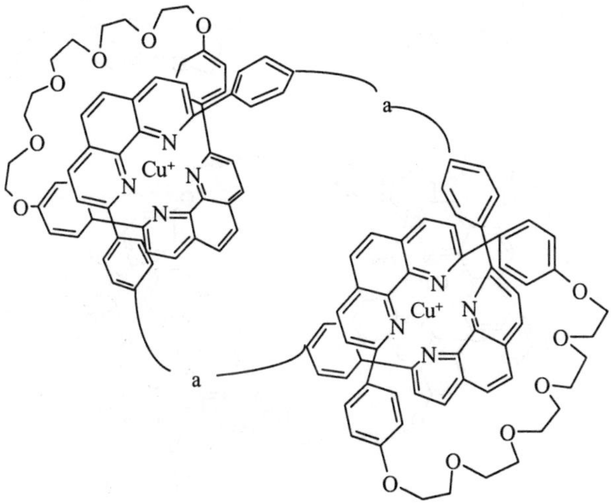

	a
6.18	-O(CH$_2$CH$_2$O)$_3$-
6.19	-O(CH$_2$CH$_2$O)$_4$-

It has resulted that the size of the central ring is important in both cyclization step and demetallation. Thus a better yield was obtained for **6.18** than for **6.19,** thus denoting that the 48-membered central ring is more easily formed than the 54-membered ring. On the other hand, the flexibility of the greatest ring favours the demetallation and the stability of the free ligand. [3]catenane has been obtained[392] in a 58% yield, as ligand in a copper(I) complex **6.20,** following an oxidative acetylenic coupling reaction to close the third macrocycle according to scheme 6.2. The [3]catenane **6.20** is a binuclear copper(I) complex, consisting of a central 66-membered cycle interlocked to three 30-membered rings.

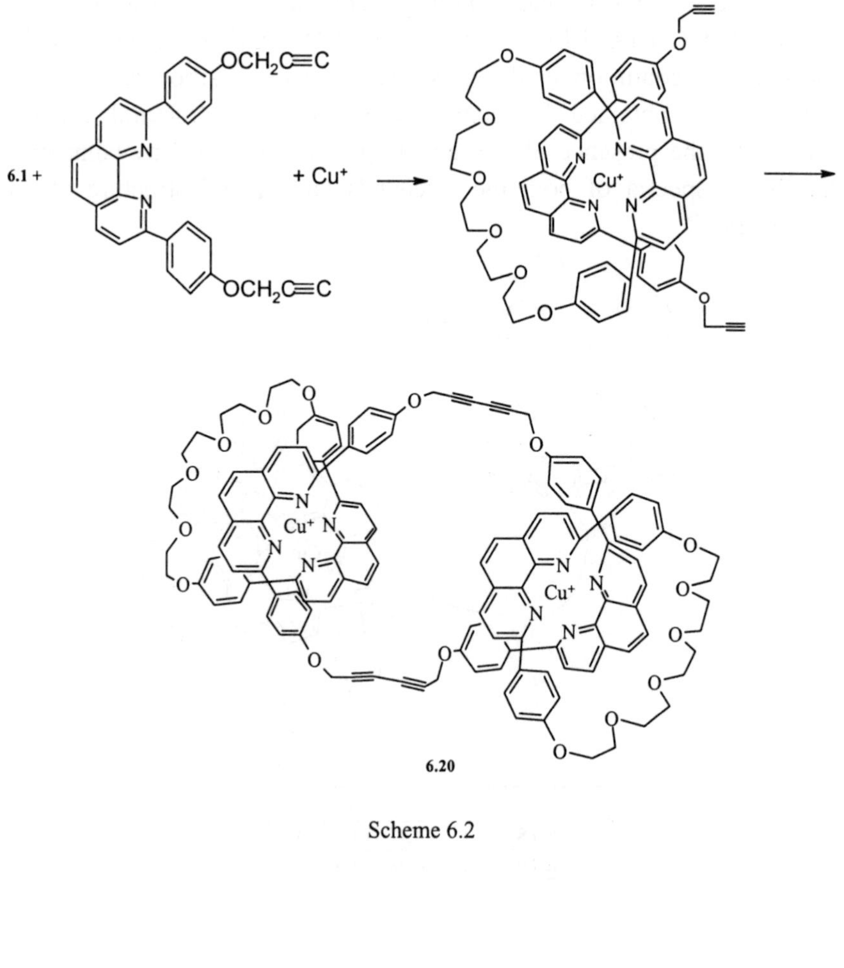

6.20

Scheme 6.2

6.2. Rotaxanes

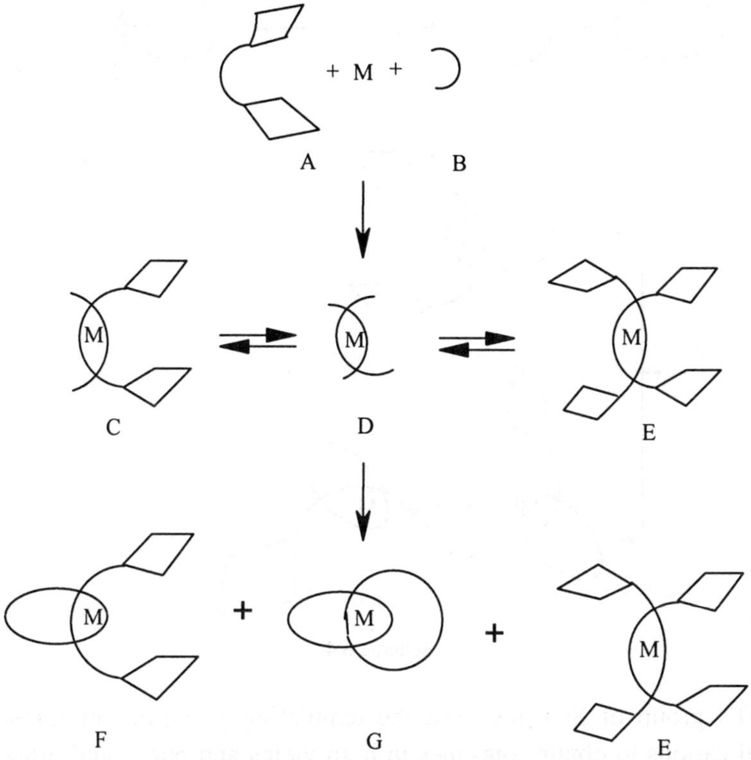

Scheme 6.3

Rotaxanes are molecular compounds composed of one or more macrocycles, trapped on the thread on the thread-like portion of a dumbbell component by large blocking groups or stoppers at its two ends. They are mechanical isomers[393] of the constituents although they are topological isomers. The rotaxanes can be regarded as intermediates to catenanes or as a catenane for which one of the rings would be infinitely large. Until earlier 80' the purely statistical low-yielding slipping methods were used.

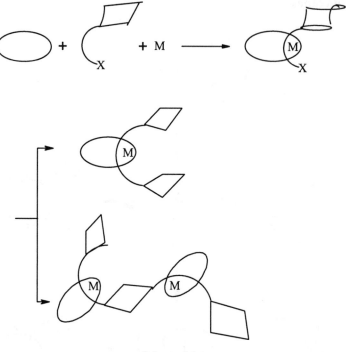

Scheme 6.4

The group of Sauvage used the templating properties of transition metal cations to obtain rotaxanes in high yields and purity and proposed two possible strategies. The first one involves the construction of a ring around the metal ion. This is a statistical step controlled by the kinetic stability of the formed species as it can be noticed in Scheme 6.3. A second strategy involves the threading of a molecule inside a pre-synthesised cycle and then its blocking by stoppers (Scheme 6.4). As it can be noticed, this second strategy is related to that used to obtain catenanes or knots. The authors[394] described the synthesis of [2]-rotaxanes (consisting of one ring and one thread) and [3]-rotaxanes (consisting of two rings and one thread) as an unexpected product following the strategy in Scheme 6.4.

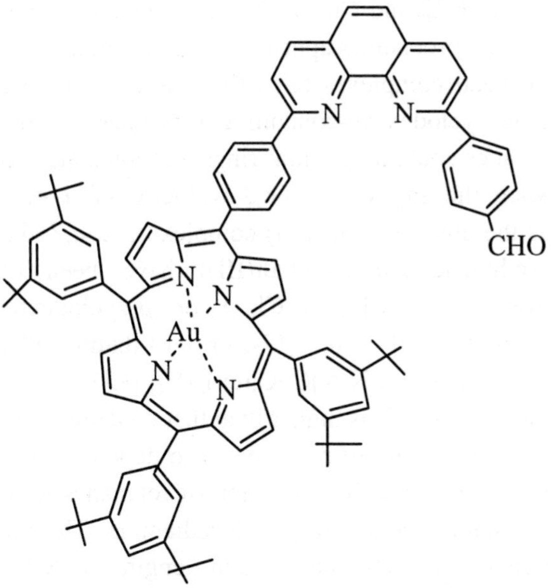

6.21

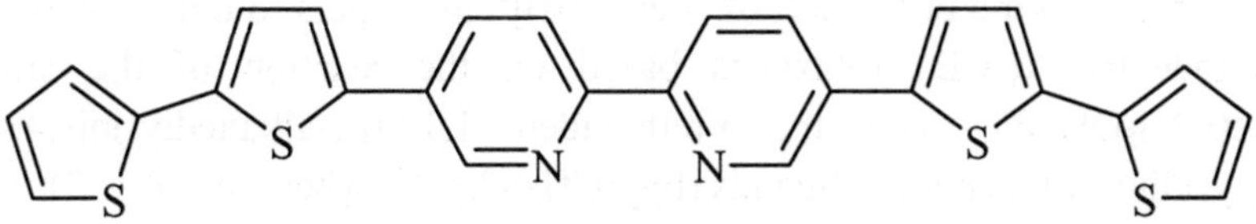

6.22

The Cu(I) complex of **6.1** played the role of the macrocyle whereas a rigid held gold(III) porphirins **6.21** acted as stoppers. The demetallation leads to a species in which a molecular ring is entrapped by the bulky end groups of the phenantroline-bridged bi-porphirin through it is threaded. However, it can be remetallated with Zn^{2+} when the [2]rotaxane is restored as in the template synthesis. Porphirins have frequently been used as building blocks for constructing novel structures and topologies such as rotaxanes

Due to a practical interest, namely to obtain new sensory and conduction functions, polymetallorotaxanes have been obtained by electrochemical polymerization of metallorotaxanes.[395] For example, metallorotaxane **6.23** has been obtained using Sauvage's macrocycle **6.1**, threading element **6.22** and Cu^+ or Zn^{2+} as template. Compared with the rotaxane, the metallorotaxanes appear to be associated.

Transition metal complexes have been used as stoppers to obtain a series of stable cyclodextrin-containing rotaxanes in good yields. For example, rotaxanes containing chiral rings and rotaxane containing metal complexes where the ring were *a-* or *β*-cyclodextrin (*a-* or *-β* -CDX) and the chain was the dimeric cobalt(III) complexes, (μ-*a,?* -diaminoalkane)-bis[chlorobis(ethylenediamine)-cobalt(III)] have been obtained. The reaction between *cis*-[CoCl2(en)$_2$]Cl, *a-* or *-β*-cyclodextrin (*a-* or *- β* CDX) in the presence of 1,10, 1,12-, or 1,14-diaminoalkanes gave the rotaxanes [2]-[[(en)2ClCo(N-N)CoCl(en)$_2$]X$_4$]-[CDX].[396] The yield of rotaxanes containing -CDX is higher than that containing -CDX due to its narrowed cavity. The yield of rotaxanes also depends on the methylenic chain length of N-N. Thus the formation of rotaxanes was not observed for N-N = 1,8-diaminooctane or *p*-xylenediamine. The yields increases with the increase of methylenic chain length of N-N up to 1,12-diaminododecane, then with 14-methylenic chain, the yield decreased. This was explained in terms of steric hindered phenomena. Macartney *et al.*[397] reported the utilization of [Fe(CN)$_5$)]$^{3-}$ as stopers to obtain stable (-cyclodextrin ((-CD) rotaxanes based on the reaction of the labile [Fe(CN)$_5$OH$_2$)]$^{3+}$ with the prethreaded 1,1''-((,(-alkanediyl)bis(4,4'-bipyridinium) dicationic ligands (bpy(CH$_2$)nbpy^{2+}, where, n = 8 –12).

Starting from the ring **6.1** and the molecular string **6.24**, containing two different coordination states, a 2-alkyl-9-phenyl-1,10-phenantroline bidentate chelate and a terdentate ligand 2,2',6',2''-terpyridine(terpy), and using Cu(I) as template, a pseudorotaxane structure which undergoes molecular motions by reducing or oxidizing the complexed copper center have been obtained.[398] As for the above mentioned [2]-catenates, the process is based on the difference of preffered coordination number for the two different redox states of the metal: 4 for Cu(I) and 5(or 6) for Cu(II). Depending on the copper oxidation state, a gliding of the ring from one part of the string to the other occurs.

A doubly threaded complex has recently been obtained by copper(I)-induced assembly of two self-complementary units and its structure was confirmed by X-ray analyses.[399]

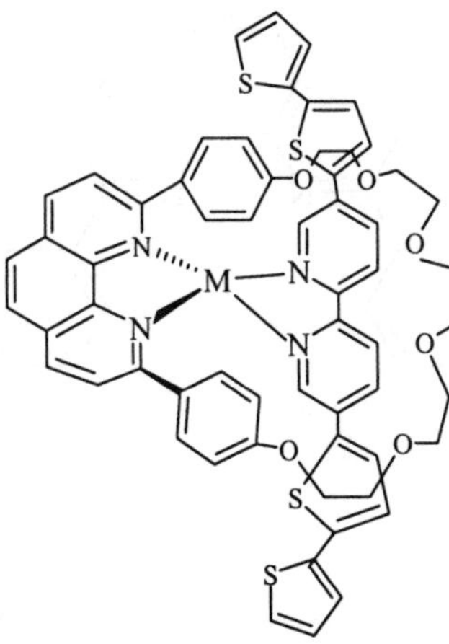

6.23

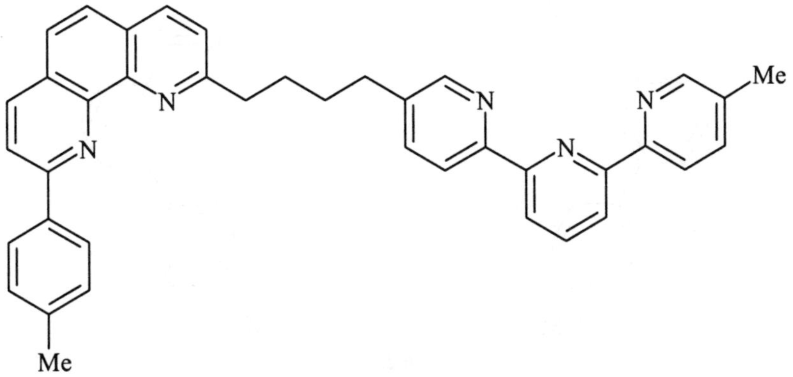

6.24

The strategy for making rotaxanes and catenanes, using two different types of transition metals, both acting as assembling and templating species but utilizing different coordinating fragments of the organic back-bone precursors have been described.[400, 401] The first metal centre will gather and orient two ligands, leading to the intertwined system which has been used extensively in the past for building interlocking rings. The second metal will effect a clipping reaction on the intertwined precursor and thus freeze the intertweaving or the interlocking situation. This step is very important in that concern the product, a rotaxane or catenate. The careful removing of the first templating metal ion preserves the structure.

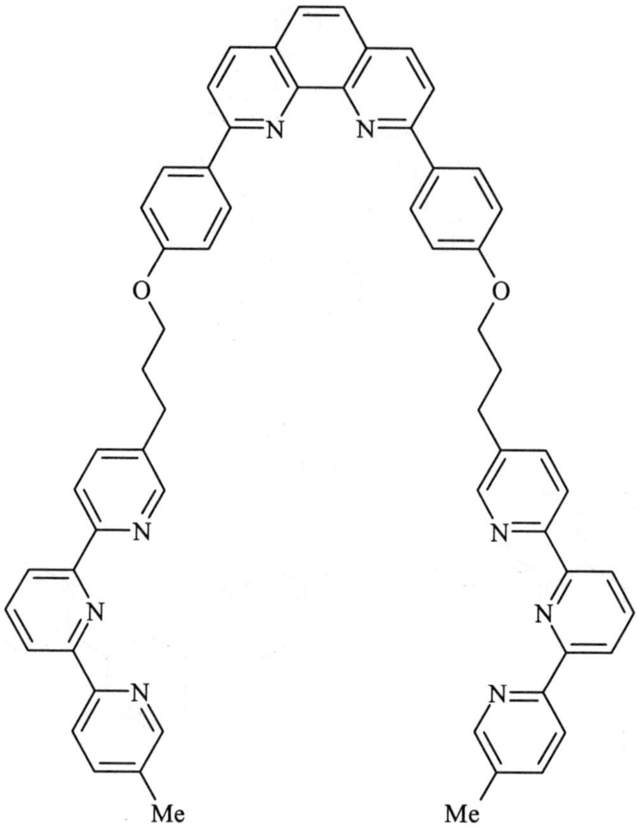

6.25

Practically, the regiospecific threading step involves the ring **6.1**, the threading molecule **6.25** and [Cu(MeCN)$_4$]BF$_4$ and lead to the intermediate **6.26**, in accordance with the preferred tetrahedral coordination for copper(I) complexes. The free terpy moieties have been further used for the preparation of a catenane **6.27** and a copper complexed rotaxane **6.28**.

6.26

6.27

The macrocyclisation coordination reaction occurs with $Ru(DMSO)Cl_2$ whereas the use of $Ru(terpy)(acetone)_3(BF_4)_2$ leads to the rotaxane. Attempt to realize copper(I) directed threading of molecular strings into more than one coordinating ring resulted in[402,403] construction of rotaxanes in which two rings are threaded by acyclic subunits.

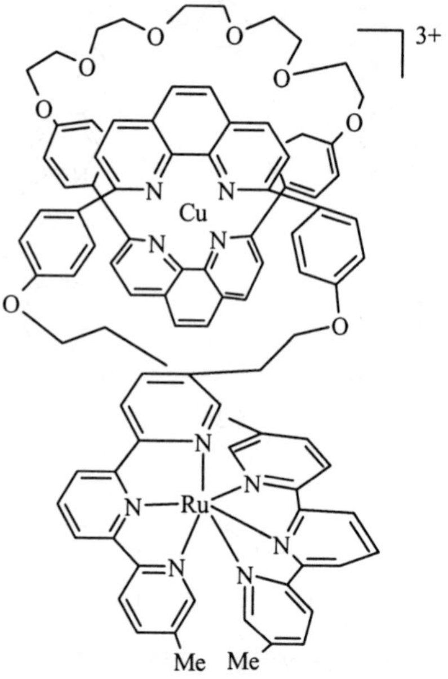

6.28

Taking in mind that the copper(I)-phenantroline dichelates are more stable than the monochelate ones, Chambron *et al.* utilized molecular strings **6.29** – **6.33** which contain two, respectively three phenantroline units separated by more or less rigid spacers. As expected, the binuclear complexes where obtained when the spacer was to rigid or to short to fold up and form a mononuclear complex, as in the case of **6.29** and **6.30**. It has been found that the more flexible spacer $-(CH_2)_6-$ determines the formation of a mixture of complexes.

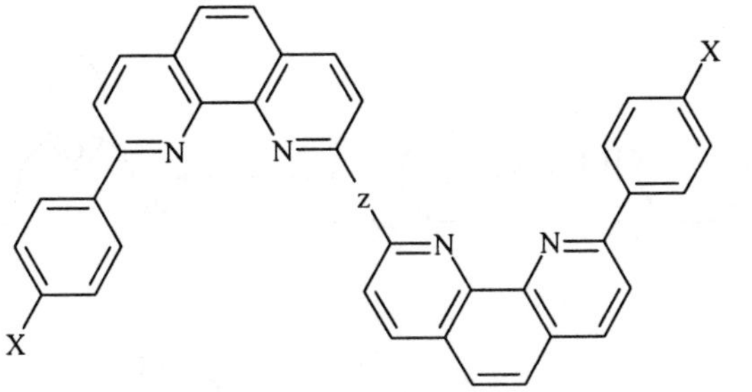

	Z	X
6.29	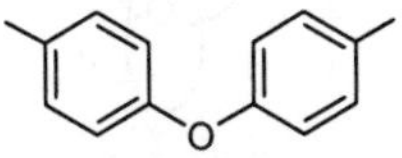	OMe
6.30	-(CH₂)₄-	OMe
6.31	-(CH₂)₄-	CHO
6.32	-(CH₂)₄-	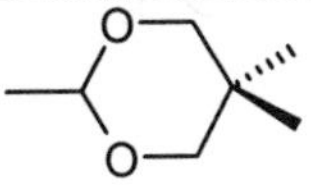

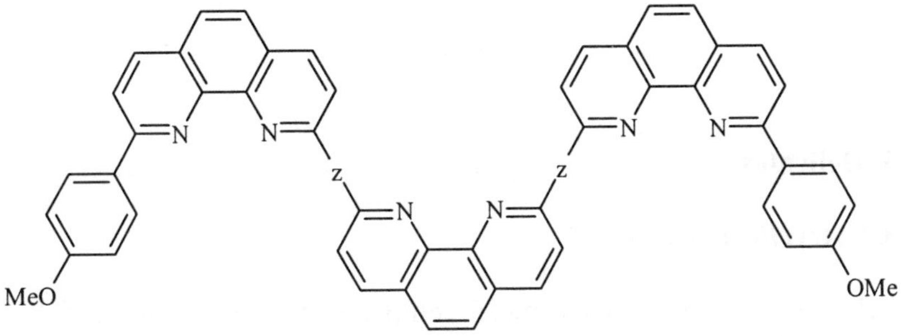

6.33

[3]rotaxanes and [5]rotaxanes have been obtained[404] following the strategy depicted in Scheme 6.6 and utilizing the ring **6.1**, strings of the type **6.31** and **6.32** and porphyrins as stopers. The strings of the [5]rotaxanes contain three pophyrins and is of approximately 60 Å long. It has been demonstrated that the driving force behind this reaction is the coordination of all 2,9-diphenyl-1,10-phenantroline chelates to copper(I) centres.

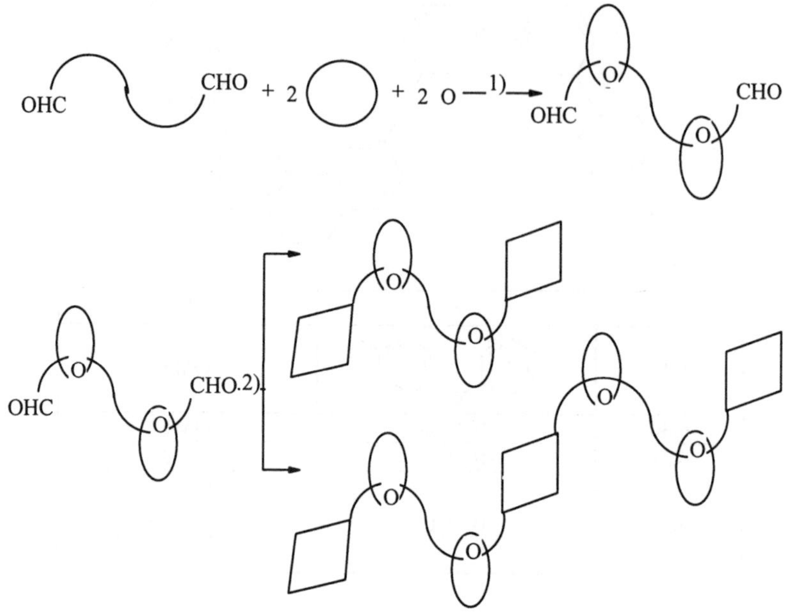

Scheme 6.5

6.3. Helicates

6.3.1. Acyclic helicates

Helicates are polynuclear metal complexes that form double[405, 406] or triple[407,408,62] helices by self-assembly of two or three ligand strands around suitable metal ions lying on the helical axis. Quadruple stranded helicates have also been obtained and classified as saturated and unsaturated[409], helicates. Polymeric metal complexes with infinite-chains having helical structures have been reported.[410, 411, 412] The factors which determine and control the helical architectures formations received considerable attention and they can be resumed as follows: *i.* the preferred coordination geometry of the metal ions[413, 414] and, *ii.* the nature of the ligands.[415, 416] The number of metal ions determines the dimension of the helicates. Thus, complexes containing from two to five metal centres are di- to pentahelicate species. Multimetallic helicates with metal ions that are two (linear), four- (tetrahedral), six- (octahedral), or nine-coordinated (tricapped trigonal prism) or combination of these geometries[417, 418,419, 420, 421] have been obtained by appropriate combination of metal ions with ligands capable of accommodating imposed geometrical requirements.

The simplest double-helical metal complexes have been formed around metal ions that favour tetrahedral coordination geometry, such as Cu(I) and Ag(I) with 2,2'-bipy strands. Many double-helical structures have been reported for tetrahedral bimetallic complexes of Cd(II), Cu(II), Fe(II), Mn(II), Ni(II). Following the simplest bipyridine, series of oligopyridine ligands with preorganizing properties were developed[422] and Constable *et al.* have demonstrated the spontaneous assembly of double-helical complexes containing these types of ligands.[423] Stable binuclear metal complexes $[Cu_2(Ph_2tpy)_2]^{2+}$, Ph_2tpy = 6,6"-diphenyl-2,2':6',2"-terpyridine were obtained and the crystallographic analyses have shown their double helical structures. The role of the phenyl substituent in the stabilisation with respect to oxidation to copper(II) was underlined. It has been demonstrated that formation and stability is not affected by the substituents in the case of the complexes of quaterpyridine (qtpy) $[M_2(qtpy)_2]^{2+}$, M = Cu(I), Ag(I), but they control the pitch of the helix.[424] Bi- and tri-nuclear complexes have been obtained from quinquepyridine[425, 426] and sexipyridine[427,428] with first-, second- and third-row transition metal cations. Crystallographic studies of copper, cadmium, and nickel complexes revealed the double helical arrangement of the ligands about metal ions. Single stranded helical structures have been reported for diruthenium(II) and europium(III) complexes with 2,2':6',2":6",2"':6"',2""-quinquepyridine, respectively, 2,2':6',2":6",2"':6"',2"":6"",2"""-sexipyridine. It has been noticed that the obtained structure depends on the nature of counterion. For example, a channel or a single cyclic helix structure depending upon whether the counterion is perchlorate or tetrafluoroboate results when tridentate ligand 2-pyrazinecarboxamide was used. Similar structures have been reported when the ligands were 4,4'-bipyridyl, 4-cyanopyridine piperazine and pyrazine. Ligands, **6.34-6.36** in which pyridine are joined through their 2,6-position, wrap spontaneous around some first row metal ions or Ag(I), resulting in double-stranded polymetallic helical complex.

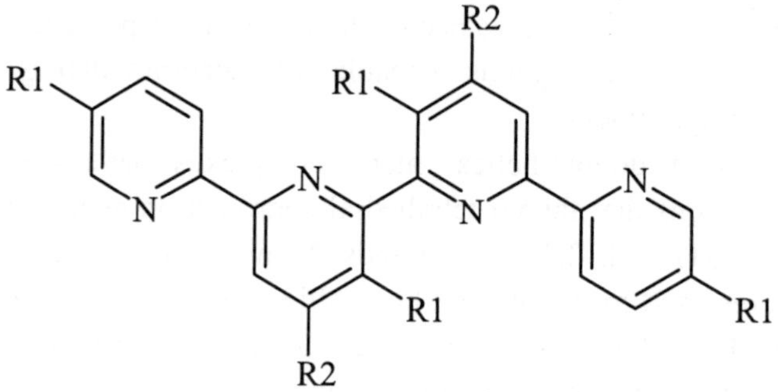

	R1	R2
6.34	CH_3	H
6.35	H	SCH_3
6.36	H	H

It has been established that the preorganizing abilities of the ligands, the structural particularities as well as the electronic communication between the metallic centres are strongly affected by the substituents attached to the pyridine rings, the number of the pyridine rings and even if this number is an odd or an even one. For example, polymetallic double-stranded helicates have been obtained from functionalized quinquepyridine (4',4'''-bis(alkylthio)-2,2':6',2'':6'',2''':6''',2''''-quinquepyridine) and sexipyridine (4',4''''-bis(alkylthio)-2,2':6',2'':6'',2''':6''',2'''':6'''',2'''''-sexipyridine) ligands and metal ions Fe(II), Co(II), Ni(II)Cu(II) and Zn(II) at ambient temperature. In all the cases bimetallic species have been obtained except the trimetallic species obtained for Cu(I). These complexes show to be mixed-valence helicate, to have metal ions in the same tetrahedral or octahedral environment or a combination of tetrahedral/octahedral or tetrahedral/linear (for Cu(I)) geometries.

The polydentate ligands involved in the self-assembled helicates should not be able to bind all their coordination sites to a single metal ion. This condition can be achieved by a mismatch between the geometry of the ligand binding sites and preferred geometry of the metal ion or by a judicious choose of spacer. First, steric factors favour the formation of tetrahedral Cu(I) complexes with the ligands **6.37 – 6.40** whereas the ligand **6.41** leads to the double –helicate architecture with octahedral core of Ni(II) or Fe(II)[429].

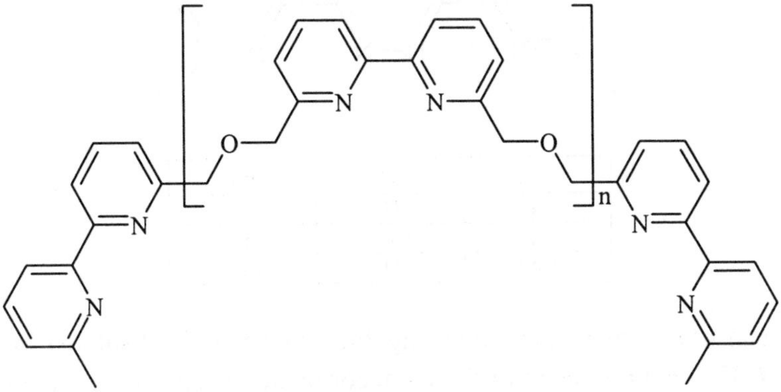

	n
6.37	0
6.38	1
6.39	2
6.40	3

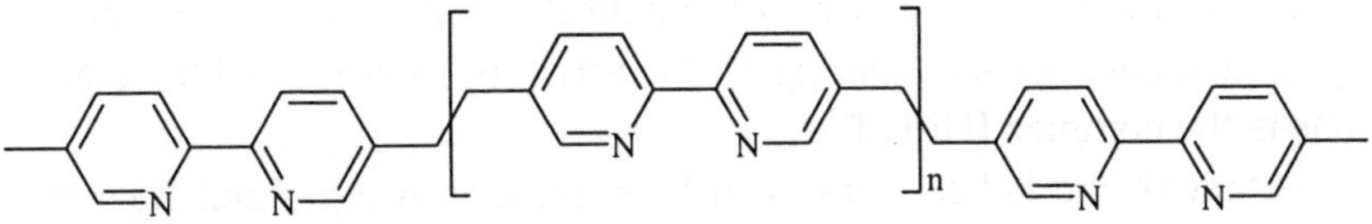

6.41

Suitable spacers hinder the formation of mononuclear species. Oligopyridines containing more or less flexible spacers have been designed. First, Bush and Stratton[430] recognised the triple-helical structure of the binuclear complexes $[M_2(6.42)_3]^{4+}$ (M = Co, Fe, or Ni). Binuclear Cu(I) and Ag(I) complexes containing ligands **6.42 – 6.44** in which the two pyridine fragments have the freedom to rotate about the central N-N bonds have recently been reported.[431] The complexes have a triple helical structure, stable both in solution and in solid state as indicated ^{1}H NMR, ESI mass spectrometry and X-ray analysis.

Oligobipyridine ligands containing spacers like CH_2OCH_2, CH_2CH_2 of type **6.41**, or CH_2SCH_2[432] between bipyridine groups have been designed. More rigid spacers, like 1,3-phenylene, between 2,2'-bipyridine groups have also been reported.[433, 434]

	R1	R2
6.42	H	H
6.43	CH$_3$	H
6.44	H	CH$_3$

Lehn *et al.* have shown that oligobidentate 6,6'-disubstituted bpy of type **6.37 – 6.40,** bonded to the tetracoordinated copper(I) or silver(I)[435] ions, lead to double-helical complexes which contain three, four and five metal ions. ^{1}H NMR and crystallographic data have proved that all metal ions that tetrahedral coordinate two bipyridines act as templates.

The substituents hinder the binding of metal ions in an octahedral manner. Lehn *et al.*[436] shifted the substituents from a 6,6'- to a 5,5'-positions and thus a trinuclear triple-helical complex have been obtained by spontaneous self-assembling of three tris(bpy) strands and three Ni(II) ions in the complex [Ni$_3$L$_3$]$^{6+}$.

Although the helical sense can be controlled using chiral ligands,[437] achiral ligands can also be used. When achiral ligands are employed, they usually give racemates although there are a few exceptions where spontaneous resolution upon crystallisation occurs. When chiral ligands are employed[438] they can be used as either enantiopure ligands or racemic mixtures. Siegel *et al.*[439] obtained a double helix in racemic modification by spontaneously wrap around Cu(I) ion of **6.39** in which the helical sense of the copper-ligand helix is controlled by the chirality of the spacer at one end. Enantiopure binuclear silver(I) double helicates have been prepared in a simple one-pot reaction using ligands which contains as spacers (R),(+) and (S),(-) enantiomers of 1,1'-binaphthalene-2,2'-diamine, pyridine-2-carbaldehyde. It resulted that the chirality of the spacer groups and, as a consequence, the ligands is preserved during the hard conditions of the synthesis.

Saturated quadruple stranded helicates were obtained by employing square-planar metal centres and oligomonodentate bridging ligands. For example, **6.45** has been obtained[440] when $[PdCl_2(PPh_3)_2]$ and $[PdI_2(py)_2]$ was treated with the ligand 1,4-bis(3-pyridyloxy)benzene in 1:2 molar ratio. The crystal X-ray structure shows that the structural unit contains a full M_2L_4 helical cage in which each square-planar palladium atom coordinate to the four bridging ligands and the dimensions of the cage are mainly defined by the Pd - Pd separation of 8.8402(8) Å.

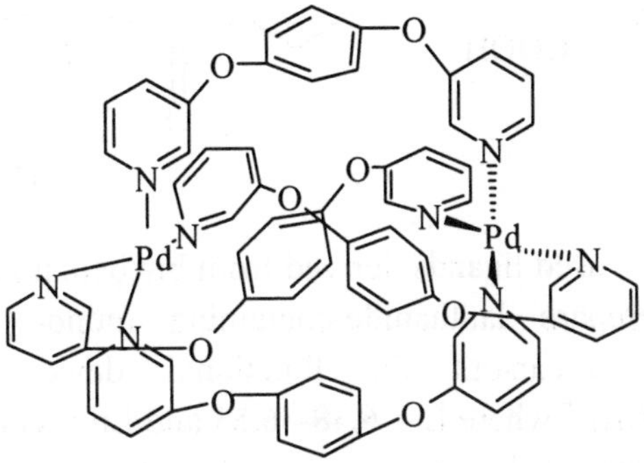

6.45

Although oligopyridines are the most studied ligands capable to promote self-assembly of mononuclear and multinuclear helical complexes, other ligand molecules, containing pyridine patterns like 1,10-phenantroline,[441] substituted pyridine like bis(benzimidazolyl)pyridines,[442] (2-pyridyl)-pyrazol-1-yl,[443] pyridine-2,6-dicarboxamide or pyridine-2,6-dicarboxylate have been used. Also, the synthesis and the X-ray structure of the trinucler, trigonally-coordinated Cu(I) complex $[Cu_3(\{PE\}_2P)_3][PF_6]_3$, where $\{PE\}_2P$ is 2,6-bis(2-pyridilethynyl)pyridine was described.[444]

Trivalent 4f lanthanide ions has been shown to control the self-assembly processes to give helices with three strands consisting of pyridine containing tridentate binding units.[445] In these systems, the nine-coordinate pseudo-tricapped trigonal prismatic coordination site suitable for these metal-ions was provided and 2,2':6'2" terpyridine is the archetype of the semi-rigid bent tridentate binding unit, despite the poor stability[446] of $[Ln(terpy)_3]^{3+}$. The presence of oxygen donors on the pyridine side arms or of some heterocyclic groups improves the stability of the triple-helical complexes. The triple-helical complexes[447] [Ln(6.46-$2H)_3$]$^{3+448}$ and [Ln(L)$_3$]$^{3+449}$ (L = 6.47, 6.49) have been obtained and, along the lanthanide series, intramolecular steric constraints which induce structural variation were observed.[450]

6.46   **6.47**

Piguet *et al.*[451] used ligands derived from bis(benzimidazolyl)pyridine in order to prepare lanthanide-containing mono- and bi-metallic supramolecular precursors for functional devices. Triple-helical complexes [Ln(L)$_3$]$^{3+}$ where L = 6.48– 6.53 and Ln = Gd, Tb or Eu, have been obtained and it has been established that their structural and photophysical properties are influenced by the substituents.[452] Thus, the substitution of the benzimidazole rings with bulky R^1 groups controls the stability and structure of the triple helix while substitution of the benzene rings as in 6.52 and 6.53, affects the electronic properties by shifting $^1\pi\pi^*$ toward lower energies. Crystal structures of [Eu(L)$_3$]$^{3+}$, L = 6.48 or 6.52 have been resolved and it has been shown that the three ligands are meridionally three-coordinated to the metal ion and wrapped about a pseudo-C_3 axis leading to a triple-helical structure.

Heterometallic triple-stranded helicates[453] [LnZn(6.51)$_3$]$^{5+}$, (Ln = Ce - Yb) and [LnCo(6.51)$_3$]$^{6+}$, have been obtained and their structure in solution was assigned. Mixing the ligand 6.57 with stoichiometric amounts of Ln(ClO$_4$)$_3\cdot n$H$_2$O (n = 6 - 8) in dichloromethane/acetonitrile, [Ln$_2$(6.57)$_3$]- (ClO$_4$)$_6\cdot x$H$_2$O$\cdot y$CH$_3$CN [Ln = La (x = 2, y = 1); Eu (x = 2, y = 0); Eu (x = 9, y = 0); Gd (x = 2, y = 1); Tb (x = 3, y = 1); Lu (x = 5, y = 0.5) were obtained as powders and [Eu$_2$(6.57)$_3$](ClO$_4$)$_6\cdot$9CH$_3$CN as crystals whose structure was resolved. It has been shown that the ligands 6.57 acts as bis(tetradentate) unit bound to the europium atoms and wrapped around the helical axis thus leading to a triple-helical structure with a pseudo-C_3 axis passing through the europium(III) ions. The Eu atoms are separated by 8.876(3) Å and the coordination sites can be described as tricaped trigonal prisms. The helice is twisted mainly due to benzimidazole rings connected by the CH_2 bridges.

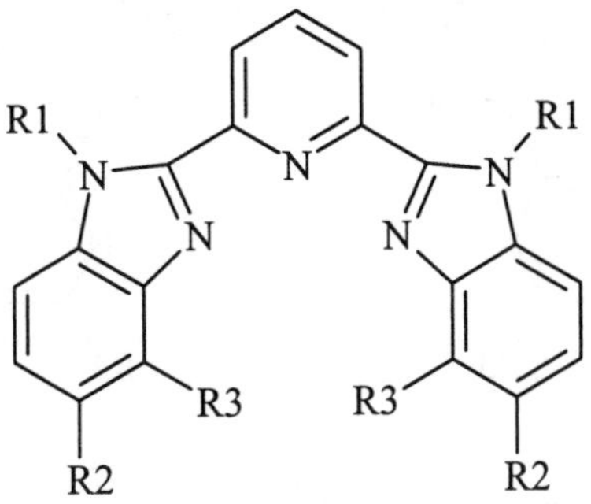

	R^1	R^2	R^3
6.48	H	H	H
6.49	Me	H	H
6.50	Et	H	H
6.51	$C_6H_3(OMe)_2$-3,5	H	H
6.52	Et	Me	H
6.53	Et	H	Me

Transition metal ions Co(III) and iron(III) form also binuclear metal complexes $[Co_2(\mathbf{6.54})_3]^{6+}$ and, respectively, $[Fe_2(\mathbf{6.54})_3]^{4+}$, with triple helical structures. Resolution into enantiomers and thermal spin crossover was observed for the iron compound.[454] It has been shown that ligand **6.62**, containing both bi- and tridentate binding units reacts with a mixture of Fe^{II} and Ag^I ions to give the self-assembled heterobinuclear double-helical complex $[FeAg(\mathbf{6.62})_2]^{3+}$ where the Fe^{II} ion is pseudooctahedrally coordinated by the two tridentate units and Ag^I lies in the pseudoterahedral site created by the two remaining bidentate units.[455] Symmetrical ligands **6.57**, **6.58**[456] and **6.59**[457] and the corresponding homodimetallic helicates $[Ln_2(\mathbf{6.57}$ or $\mathbf{6.58})_3]^{6+}$ and $[Ln_2(\mathbf{6.59}\text{-}2H)_3]$ stable also in acetonitrile and water have been obtained. It has been established that halogen substitution[458] in **6.60** and **6.61** does not influence the overall wrapping of the ligands strands around the two lanthanide ions in $[Ln_2(L)_3]^{6+}$, (L= **6.60** or **6.61**, Ln= La, Eu, Gd, Tb, Lu) and their structure is similar to helicats with **6.58**.

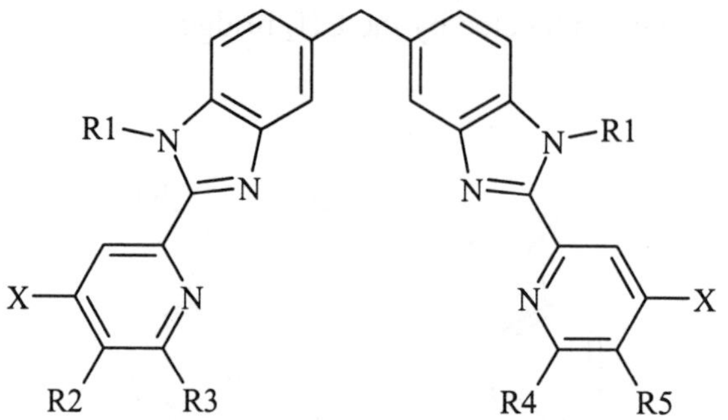

	R^1	R^2	R^3	R^4	R^5	X
6.54	Me	Me	H	H	Me	H
6.55	Me	H	Me	Me	H	H
6.56	Me	H	CONEt$_2$	H	H	H
6.57	Me	H	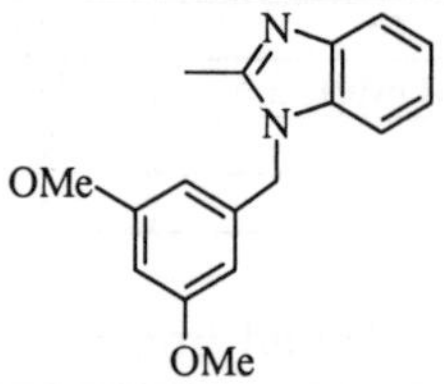 OMe / OMe (benzimidazole)	R^3	H	H
6.58	Et	H	CONEt$_2$	R2	H	H
6.59	Et	H	COOH	R2	H	H
6.60	Et	H	CONEt$_2$	CONEt$_2$	H	Cl
6.61	L^F	H	CONEt$_2$	CONEt$_2$	H	Br
6.62	L^G	H	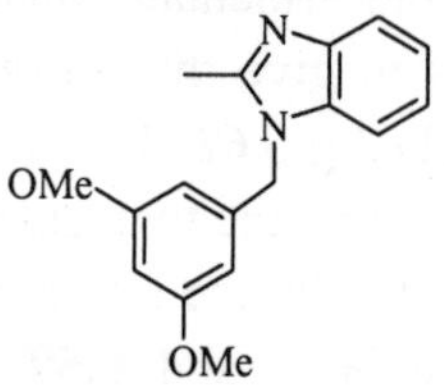 OMe / OMe (benzimidazole)	Me	H	H

Ligands which incorporate oxygen[459, 460] or sulfur donor atoms[461] have been obtained. For example, binuclear complexes $(NEt_4)_4[Ni^{II}_2(\textbf{6.63})]$, $(NEt_4)_2[Ni^{III}_2(\textbf{6.63})]$ and $Na_4[Ni^{II}_2(\textbf{6.64})]$ where the ligands 1,2-bis(2,3-dimercaptobenzamido)ethane, (H_4-**6.63**), and 1,2-bis(2,3-dimercaptophenyl)ethane, (H$_4$-**6.64**), contain benzenedithiolate donors have been obtained and their structural analyses show that the mononuclear units are linked in a double-stranded fashion by the carbon backbone. Achiral acyclic thioether-oligopyridine and thioether –oligopyrazine ligands were used to obtain helix self-assemblies. It has been established that these ligands, upon interaction with silver(I) folded, through strong intraligand π-stacking interactions, in such a way as to exhibit planar chirality. [65]

Oxygen donor ligands in combination with hard metal ions have also resulted in helicates. Cathechol ligands with pure alkyl chains, *p*-phenylene or amide spacers[462, 463] between the chelating units have been used. Aromatic spacers based on 1,5-diaminonaphthalene[464] or 2,6-diaminoantracene[465] was used to obtain more rigid ligands. The number of catechol chelating units determines the number of metal centres.

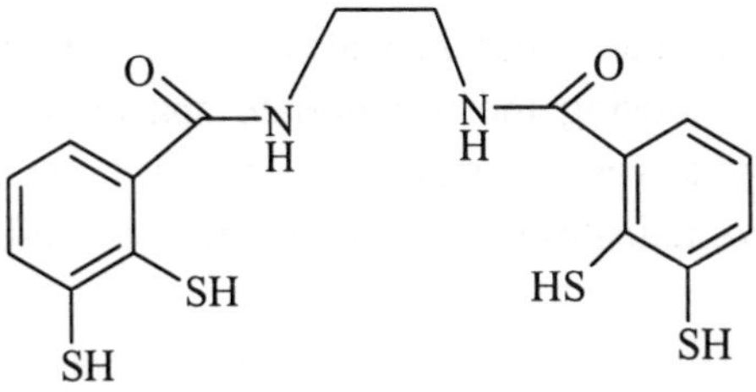

6.63 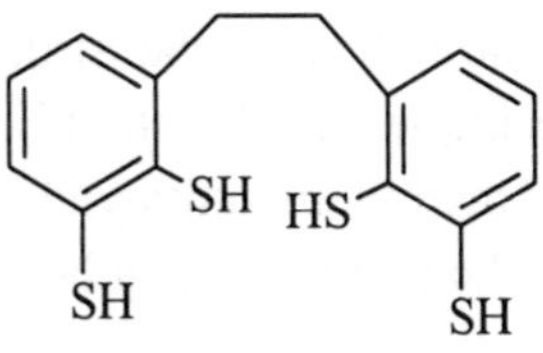　　　　　　　　　　　**6.64**

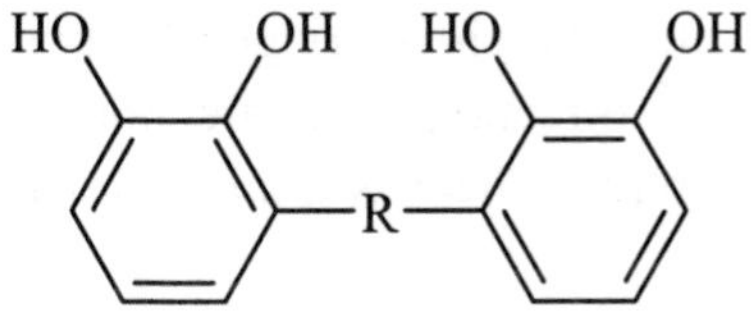

	R
6.65	(CH$_2$)$_n$
6.66	*p*-C$_6$H$_4$
6.67	*p*- C$_6$H$_4$-C$_6$H$_4$
6.68	*p*-CONH-C$_6$H$_4$-NHCO
6.69	*p*-CONH-(CH$_2$)$_4$-NHCO

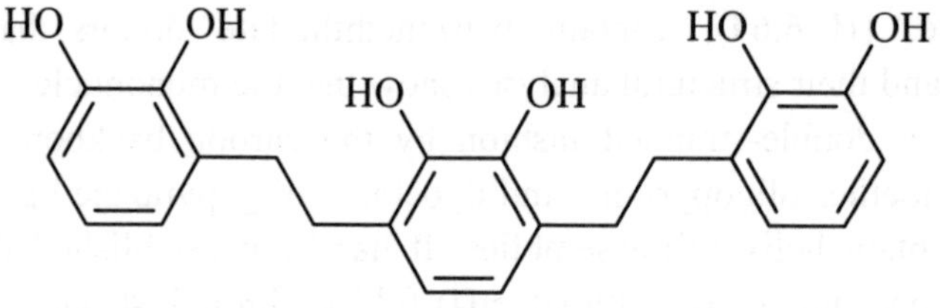

6.70

Thus, binuclear or trinuclear metal complexes have been obtained when the ligands contain two (**6.65 – 6.69**), respectively, three (**6.70**), cathechol moieties. Soluble binuclear metal complexes have been selectively obtained but only for $Na_4[(\mathbf{6.69})_2(MoO_2)_2]$ a double-stranded helical structure in solid state as well as in solution was established. For both binuclear and trinuclear iron(III) or titanium(IV) structural analyses shown a triple-stranded helical structure. It has been shown that it is possible to obtain selectively either the chiral helicates or the achiral *meso*-helicate type structures by choosing the appropriate spacer. For example, for the alkyl-bridged derivatives, a ligand with an odd number of methylene units yields the *meso*-helicate while an even number leads to the corresponding helicate.

Albrecht et al.[466] used amino acids as spacers into dicatechol ligands. Complexation of phenylalanine-bridged dicathechol ligand, **6.71**, with titanium(IV) leads to double-stranded coordination compound $Na_2[(\mathbf{6.71})_2Ti_2(OCH_3)_2]$. The X-ray structure shows a central four-member $[Ti(OCH_3)_2]$ unit, which is bridged by two ligands **6.71** which are oriented in opposite directions and are bound in a "side-by-side" fashion to the metal centres. There is a *meso*- relation between the two complex units.

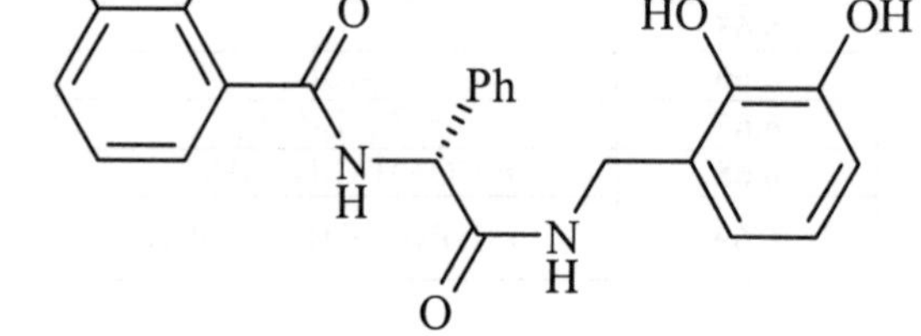

6.71

The exclusive formation of a predesigned cluster requires ligand rigidity, kinetically labile metal-ligand interactions and thermodynamically stability. Raymond *et al.*[467,468] obtained a series of three bis(catecholamide) ligands and their triple helical complexes $K_6[Fe_2L_3]$, L = **6.72** – **6.74**. The authors demonstrate that molecular resolution can be achieved based solely on the distance between two coordination sites.

Triple helical structures have been reported for the complexes $[M_2(p\text{-}XBP)_3(NO_3)_6]$ (M = Pr, Nd, Sm, Eu) and p-XBP =N,N'-xylene-1,4-diylbis(pyridin-2-one). The X-ray structure of neodymium complex shows that each metal ion is nine-coordinate and bonds to three bidentate nitrates and to one oxygen atom of each of the three p-XBP ligand.

Tripodlike molecules **6.75** and **6.76** have been used[469] to obtain triple-stranded helices which incorporate two octahedral iron(III) ions. The presence of the chiral amino acids and hydroxamate chromophores allows the determination of the absolute configuration around each metal ion and thereby the helicity of the overall structure.

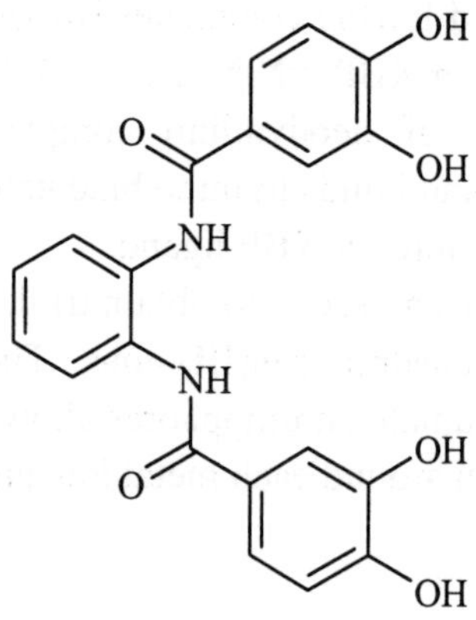

6.72 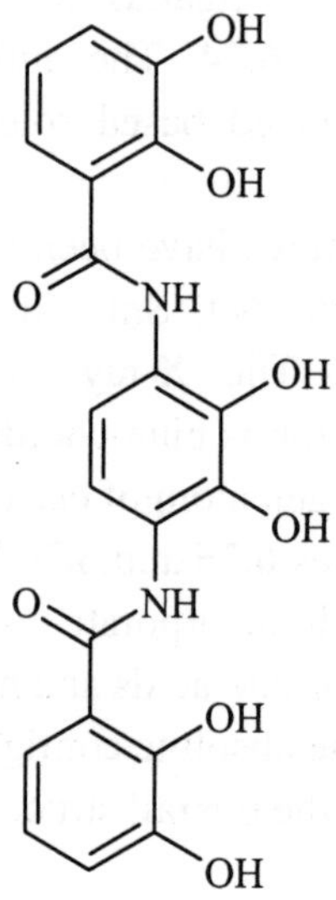   **6.73** 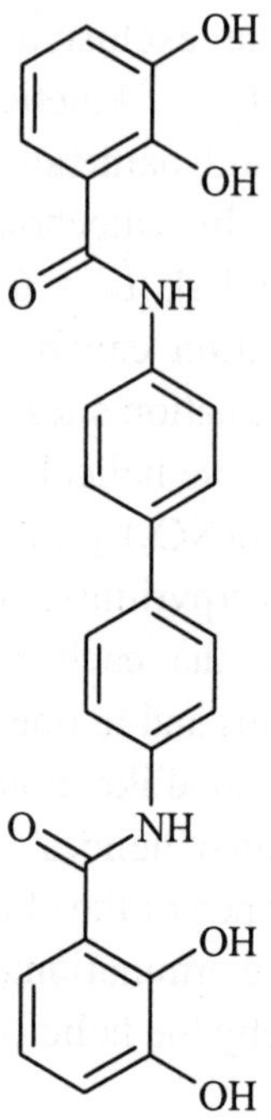   **6.74**

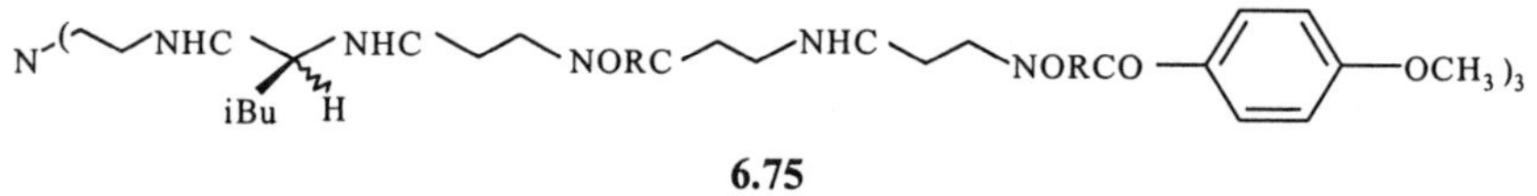

6.75

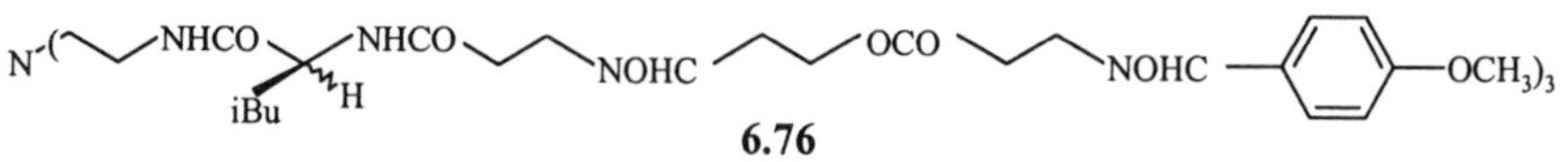

6.76

6.3.2. Circular helicates

Acyclic helicates are the most common. However, some circular helicates have been obtained.[470] In circular helicates each ligand extends over three adjacent metal centres and the ligand stands wrap around each other, thus generating a double-helical structure. The involved ligands contain bipyridine binding sites and the nature of the complex is determined by the flexibility of the linkers between these sites and of the counter ion. Circular helicates were obtained as iron(II) complexes with ligands **6.38** and **6.41**.

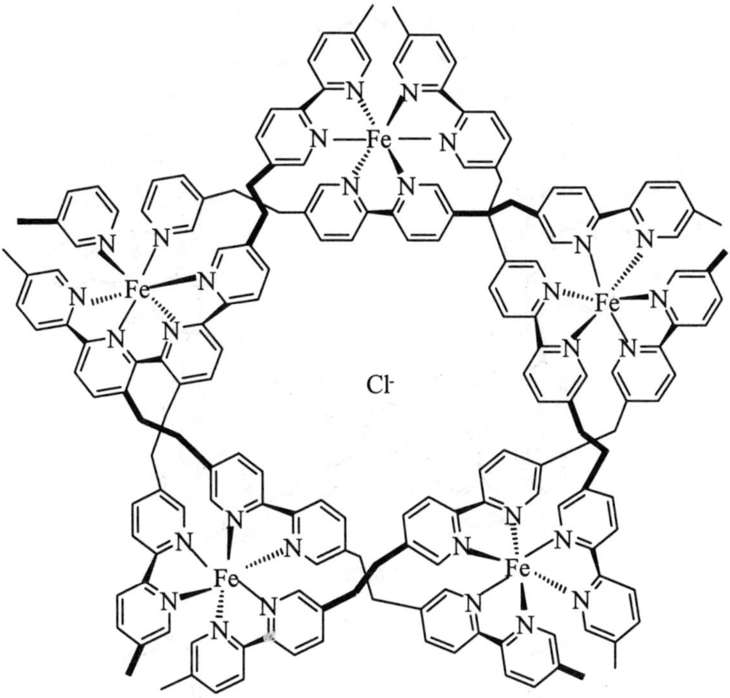

6.77

Pentanuclear, **6,77** or hexanuclear species, **6.78,** were obtained using **6.37** as ligand and, depending if the counter ion was Cl⁻ or, respectively, SO_4^{2-}, BF_4^-, and SiE_6^{2-}. It was established that the pentanuclear complex **6.77** presents a pentagonal shape with a cavity radius of 1.75 Å which can explain the weak binding of Cl⁻ inside. A tetranuclear circular helicate has been obtained when ligand to metal ratio was 7 : 1. As in the previous complexes, each Fe(II) ion is coordinated to three bipy subunits, belonging to three different ligands, thus describing pseudo-octahedral coordination geometry. The elongation of the spacer, move the metal ions at a distance of 10.62 Å which is 2.2 Å longer than in **6.77** and creates an inner cavity with a radius of 1.85 Å.

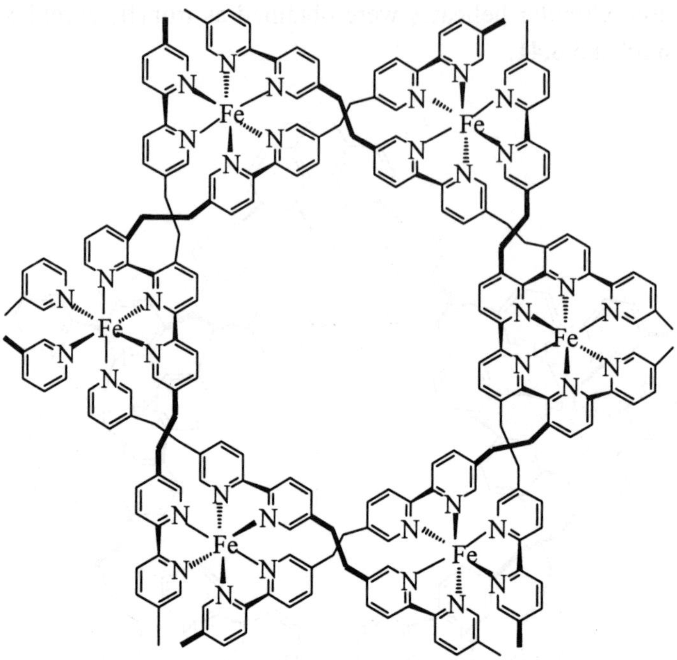

6.78

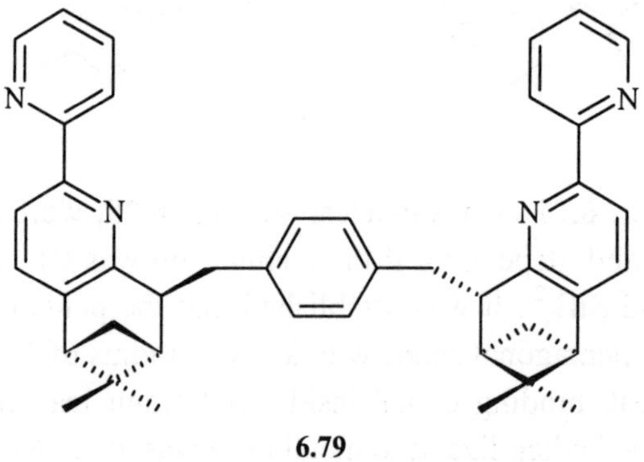

6.79

A hexanuclear silver(I) complex containing enantiopure ligand based on (-)-5,6-pinene bipyridine, **6.79**, has recently been obtained and the circular helical structure has been elucidated.[471] It has been demonstrated that at low temperature and high pressure, the hexanuclear circular helicate forms a tetranuclear circular helicate[472] in a process characterised by $K^{298} = (8.7 \pm 0.7) \times 10^{-5}$ mol kg^{-1}, $\Delta H° = -15.65 \pm 0.8$ kJ mol^{-1}, $\Delta S° = -130.2 \pm 3$ J mol^{-1}K^{-1}.

6.4. Knots

The strategy for making knots, IV in Scheme 6.6, is based on the three-dimensional template effect of transition metals,[473] which are able to gather and interlace coordinating molecular strings prior to the ultimate cyclization step. The ligands are strands with appropriate ends and containing two coordinatig sites separated by spacer which confers chemical robustness and ease of introduction the following steps. For an efficient preparation of knotted systems, the construction of double-stranded helical complexes, II, from metals and the strands, I, is the first essential requirement as it results from Scheme 6.6.

Following template syntheses, the trefoil knots **81** and **82**, respectively, were obtained using the hydrolised form of **30** as the strands which contains two phenantroline groups, linked by an enough long spacer, and copper(I) as template.

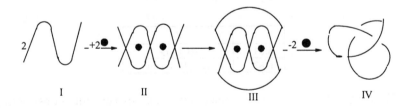

Scheme 6.6.

The helical complex precursor of type **6.80,** is formed in the first stage, which further was treated with biiodide to lead the bicopper(I) trefoil knot. X-ray structure of the knot has been resolved[474] and further, it has been shown that the chiral bicopper(I) knotted compound crystallises as a pure enantiomer. Since the first report, a lot of works have been done in which various factors of the precursor were changed with the aim to realise a really preparative method for knotted molecules.

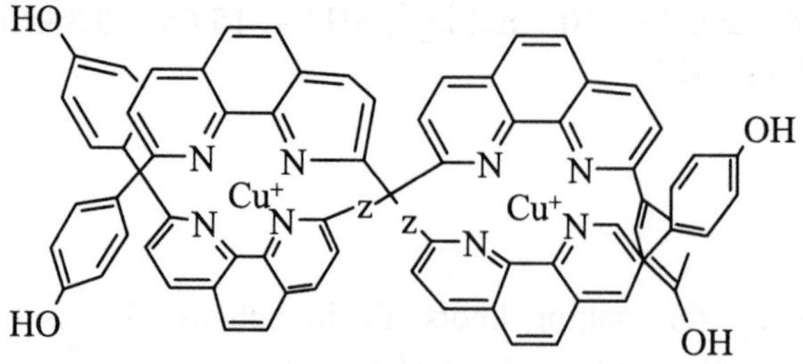

6.80

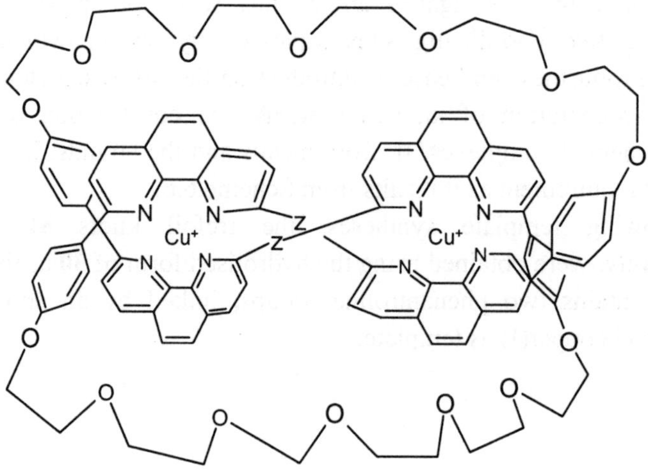

	Z
6.81	-(CH₂)₄-
6.82	1,3-phenylene

However, the use of 1,3-phenylene as spacer for constructing binuclear helical complex was found to be more beneficial.[475] Thus, the preparation of the double-stranded helix is almost quantitative and it can be separated as crystals suitable for an X-ray investigation. It has been shown that the precursor has a Cu-Cu distance of 4.76, shorter than that of the previously obtained knots of 6.3 or 7Å which affords a favourable condition to the successful continuation of the synthesis. The nature of the trefoil knot was established on the ¹NMR and FAB-MS spectroscopy data and on electrochemical measurements, which show strong electronic interaction between both metal atoms. The knotted molecules allowed the separation of the pure enantiomers[476, 477] by using enantiomerically pure chiral counterion (S)-(+)-BNP⁻ (BNP⁻ is 1,1'-binaphthyl-2,2'-diyl phosphate, potassium salt).

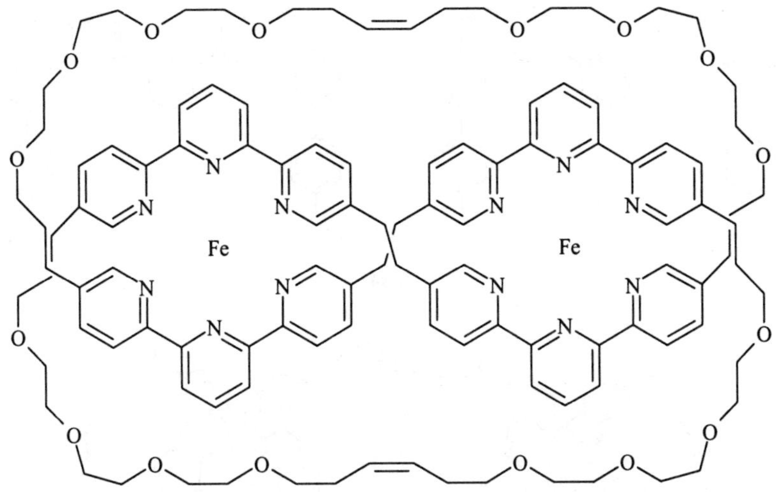

6.83

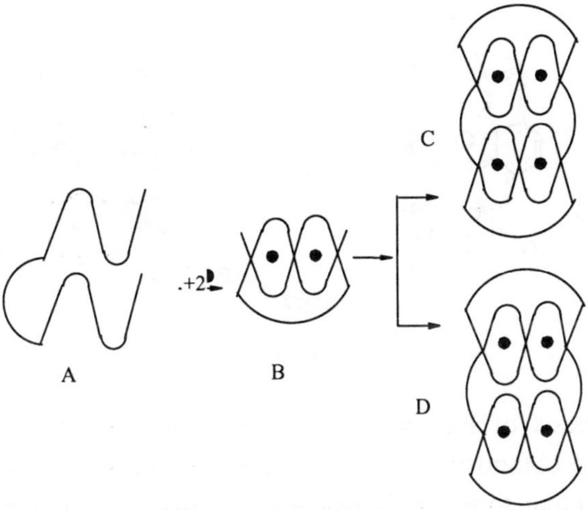

Scheme 6.7

The ring-closing metathesis, already efficient in the preparation of [2]catenanes has been successfully utilised to obtain the copper(I) trefoil knot. Following this approach a trefoil knot **6.83** has been obtained using iron(II) as highly efficient template and 2,2':6',2''-terpyridine as coordinating fragments. The knot contains two pseudooctahedral coordination sites of the very stable $[Fe(terpy)_2]^{2+}$ type.

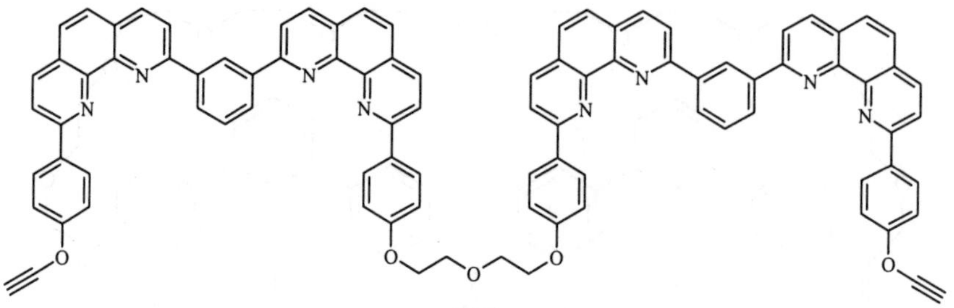

6.84

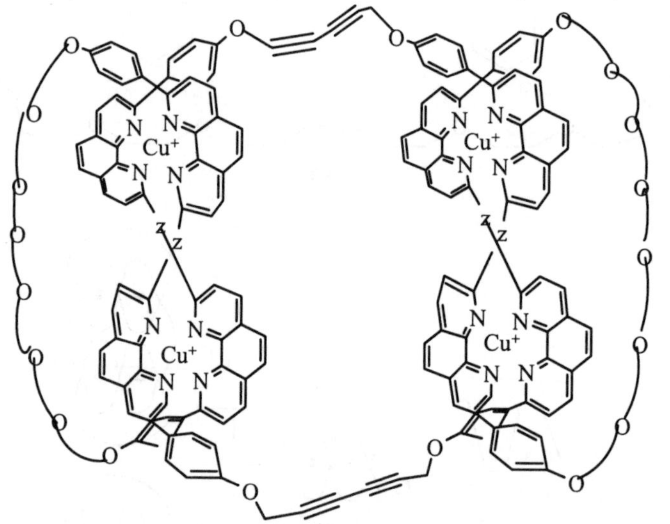

6.85

The synthesis of the molecular trefoil knot was extended[478] to the preparation of composite knots as it is shown in Scheme **6.7**. The long flexible molecular thread **6.84** reacted with two equivalents of copper(I) when a mixture of the binuclear half-closed helical complexes have been obtained. A cyclodimerization step afforded a mixture from which the composite knots $K_{(-)}$-$K_{(-)}$, **6.85**, belonging to the topological forms D, were chromatographically separated in very low amounts.

Siegel *et al.*[479] proposed recently an alternative representation of the trefoil knots, which consist of a D3-symmetric core, embedded within. In this respect, the synthesis and the stereochemical control involve both template and a bascule effects of a copper(I) ion and, respectively, of a ligand **6.87**. The X-ray crystal structure of the deep red colour Cu$_3$(**6.87**) shows that the bipyridine arms from neighbouring units cross to form the tetrahedral metal binding sites. It was also established that the initial metal binding promotes subsequent metal bindings, thus showing a positive cooperative effect.

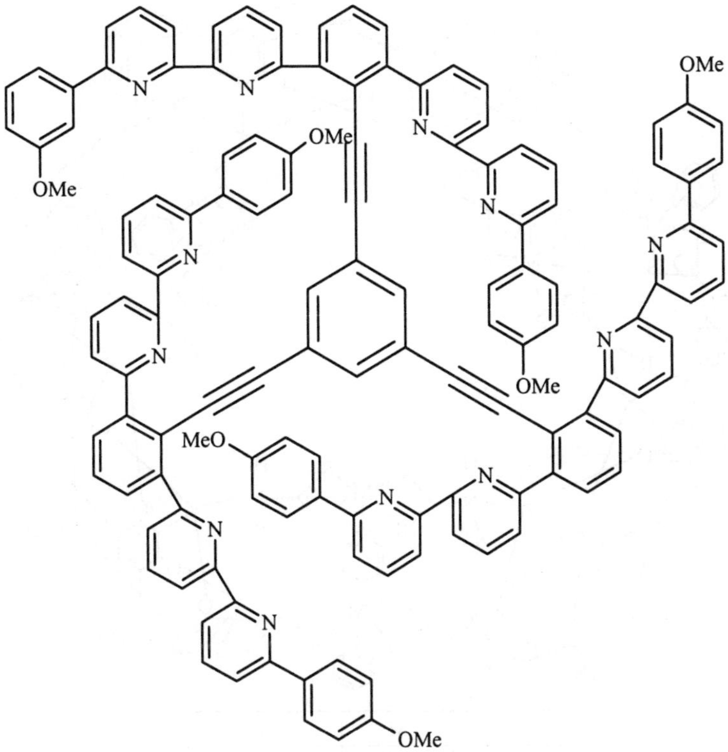

6.86

6.5. Macrocycles and Cages

The use of metal-directed self-assembly techniques provides a facile route to novel molecular-sized macrocycles and cages capable of binding charged and neutral guest substrates.[480] The process occurs in quantitative yields directly from the preorganized precursors. The ligands containing specific coordination sites like dithiocarbamate (*dtc*),[481] pyrazole[482] or pyridine[483] have been used to obtain copper(II)-macrocycles.

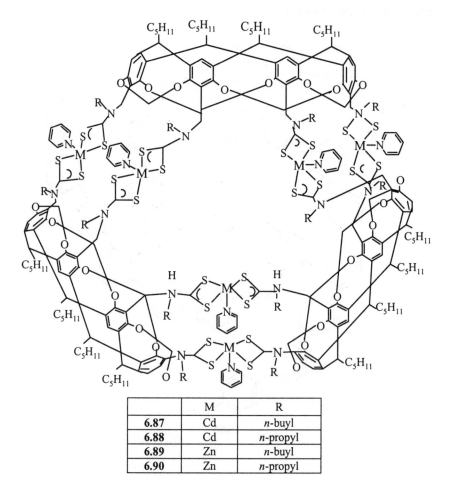

	M	R
6.87	Cd	*n*-buyl
6.88	Cd	*n*-propyl
6.89	Zn	*n*-buyl
6.90	Zn	*n*-propyl

For example copper(II) directed macrocycles which incorporate aromatic rings with *dtc* binding sites have been reported and, further, nano-sized architectures **6.87** – **6.90** have been constructed by cadmium(II) and zinc(II) directed assembly of a trimer of *dtc*–functionalized resorcarene ligands. These are calice-shaped molecules, calixarenes, and serve as excellent hosts for the binding of organic guest molecules through non-covalent interactions. It has been shown[484] that the cavity is circular with a diameter of ca. 16-17 Å and therefore suitable for the binding of spherically shaped molecules such as C_{60}. These loop-shaped hosts strongly binds fullerenes through favourable interactions between the π-electron system of the calixarene aromatics and the surface of the carbon sphere.

Of particular relevance are the works of Fujita's group. The authors reported the assembly of large multidentate organic ligands into large two- and three-dimensional structures through metal-coordination. The process consists in the incorporation of coordination angles of transition metals into metal-organic frameworks. Due to the preferred square-planar geometries, *cis*-protected square-planar Pd(II) or Pt(II)[485] complexes were currently used to incorporate 90° coordination angles.[486] Organic molecules having different shapes describe panels corresponding to regular polygons: triangles, squares and pentagons, which - arranged in space such that the edges, vertices and three coordinate directions, are equivalent. The most organic molecules contain pyridine units as bridging points. The simplest ligand, 4-4'bipyridine has a rod-like shape, and was used to construct a *2D* square panel **6.91**.[487]

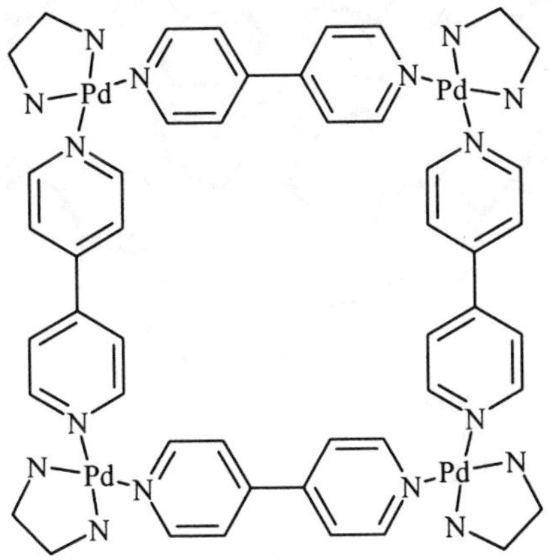

6.91

Pyridine containing organic panels with different number (from three to six) and positions of the binding sites which can be assembled in octahedral, square pyramids, tetrahedra and hexahedra by palladium(II) complexes have been synthesized. For example, molecular panelling of **6.92** form octahedral *3D* complex with adamantates like cage M_6L_4 (structure **6.93**[488]), with 2-5 nm in diameter. In this large central void guest molecule – adamantly carboxylates can be accommodated.

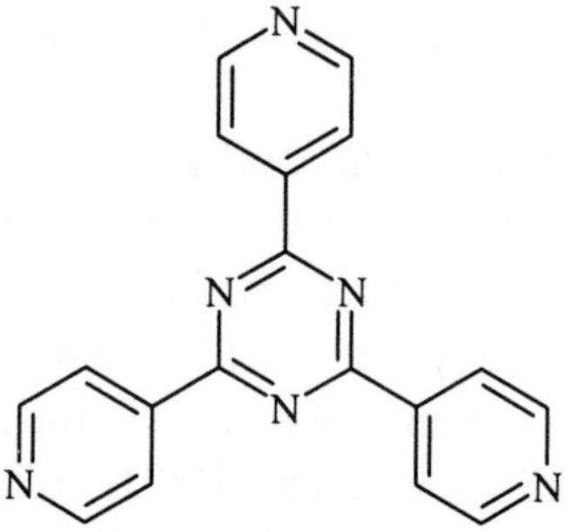

6.92

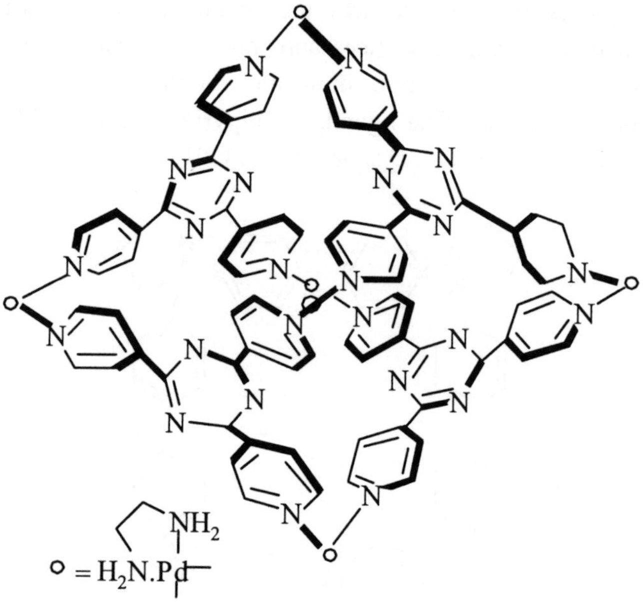

6.93

A similar Pt cage assemble was obtained and it has been established that it is more stable under extreme conditions compared to the Pd analogue which decomposes in acidic, basic or nucleophilic media or even the guest is removed. This was explained by the irreversible nature of Pt(II)-pyridine bond.

The formation of nanoscale cage-like complexes **6.94** composed of two resorcin[4]arene derivatives having pendent pyridine groups and four Pd(II) or Pt(II) square planar precursors have recently been described.[489]

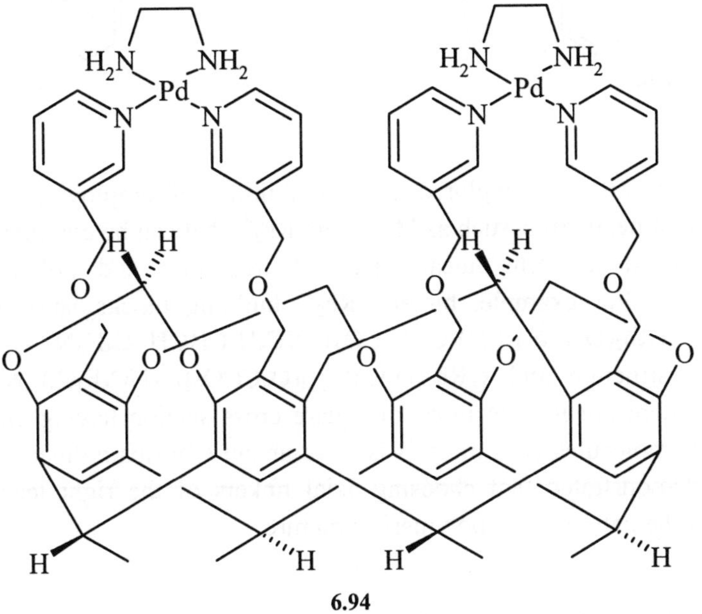

6.94

The pyridine groups act as a guest molecule: they are encapsulated inside cage due to the strong cation-π interactions.

The incorporation of different metal components or two or more than two different types of bridging ligands are much rarer.[490] Cage-type molecules that comprise two different types of ligands, have been prepared in a simple one-step process by refluxing (Otf)Re(CO)₅, **6.95** and, **6.96** or, respectively, **6.92**, in 47 and 41 % yield, respectively. (Otf = triflate).

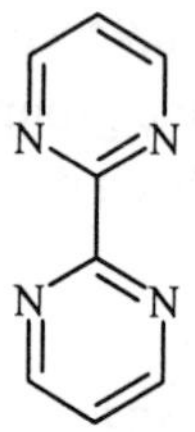

6.95 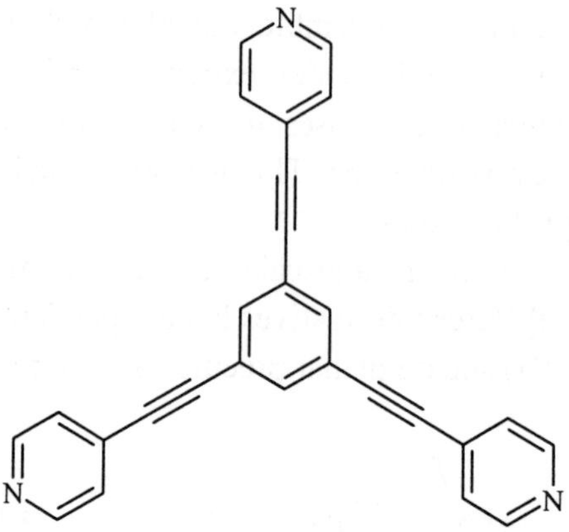   **6.96**

Cotton *et al.*[491] explored the possibilities of employing strongly bonded bimetal units, such as Mo_2^{4+} and Rh_2^{4+} that can be complexed and then linked to form linear square,[492] triangular and polyhedral,[493] structures. For example, bimetal Rh_2^{4+} building blocks have used to obtain $\{[Rh_2(DAniF)_2]_2(O_2CCH_2CO_2)_2(NC_5H_4CHCH\ C_5H_4N)_2\}_n$ with a tubular structure and $\{[Rh_2(DAniF)_2]_4(O_2CCO_2)_4\ (NC_5F_4C_5F_4N)_4\}_n$ in which there are infinite tubes of square cross section having entrained CH_2Cl_2 molecules (DAniF = N,N'-di-p-anisyl-silformamidinate). It has been demonstrated that choosing axial linkers of the right length can control the nature of the polymeric structure.

6.6. Racks, Ladder and Grids

Linear structures of [*n*]-rack type and two-dimensional arrays of [2×*n*]-ladder and [*m*×*n*]-grid type structures have been first reported by Lehn's group. These architectures have been generated from polybipyridine ligand systems and transition-metal ions of adequate coordination geometries. Ladder architectures **6.97** and **6.98** have been obtained[494] from the corresponding oligobipyridine ligands and bipyrimidine in the presence of Cu(I) ions. It can be noticed that the products are generated spontaneously from a mixture of two different ligand species.

6.97

6.98

Rigid, linear polytopic ligands containing bidentate chelate sites can be able to self-assemble into square – matrix arrays of metal centres with ions of tetrahedral coordination geometry. The metal ions are generally copper(I) and silver(I) whereas the used ligands contain pyridine, bipyridine and pyridazine groups. An inorganic 3x3 grid consisting of an array of nine silver(I) metal ions and six tritopic ligands 6,6'-bis[2-(6-methylpyridyl)]-3,3'-bipyridazine have been obtained[495] and, its structure was determined by X-ray crystallography.

It has been established that all silver ions are in distorted tetrahedral environment and an average Ag...Ag distance of approximately 3.72 Å.

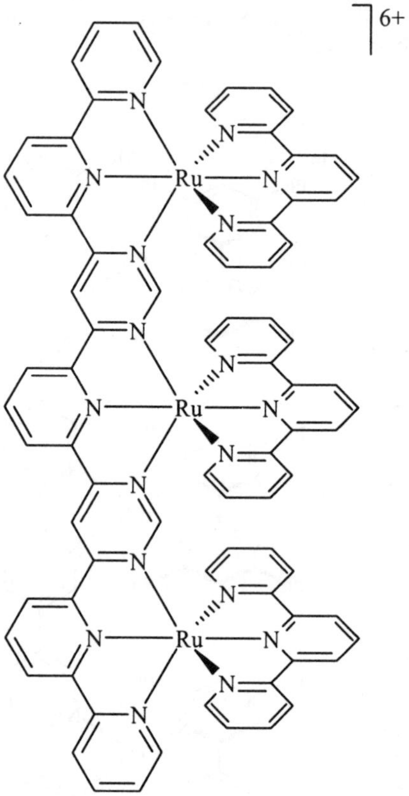

6.99

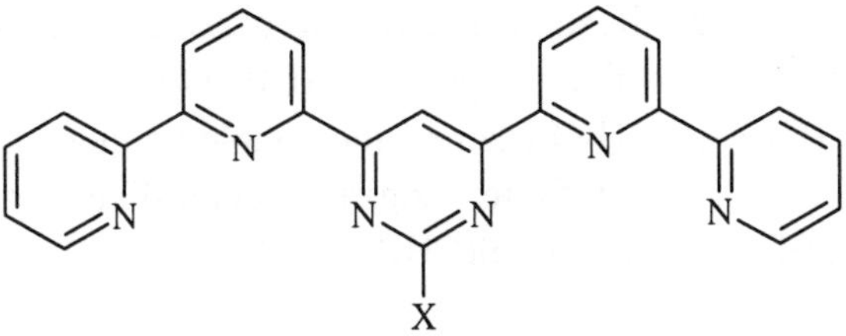

6.100

Recently, a [4×5]-grid type species has been obtained[496] from the reaction of silver(I) with the pentatopic ligand 6,6'''-bis(6-methyl-pyridin-2-yl)-[3,3':6',6'':3'',3''']quarter-pyridazine. It consists of two [4×5]-Ag$_{10}$ rectangular subgrids located on opposite sites of an array of parallel ligands.

Octahedral metal ions can assemble ligands containing tridentate coordination sites like 2,2':6'2''-terpyridinyl (tpy). For example, ligands of the type **6.99** and **6.100** containing two and three binding sites, respectively, can be assembled by Ru(II) ions into rack complexes of the type **6.101**.[497]

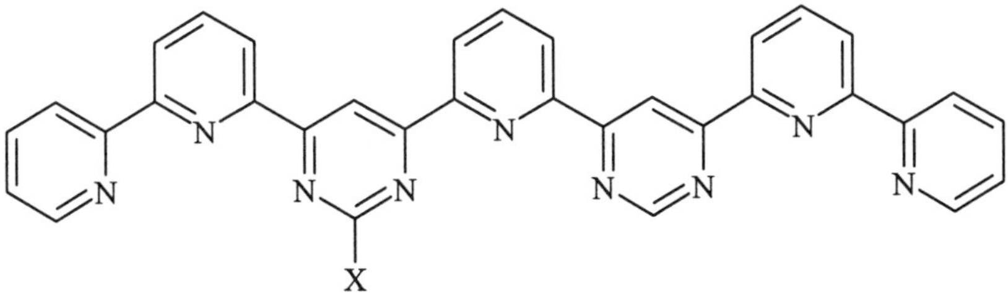

6.101

In contrast, cobalt(II) forms a tetranuclear complex, [Co$_4$(**6.99**)$_4$]$^{8+}$ for which, X-ray analyses shown that the four cobalt cations are arranged in a [2×2]-grid-type structure of distorted D_{2d} symmetry.[498] Similar ligands lead to [2×2]-, [3×3]- and [4×4]-grid type structures containing four, nine and sixteen lead(II) ions, respectively, in octahedral coordination sites.[499]

As for the above presented architectures, the spacers between the binding sites affect the nature of the product. Double bond in the spacer moyety may lead to a more rigid ligand which further has a large effect on the construction of supramolecular framework. It has been shown that the flexible tetradentate ligand **6.102** lead to a *2D* grid network structure for complexes of the type $[Cu(L)]NO_3 \cdot H_2O$ and $[Cu(L-2H)](AcOH)_2 \cdot (C_2H_5)_2O\ 4H_2O$ whereas the double bounds in the spacer of **6.103**, lead to an infinite zigzag chain-like structure for $[Cu(L)(CH_3CN)]NO_3$ and $[Cu(L)](CH_3CN)ClO_4$.[500]

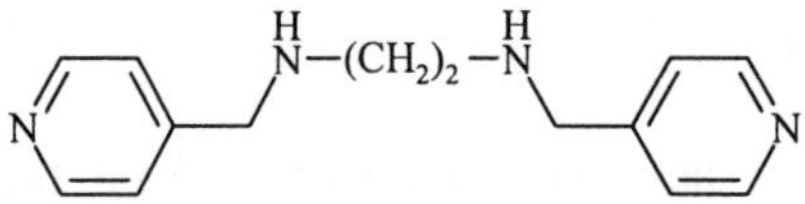

6.102 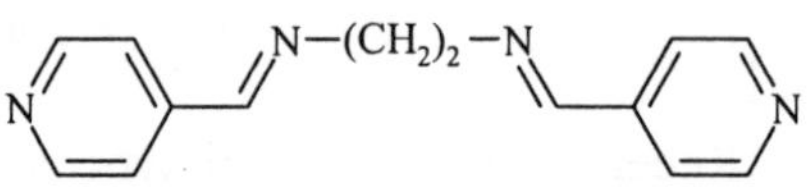   **6.103**

References

1 Ettling, C. (1840). Untersuchungen über das ätherische Oel der Spiraea Ulmaria und die salicylige Säure, *Ann.*, 35, pp. 241-276.

2 Schiff, H. (1869). Untersuchungen über Salicylderivate. I. Amide des Salicylaldehyds, *Ann.*, 150, pp. 193-200.

3 Einchorn, G. L and Latif, R. A (1954). The metal complexes of tris-anhydro-*o*-aminobenzaldehyde. *J. Am. Chem. Soc.*, 76, pp. 5180-5182.

4 Thompson, M. C. and Busch, D. H. (1964). Reactions of coordinated ligands. IX. Utilization of the template hypothesis to synthesize macrocyclic ligands *in situ. J. Am. Chem. Soc.*, 86, pp. 3651-3656.

5 Thompson, M. C. and Busch, D. H. (1962). Reactions of coordinated ligands. II. Nickel(II) complexes of some novel tetradentate ligands. *J. Am. Chem. Soc.*, 84, 1762-1763.

6 Dietrich, B., Viout, P. and Lehn, J. -M. Macrocyclic Chemistry, Weinheim, 1993.

7 Blin, E. L. and Busch, D.H. (1968). Reactions of coordinated ligands. XV. Demonstration of the kinetic coordination effect. *Inorg. Chem.*, 7, pp. 820-824.

8 Curtis, N. F. (1968). Macrocyclic coordination compounds formed by condensation of metal-amine complexes with aliphatic carbonyl compounds. *Coord. Chem. Rev.*, 3, pp. 3-47.

9 Thompson, M. C. and Busch, D. H. (1964). Reactions of coordinated ligands.VI. Metal ion control in the synthesis of planar nickel(II) complexes of α-diketo-bis-mercaptoimines. *J. Am. Chem. Soc.*, 86, pp. 213-217.

10 Busch, D. H. and. Stephenson, N. A. (1990). Molecular organization, portal to supramolecular chemistry. Structural analysis of the factors associated with molecular organization in coordination and inclusion chemistry, including the coordination template effect. *Coord. Chem. Rev.*, 100, pp. 119-154.

11 Ahrland, S., Chat J. and Davies, N. R. (1958). The relative affinities of ligand atoms for acceptor molecules and ions. *Quart. Rev. Chem. Soc.*, 12, pp. 265- 276.

12 Pearson, R. G. (1963). Hard and soft acids and bases. *J. Am. Chem. Soc.*, 85, pp. 3533-3539.

13 Hubin, T.J. and Busch, D.H. (2000). Template routes to interlocked molecular structures and orderly molecular entlargements. *Coord. Chem. Rev.*, 200-202, pp. 5-52.

14 Hancock, R.D. and Marsicano, F. (1976). The chelate effect : a simple quantitative approach. *J. Chem. Soc., Dalton Trans.*, pp. 1096-1998.

15 Adamson, A.W. (1954). A proposed approach to the chelate effect. *J. Am. Chem. Soc.*, 76, pp. 1578-1579.

[16]   Hancock, R. D. and Martell, A. E. (1989). Ligand design for selective complexation of metal ions in aqueous solution. *Chem. Rev.*, 89, pp. 1875-1914.

[17]   Izatt, R. M. and Christensen, J. J. (1978). Synthetic Multidentate Macrocyclic Ligands, Academic Press, New York.

[18]   Christensen, J. J. and Izatt R. M. (1987). Synthesis of Macrocycles, the Design of Selective Complexing Agents. Wiley, New York.

[19]   Cabbines, D. K. and Margerum, D. W. (1969). Macrocyclic effect. On the stability of copper(II) tetramine complexes. *J. Am. Chem. Soc.* , 91, pp. 6540-6541.

[20]   Kodama, M. and Kimura, E. (1978). Equilibria complex formation between several bivalent metal ions and macrocyclic tri- and penta-amines. *J. Chem. Soc., Dalton Trans.*, pp. 1081-1085.

[21]   Fabbrizzi, L., Micheloni, M., Paoletti, P., Poggi A. and Lever, A.B.P. (1981). High-spin nickel(II) complexes of penta-aza macrocycles: characterization, electronic spectra and thermodynamic properties. *J. Chem. Soc., Dalton Trans.*, pp. 1438-1441.

[22]   De Sousa Healy, M. and Rest, A. J. (1978). Template reactions. *Adv. Inorg. Chem. Radiochem.*, 21, pp. 1-40.

[23]   Henrick, K., Lindoy, L. F., McPartlin, M., Tasker, P. A. and Wood, M. P. (1984). Specification of the bonding cavities available in metal-binding sites: a comparative study of a series of quadridentate macrocyclic ligands. *J. Am. Chem. Soc.*, 106, pp.1641-1645.

[24]   Martin, L. Y., DeHayes, L. J., Zompa, L. J. and Busch, D. H. (1974). The relationship between metal-donor distance and ring size in macrocyclic complexes. *J. Am. Chem. Soc.*, 96, pp. 4046-4048.

[25]   Chantson T. E. and Hancock, R. D. (1995). The best-fit size and geometry of metal ions for coordination with nitrogen-donor macrocycles. *Inorg. Chim. Acta*, 230, pp. 165-167.

[26]   Thöm, V. J., Mc Dougall, G. J., Boeyens, J. C. A. and Hancock, R. D. (1984). Origin of the high ligand field strength and macrocyclic enthalpy in complexes of nitrogen-donors. *J. Am. Chem. Soc.*, 106, pp. 3198-3207.

[27]   Anderson, S., Anderson, H. L. and Sanders, J. K. M. (1992). Scavenger templates: synthesis and electrospray mass spectrometry of a linear porphyrin octamer. *Angew. Chem. Int. Ed. Engl.*, 31, pp. 907-910.

[28]   Uhleman, E. and Plath M. (1969). Metallselektivität von *o*-Aminoarylketonen und ihrer Schiffschen Basen mit 1,2-diaminen, *Z. Chem.*. 9, pp. 234-235.

[29]   Busch, D. H., Jicha, D. C., Thompson, M. C., Wrathall, J. W. and Blinn, E. (1964). Reactions of coordinated ligands. VIII. The reactions of alkyl halides with mercapto groups in transition metal complexes of mercaptoamines, *J. Am. Chem. Soc.*, 86, pp. 4642-3650.

[30]   Elder, M. S., Pinz, G. M., Thornton, P. and Busch, D. H. (1968). *Inorg. Chem.*, 7, 2426 (1968).

[31]   Lindoy, L. F. and Busch, D. (1974). Metal ion controlled syntheses of novel five-coordinate zinc and cadmium complexes containing helical coordination geometry and their template reaction to form complexes of a pentadentate macrocyclic ligand, *Inorg. Chem.*, 13, pp. 2494-2498.

[32] Lindoy, L.F. (1975). Transition metal complexes of synthetic macrocyclic ligands, *Chem.Soc. Rev.*, 4, pp. 421-441.

[33] Laurie, S. H. (1967). Coordination complexes of amino acids: preparation and properties of some copper(II) and copper(I) complexes containing mixed bidentate ligands, *Austral. J. Chem.*, 20, pp. 2597-2608.

[34] Nunez, L. J. and Eichorn, G. L. (1962). On the mechanism of formation of the metal complexes of Schiff bases, *J. Am. Chem. Soc.*, 84, pp. 901-906.

[35] Baker, D., Dudley, G. W. and Oldham, C. (1970). Metal-triketone complexes. Part II. Copper complexes exhibiting antiferromagetism, J. Chem. Soc., A, pp. 2608-2612.

[36] Fenton, D. E. and Lintverdt, R. L. (1978). Binuclear copper(II) complexes as mimics for type 3 coppers in metalloenzymes. 1. The importance of cooperative interactions between metals in the reversible multielectron transfer in bis(1,3,5-triketonato)dicopper(II) complexes, *J. Am. Chem. Soc.*, 100, pp. 6367-6375.

[37] Okawa, H. (1970). Copper(II)complexes of 3-formyl-5-methylsalicylaldehyde and its Schiff bases with alkyl amines, *Bull. Chem. Soc. Jpn*, 43, pp. 3019-3020.

[38] Mandal, S.K. and Nag, K. (1983). Binuclear metal complexes. 1. Synthesis, characterization and electrochemical studies of dicopper(II) complexes with 4-methyl-2,6-di(acyl/benzoyl)phenol, *Inorg. Chem.*, 22, pp. 2567-2572.

[39] Balasubramanian, S. (1987). Macrocyclic dicarbinolamine complexes of nickel(II) with planar $N_4(N_2)$ ligands: synthesis and spectral and electrochemical properties, *Inorg. Chem.*, 26, pp. 553-559.

[40] Haque, Z. P., McPartlin, M. and Tasker, P. A. (1979). Macrocyclic ligand synthesis. Isolation of a dicarbinolamine complex from zinc(II)-promoted cyclization reaction, *Inorg. Chem.*, 18, pp. 2920-2921.

[41] Guerriero, P., Vigato, P. A., Fenton, D. E. and Hellier, P. C., (1992). Synthesis and application of macrocyclic and macroacyclic Schiff bases, *Acta Chem. Scand.*, 46, pp. 1025-1046.

[42] Eggleston, D. S. and Jackels, S. (1980). Tetrasubstitued [14]-1,3,8,10-tetraeneN$_4$ macrocyclic complexes: synthesis, organic precursor, and template reaction mechanism, *Inorg. Chem.*, 19, pp. 1593-1599.

[43] Mendoza-Diaz, G., Ruiz-Ramirez, L. and Moreno-Esparza, R. (2000). Template reaction of [Ni(ben)$_3$](ClO$_4$)$_2$ with acetone. Crystal structure of 1,11-bis-benzyl-5,7,7-trimethyl-1,4,8,11-tetra-aza-undeca-4-ene   nickel(II)   perchlorate, *Polyhedron*, 19, pp. 2149-2154.

[44] Fujiwara, M., Kinoshita, S., Wakita, H., Matsushita, T. and Shono, T. (1987). Template synthesis of copper(II) metal complexes with tetraaza macrocycle in solid phase, *Chem.Lett.*, pp. 1323-1326.

[45] Curtis, N. F. and House, D. A. (1961). Structure of some aliphatic Schiff base complexes of nickel(II) and copper(II), *Chem. Ind.*, 42, pp. 1708-1709.

[46] Konefal, E., Loeb, S. J., Stephan, D. W. and Willis, C. J. (1984). Coordination modes of polydentate ligands. 1. Template synthesis of complexes of Ni^{2+}, Cu^{2+}, and Co^{2+} with pentadentate and hexadentate ligands: structure of a highly distorted six-coordinate Co^{2+} complex, *Inorg. Chem.*, 23, pp. 538-545.

[47]   Torzilli, M. A., Colquhoun, S., Kim, J. and Beer, R. H. (2002). Structural and ^{1}H NMR spectroscopic characterization of bis(N-isopropylsalicyldiminato)iron(II), *Polyhedron*, 21, pp. 705-713.

[48]   Costes, J. -P, Dahan, F., Fernandez Fernandez, M. B., Fernandez Garcia, M. I., Garcia Deibe, A. M. and Sanmartin, J. (1998). General synthesis of 'salicylaldehyde half-unit complexes': structural determination and use as synthon for the synthesis of dimetallic or trimetallic complexes and of 'self-assembling ligand complexes', *Inorg. Chim. Acta*, 274, pp. 73-81, and references therein.

[49]   Root, C. A., Hoeschele, J. D., Cornman, C. R., Kampf, J. W. and Pecoraro, V. L. (1993). Structural and spectroscopic characterization of dioxovanadium(V) complexes with asymmetric Schiff base ligands, *Inorg. Chem.*, 32, pp. 3855-3861, and references therein.

[50]   Liu, C. -M., Xiong, R. -G., Zuo, J. -L. and You, X. -Z. (1996). Synthesis and crystal structure of a novel zinc(II) complex [Zn(pbp)$_2$](ClO$_4$)$_2$, *Polyhedron*, 15, pp. 2051-2055.

[51]   Gourbatsis, S., Perlepes, S. P., Butler, I. S. and Hadjiliadis, N. (1999). Zinc(II) complexes derived from the di-Schiff-base ligand N,N?-bis[1-(pyridin-2-yl)ethylidene]ethane-1,2-diamine (L$_A$) and its hydrolytic-cleavage product N-[1-pyridin-2-yl-ethylidene]etane-1,2-diamine (L): preparation, characterization and crystal structure of the 5- coordinate species [ZnLCl$_2$], *Polyhedron*, 18, pp. 2369-2375.

[52]   Goto, M., Ishikawa, Y., Ishihara, T., Nakatake, C., Higuchi, T., Kurosaki H. and Goedken, V.L. (1998). Iron(II) complexes with novel pentadentate ligands via C-C bond formation between various nitriles and [2,4-bis(2-pyridylmethyl-imino)pentane]iron(II) perchlorate: synthesis and structures, *J. Chem. Soc.,Dalton Trans.*, pp. 1213-1222.

[53]   Nelson, S. M., Esho, F. S. and Drew, M. G. B. (1982).). Metal-ion controlled reactions of 2,6-diacetylpyridine with 1,2-diaminomethane and 2,6-diformylpyridine with o-phenylenediamine and the crystal and molecular unidentate o-phenylenediamine, *J. Chem. Soc., Dalton Trans.*, pp. 407-415.

[54]   Constable, E. C. and Holmes, J. M. (1987). The preparation and coordination chemistry of 2,6-diacetylpyridine bis(6-chloro-2-pyridylhydrazone), *Inorg. Chim. Acta*, 126, pp. 187-193.

[55]   Duncan, C. A., Copeland, E. P., Kahwa, I. A., Quick, A. and Williams, D. J. (1997). Unusual formation and crystal structure of a new stable dinuclear vanadium(V) amino-imino-acetal, *J. Chem. Soc., Dalton Trans.*, pp. 917-919.

[56]   Hernandez-Molina, R., Menderos, A., Dominiguez, S., Gili, P., Ruiz-Pérez, C., Castiñeiras, A., Solanas, X., Lloret, F. and Real, J. A. (1998). Different ground spin states in iron(III) complexes with quadridentate Schiff bases: synthesis, crystal structures, and magnetic properties, *Inorg. Chem.*, 37, pp. 5102-5108.

[57]   Carlisle, W. D., Fenton, D. E., Mulligan, D. C., Roberts, P. B., Vigato, P. A. and Tamburini, S. (1987). Metal complexes of binucleating ligands derived from 2,6-diformyl- and 2,6-diacetyl-4-metylphenol, *Inorg. Chim. Acta*, 126, pp. 233-235.

[58]   Cheng, P., Liao, D., Yan, S., Jiang, Z., Wang G., Yao, X. and Wang, H. (1996). Magnetic interaction in the keto and enol forms of binuclear copper(II) complexes

with a Robson-type ligand. X-ray crystal structure of [Cu$_2$(HL)(μ-N$_3$)(H$_2$O)(C$_2$H$_5$OH)(ClO$_4$)], *Inorg. Chim. Acta*, 248, pp. 135-137.

[59] Brubaker, G., Latta, J. and Aquino D. (1970). Nickel(II) complexes with sulfur containing Schiff base ligands, *Inorg. Chem.*, 9, pp. 2608-2610.

[60] Kaasjager, V. E., van den Broeke, J., Henderson, R. K., Smeets, W. J. J., Spek, A. L., Driessen, W. L., Bouwman, E. and Reedijk, J. (2001). Synthesis and X-ray crystal structures of dinuclear and a trinuclear nickel complex with mixed N, O, S ligand environments, *Inorg. Chim. Acta*, 316, pp. 99-104.

[61] Mohan, M., Sharma, P., Kumar, M. and Jha, N. K. (1986). Metal complexes of 2,6-diacety-lpyridine bis(thiosemicarbazone): their preparation, characterization and antitumour activity, *Inorg. Chim. Acta*, 125, pp. 9-15.

[62] Guerriero, P., Ajo, D., Vigato, P. A., Casellato, U., Zanello, P. and Graziani, R., (1988). Transition metal complexes with 'and-off' compartmental Schiff bases containing sulphonic or phosphonic groups, *Inorg. Chim. Acta*, 141, pp. 103-118.

[63] Martin, J. W. L., Johnson, J. H. and Curtis, N. F. (1978). Complexes of 2,4,4-trimethyl-1,5,9-triazacyclododec-1-ene with cobalt(II), nickel(II), and copper(II); X-ray structure determination of di-isothiocyanato(2,4,4-trimethyl-1,5,9-triazacyclododec-1-ene)nickel(II), *J. Chem. Soc., Dalton Trans*, pp. 68-76.

[64] Curtis, N. F., Curtis, Y. M. and Powell, H. K. J. (1966). Transition-metal complexes with aliphatic Schiff bases. Part VIII. Isomeric hexamethyl-1,4,8,11-tetra-azacyclotetradecadienenickel(II) complexes formed by reaction of trisdiaminoethanenickel(II) with acetone, *J. Chem. Soc., A*, pp. 1015-1018.

[65] Bailey, M. F. and Maxwell, I. E. (1972). Crystal and molecular structure of a nickel(II) macrocyclic di-imine-diamine complex: rac-5,7,7,12,14,14-hexamethyl-1,4,8,11-tetra-azacyclotetradeca-4,11-dienenickel(II) perchlorate, *J. Chem. Soc., Dalton Trans*, pp. 938-944.

[66] Martin, J. W. L., Timmons, J. H., Martell, A. E. and Willis, C. J. (1980). Synthesis and characterization of copper(II) and nickel(II) complexes of novel 18-membered tetraaza macrocyclic ligands, *Inorg. Chem.*, 19, pp. 2328-2331.

[67] Timmons, J. H., Rudolf, P., Martell, A. E., Martin, J. W. L. and Clearfield, A. (1980). Crystal and molecular structures of complexes of two isomeric 18-membered tetraaza macrocyclic ligands having the empirical formula [CuC$_{20}$H$_{40}$N$_4$](ClO$_4$)$_2$. Effects of chelate ring size and double-bond placement on coordination geometry about copper(II), *Inorg. Chem.*, 19, pp. 2331-2338.

[68] Black, D. St. C., Chaichit, N., Gatehouse, B. M. and Moss, G. I., (1987). Metal template reactions. XXV. N-acyl isatin precursors for the synthesis of malonamido macrocyclic metal complexes, *Aust. J. Chem.*, 40, pp. 1745-1754.

[69] Black, D. St. C. and Moss, G. I. (1987). Metal template reactions. XXIV. Synthesis of macrocyclic amide and ester complexes via substituted 1,1?-oxalylbisisatins and the deesterification of some ester complexes, *Aust. J. Chem.*, 40, pp. 143-155;

[70] Black, D. St. C. and Moss, G. I. (1987). Metal template reactions. XXIII. Synthesis of macrocyclic amide and ester complexes via 1,1?-oxalylbisisatin, *Aust. J. Chem.*, 40, pp. 129-142;

[71] Mederos, A., Dominiguez, S., Hernandez-Molina, R., Sanchiz, J. and Brito, F. (1999). Coordinating ability of ligands derived from phenylenediamines, *Coord. Chem. Rev.*, 193-195, pp. 857-911.

[72] Barefield, E. K., Wagner, F. and Hodges, K. D. (1976). Synthesis of macrocyclic tetramines by metal ion assisted cyclization reaction, *Inorg. Chem.*, 15, pp. 1370-1377.

[73] Cummings, S. C. and Sievers, R. E. (1970). A new and simple template synthesis of uninegative, macrocyclic, corrin-type ligands, *J. Amer. Chem. Soc.*, 92, 215-217.

[74] Martin, J. W. L., Wei, R. M. C. and Cummings, S. C. (1972). Copper(II) complexes with 13-membered macrocyclic ligands derived from triethylenetetramine and acetylacetone, *Inorg. Chem.*, 11, pp. 475-479.

[75] Martin, J. W. L. and Cummings, S. C. (1973). Square-planar nickel(II) and copper(II) complexes containing 14- and 15-membered tetraaza macrocyclic ligands, *Inorg. Chem.*, 12, pp. 1477-1482.

[76] Roberts, G. W., Cummings, S. C. and Cunningham, J. A. (1976). Synthesis and characterisation of low-spin cobalt(II) complexes with macrocyclic tetraaza ligands. The crystal structure of $[Co([14]dieneN_4)'H_2O](PF_6)_2$, *Inorg. Chem.*, 15, pp. 2503-2510.

[77] Wolf, V. L. and Jäger, E. G. (1966). Kupfer- und Nickelchelate von Diaminderivaten, *Z. Anorg. Allg. Chem.*, 346, pp. 76-91.

[78] Abid, K.K. , Fenton, D. E., Consellato, U., Vigato P.A. and Graziani, R. J. (1984). The template synthesis and crystal and molecular structure of a sexidentate Schiff base macrocyclic complex of samarium(III), *J. Chem. Soc., Dalton Trans.*, pp. 351-354.

[79] Radecka-Paryzek, W. and Luks, E. (1990). The new triaza macrocyclic complex of yttrium, *Polyhedron*, 9, pp. 475-477.

[80] Karn, J. L. and Busch, D. H. (1966). Nickel (II) complexes of the tetradentate macrocycle 2,12-dimethyl-3,7,11,17-tetraazabicyclo (11.3.1) heptadeca-1(17),2,11,13,15-pentaene, *Nature,* 211, pp. 160-162.

[81] Alcock, N. W., Balakrishnan, K. P., Moore, P. and Pike, G. A. (1987). Synthesis of pyridine-containing tetra-aza macrocycles: 3,7,11,17-tetra-azabicyclo[11.3.1]heptadeca-1(17),13,15-triene (L^1), its 3,11-dibenzyl (L^2) and 3,7,11-tribenzyl (L^3) derivatives, and their nickel(II), copper(II), and zinc(II) complexes: crystal structures of L^2HCl and $[Ni(L^2)Cl]ClO_4'H_2O$, *J.Chem. Soc., Dalton Trans.*, pp. 889-894.

[82] Prince, R. H., Stotter, D. A. and Woolley, P. R. (1974). Metal complexes of Schiff base ligands. Studies on the formation of macrocyclic condensates, *Inorg. Chim. Acta,* , 9, pp. 51-54.

[83] Curry, I. D. and Busch, D. H. (1964). The reactions of coordinated ligands. VII. Metal ion control in the synthesis of chelate compounds containing pentadentate and sexadentate macrocyclic ligands, *J. Am. Chem. Soc.*, 86, pp. 592-594.

[84] Nelson, S. M. and Busch, D. H. (1969). Seven-coordination in some mononuclear and binuclear iron(III) complexes containing a pentadentate macrocyclic ring, *Inorg.Chem.*, 8, pp. 1859-1862.

[85] Drew, M. G. B., bin Othman, A. H., McIlroy, P. D. and Nelson, S. M. (1975). Seven-co-ordination in metal complexes of quinquedentate macrocyclic ligands. Part II. Synthesis, properties, and crystal and molecular structures of some iron(III) derivatives of two 'N$_5$' macrocycles, *J. Chem. Soc., Dalton Trans*, pp. 2507-2516.

[86] Alexander, M. D., Von Heuvelen, A. and Hamilton, H. G. Jr. (1970). Manganess(II) complexes of a macrocyclic ligand, *Inorg. Nucl. Chem. Lett.*, 6, pp. 445-448.

[87] Drew, M. G. B., bin Othman, A. H., McFall, S. G., McIlroy, P. D. A. and Nelson, S. M. (1977). Seven-co-ordination in metal complexes of quinquedentate macrocyclic ligands. Part 5. Synthesis and properties of pentagonal-bipyramidal and pentagonal-pyramidal manganese(II) complexes and crystal and molecular structure of {2,15-dimethyl-3,7,10,14,20-penta-azabicyclo-[14.3.1]eicosa-1(20),2,14,16,18-pentaene}bis(isothio- cyanato)manganese(II), *J. Chem. Soc., Dalton Trans.*, pp. 438-446.

[88] Drew, M. G. B., Grimshaw, J., McIlroy, P. D. A. and Nelson, S. M. (1976). Seven-co-ordination in metal complexes of quinquedentate macrocyclic ligands. Part III. Preparation and properties of some iron(II) complexes of 2,13-dimethyl-3,6,9,12,18-penta-azabicyclo[12.3.1]octadeca-1(18),2,12,14,16-pentaene and 2,14-dimethyl-3,6,10,13,19-penta-azabicyclo[13.3.1]nonadeca-1(19),13,15,17-pentaene, *J.Chem.Soc., Dalton. Trans.*, pp. 1388-1394.

[89] Drew, M. G. B., McFall, S. G. and Nelson, S. M. (1979). Pentagonal-pyramidal cadmium(II) complexes of the quinquedentate macrocyclic ligand 2,15-dimethyl-3,7,10,14,20-penta-azabicyclo[14.3.1]eicosa-1(20),2,14,16,18-pentaene, *J. Chem. Soc., Dalton Trans.*, pp. 575-581.

[90] Drew, M. G. B., Hollis, S., McFall, S. G. and Nelson, S. M. (1978). The structure of a 7-coordinate Cd(II) complex of a macrocyclic ligand containing both bridging and uncoordinated perchlorate groups, *J. Inorg. Nucl. Chem.*, 40, pp. 1595-1596.

[91] Ferraudi, F. (1980). Photochemistry of high-spin iron(III) complexes of the macrocyclic ligands [15]pyridineN5 and [15]pyaneN5. An investigation of the charge-transfer processes, *Inorg. Chem.*, 19, pp. 438-444.

[92] Scovolle, A. N. and Reiff, W. M. (1983). Low temperature magnetic susceptibility study of the zero-field splitting in seven coordinate high-spin iron(III) complexes based on pentadentate macrocyclic nitrogen ligand. *Inorg. Chim. Acta*, 70, pp. 127-131.

[93] Cofield, M. L. and Bryan, P. S. (1986). Low temperature magnetic susceptibility and zero-field splitting in some high-spin manganese(III) compounds, *Inorg. Chim. Acta*, 112, pp. 1-4.

[94] Drew, M. G. B., Rice, D. A. and bin Silong, S. (1983). Studies in the flexibility of macrocyclic ligands. Crystal and mplecular structure of 2,13-dimethyl-3,6,9,12,18-pentaazabicyclo (12.3.1) octadeca-(18),14,16-triene-dichloro-ion (III) hexafluoro-phosphate, *Polyhedron*, 2, pp. 1053-1056.

[95] Drew, M. G. B., Hollis, S. and Yates, P. C. (1985). Studies in the flexibility of macrocycle ligans. Calculation of macrocycle cavity size by force-fields methods. Crystal and molecular structure of (CoLCl)(ClO$_4$)$_2$ and (CuL)(PF$_6$)$_2$, L= 2,13-

dimethyl-3,6,9,12,18-pentaazabicyclo[12.3.1]octadeca-118,14,16-triene, *J.Chem.Soc., Dalton Trans.,* pp. 1829-1834.

[96] Drew, M. G. B., bin Othman, A. H., McFall, S. G. and Nelson, S. M. (1977). Molecular association without co-ordination in an adduct od 1,10-phenanthroline with a manganese(II) macrocyclic complex; X-ray crystal and molecular structure, *J. Chem. Soc., Chem. Commun.,* pp. 558-560.

[97] Radecka-Paryzek, W. and Litkowska, H. (2000). Rare earth macrocyclic complexes derived from spermine, *J. Alloys Commp.,* 300-301, pp. 435-438.

[98] Alcock, N. W., Liles, D. C., McPartlin, M. and Tasker, P. A. (1974). A novel series of planar pentadentate macrocyclic ligands. X-ray structure of 10, 11,12,13-tetrahydrodibenzo[b,k]pyrido[g,f]1,4,7,10,13]penta-azacyclo-pentadecin- N^5, N^{10}, N^{13}, N^{18}- N^{19b}-di(perchlorato)manganese, *J. Chem. Soc., Chem. Commun.,* 727-728.

[99] Adam, K. R., Antolovich, M., Baldwin, D. S., Duckworth, P+. A., Leong, A. J., Lindoy, L. F., McPartlin, M. and Tasker, P. A. (1993). Ligand design and metal-ion recognition. The interaction of copper(II) with a range of 16- to 19-membered macrocycles incorporating oxygen, nitrogen and sulfur donor atoms, *J. Chem. Soc., Dalton. Trans.,* pp. 1013-1018.

[100] Radecka-Paryzek, W., Patroniak-Krzyminiewska, V. and Litkowska, H. (1998). The template synthesis and characterization of the yttrium and lanthanide complexes of new 19-membered pentadentate azaoxa macrocycle, *Polyhedron,* 17, pp. 1477-1480.

[101] Arif, A. M., Gray, C. J., Hart, F. A. and Hursthouse, M. B. (1985). Synthesis and structure of lanthanide complexes of a mixed donor macrocyclic ligand, *Inorg. Chim. Acta,* 109, pp. 179-183.

[102] Bastida, R., de Blas, A., Castro, P., Fenton, D. E., Macias, A., Rial, R., Rodriguez, A. and Rodriguez-Blas, T. (1996). Complexes of lanthanide(III) ions with macrocyclic ligands containing pyridine head units, *J. Chem. Soc., Dalton Trans.,* pp. 1493-1497.

[103] Bandin, R., Bastida, R., de Blas, A., Castro, P., Fenton, D. E., Macias, A., Rodriguez, A. and Rodriguez-Blas, T. (1994). Complexes of lanthanide ions with Schiff base macrocyclic ligands derived from 2,6-diformylpyridine, *J. Chem. Soc., Dalton Trans.,* pp. 1185-1192.

[104] Alcock, N. W., Liles, D. C., McPartlin, M. and Tasker, P. A. (1974). A novel series of planar pentadentate macrocyclic ligands. X-ray structure of 10,11,12,13-tetrahydrodibenzo[b,k]pyrido[g,f][1,4,7,10,13]penta-azacyclopenta-decin-N^5,N^{10},N^{13},N^{18}-N^{19b}-di(perchlorato)manganese, *J. Chem. Soc., Chem. Commun.,* pp. 727-728.

[105] Fenton, D. E., Murphy, B. P., Leong, A. J., Lindoy, L. F., Bashall, A. and McPartlin, M. (1987). Studies of metal-ion recognition. The interaction of Co^{II}, Ni^{II}, and Cu^{II} with new oxygen-nitrogen donor macrocycles; X-ray structures of complexes of Cu^{II} and Ni^{II} with a 15-membered O_2N_3 derivative, *J. Chem. Soc., Dalton Trans.,* pp. 2543-2553.

[106] Lodeiro, C., Bastida, R., Bértolo, E., Macias, A. and Rodriguez, A. (2003). Metal complexes with four macrocyclic ligands derived from 2,6-bis(2-

formylphenoxymethyl)pyridine and 1,7-bis(2?-formylphenyl)-1,4,7-trioxaheptane, *Inorg. Chim. Acta.,* 343, pp. 133-140.

[107] Lodeiro, C., Bastida, R., de Blas, A., Fenton, D. E., Macias, A., Rodriguez, A. and Rodriguez-Blas, T. (1998). Complexes of lead(II) and lanthanide(III) ions with two novel 26-membered-imine and – amine macrocycles derived from 2,6-bis(2-formylphenoxymethyl)pyridine, *Inorg. Chim. Acta,* 267, pp. 55-62.

[108] Vicente, M., Lodeiro, C., Adams, H., Bastida, R., de Blas A., Fenton, D. E., Macias, A., Rodriguez, A. and Rodriguez-Blas, T. (2000). Synthesis and characterization of some metal complexes with new nitrogen-oxigen donor macrocyclic ligand – X-ray crystal structures of a 26-membered reduced monoprotonated macrocycle and a 20-membered pendant-arm Schiff-base macrocyclic cadmium(II) complex, *Eur. J. Inorg. Chem.*, pp. 1015-1024.

[109] Riker-Nappier, J. and Meek, D. W. (1974). Nickel(II) complexes of two new phosphorus-nitrogen macrocyclic ligands, *J. Chem. Soc., Chem. Commun.*, pp. 442-443.

[110] Rothermel, G. L., Miao, L., Hill, A. L. and Jakels, S.C. (1992). Macrocyclic ligands with 18-membered rings containing pyridine or furan groups: preparation, protonation, and complexation by metal ions, *Inorg. Chem.* 31, pp. 4854-4859.

[111] Shakir M. and Varkey, S. P. (1994). Synthesis and structural characterization of cobalt(II), nickel(II) and copper(II) complexes of 18-membered mixed-donor macrocycles, *Polyhedron,* 13, pp. 791-797.

[112] Flangan, S., Dong, J., Haller, K., Wang, S., Scheidt, W. R.,. Scott, R. A., Webb, T. R., Stanbury, D. M. and Wilson, L. J. (1997).). Macrocyclic $[Cu^{I/II}(bite)]^{+/2+}$ (bite = biphenyldiimino dithioether): an example of fully-gated electron transfer and its biological relevance, *J. Am. Chem. Soc., 119*, pp. 8857-8868.

[113] Jackels, S. C., Farmery, K., Barefield, E. K., Rose, N. J. and Busch, D. H. (1972). Some tetragonal cobalt(III) complexes containing tetradentate macrocyclic amine ligands with different degrees of unsaturation, *Inorg. Chem.*, 11, pp. 2893-2900.

[114] Baldwin, D. A., Pfeiffer, R. M., Reichgott, D. W. and Rose, N. J. (1973). Synthesis and reversible ligation studies of new low-spin iron(II) complexes containing a planar cyclic tetradentate ligand and other donor molecules including carbon monoxide, *J. Amer. Chem. Soc.*, 95, pp. 5152-5158.

[115] Coltrain, B. K. and Jackels, S. C. (1981). Coordination chemistry of a cooper(II) tetraimine macrocycle: four-, five-, and six-coordinate derivatives and reduction transmetalation to the zinc(II) complex, *Inorg. Chem.*, 20, pp. 2032-2039.

[116] Welsh, W. A., Reynolds, G. J. and Henry, P. M. (1977). Synthesis of hydroxy-substituted macrocyclic ligand complexes of cobalt and isolation of a macrocycle precursor, *Inorg. Chem.*, 16, pp. 2558-2561.

[117] Jäger, E. G. (1968). Neutralkomplexe mit vierzehngliedrigen makrocyclischen Liganden, *Z. Chem.*, 8, pp. 30-31.

[118] Bamfield, P. (1969). Reactions of the metal complexes of 2-hydroxymethylenecyclohexanone and its derivatives with amines, *J. Chem. Soc A,* pp. 2021-2027.

[119]   Black, D. St. C. and Lane, M. J. (1970). Metal template reactions. II. Formation of macrocyclic quadridentate α-diimine nickel(II) complexes, *Aust. J. Chem.*, 23, pp. 2027-2037.

[120]   Black, D. St. C. and Lane, M. J. (1970). Metal template reactions. IV. The attempted dimerization of α, ω amino dialdehydes around nickel(II) ions, *Aust.J. Chem.*, 23, pp. 2055-2063.

[121]   Park, Y. C., Kim, S. S., Lee, D. -C. and An, C. H. (1997). Template synthesis of asymmetrical nickel(II) of substitute mono-benzo-N_4 and effect of substitute for methane sites in their complexes, *Polyhedron*, 16, pp. 253-258.

[122]   Reddy, K. H., Reddy, M. R. and Raju, K. M. (1997). Synthesis, characterization, electrochemistry and axial ligation properties of macrocyclic divalent metal complexes of acetylacetone buckled with different diamines, *Polyhedron*, 16, pp. 2673-2679.

[123]   Jäger, E. G. (1964). Ein neues Nickelchelat mit allseitig geschlossenem Ringsystem, *Z. Chem.*, 4 , pp. 437-438.

[124]   L'Eplattenier, F. A. and Pugin, A. (1975). Template -reaktionen I. Herstellung von tricyclischen und tetracyclischen metalkomplexen aus aromatischen 1,2-diaminen und 1,3-dicarbonylverbindungen, *Helv. Chim. Acta*, 58, pp. 917-929.

[125]   Lukes, P. J., McGregor, A. C., Clifford, T. and Crayston, J. A. (1992). Electrochemistry of planar cobalt(II) and nickel(II) tetraaza[14]annulene complexes, *Inorg. Chem.*, 31, pp. 4697-4699 and references therein.

[126]   Arion, V. B., Gerbeleu, N. V., Levitsky, V. G., Simonov, Yu. A., Dovorkin, A. A. and Bourosh, P. N. J. (1994). Template synthesis, structure and properties of a bis(macrocyclic) dinickel(II) complex based on a 14-membered hexaaza unit, *Chem. Soc. Dalton Trans.*, pp. 1913-1916.

[127]   Melson, G. A. and Busch, D. H. (1965). Reactions of coordinated ligands. XI. The formation and properties of a tridentate macrocyclic ligand derived from o-aminobenzaldehyde, *J. Am. Chem. Soc.*, 87, pp. 1706-1710.

[128]   Fleischer, E. B. and Klem, E. (1965). The structure of a self-condensation product of o-aminobenzaldehyde in the presence of nickel ions, *Inorg. Chem.*, 4, pp. 637-642.

[129]   Bayley, N.A., Fenton, D. E., Jackson, I. T., Moody, R. and Rodriguez de Barbarin, C. (1983). Metal ion controlled ring contraction to produce an oxazolidine-containing Schiff base macrocycle and the X-ray structure of the $Pb(NCS)_2$ complex, *J. Chem. Soc., Chem. Commun.*, pp. 1463-1465.

[130]   Fenton, D. E., and Vigato, P. A. (1988). Macrocyclic Schiff base complexes of lanthanides and actinides, *Chem. Soc. Rev,* 17, pp. 69-90.

[131]   Radecka-Paryzek, W. (1981). The synthesis and characterization of the macrocyclic and ring-opened complexes formed in the reaction of the lanthanides with 2,6-diacetyl-pyridine and hydrazine, *Inorg. Chim. Acta*, 52, pp. 261-268.

[132]   Arif, A. M., Backer-Dirks, J. D. J., Gray, C. J., Hart, F. A. and Hursthouse, M. B. (1987). Syntheses, X-ray structures and properties of complexes of macrocyclic hexaimines with lanthanide nitrates, *J. Chem. Soc., Dalton Trans.*, pp. 1665-1673. and references.

133 Aruna, V. A. J. and Alexander, V. (1996). Macrocyclic complexes of lanthanides in identical ligand frameworks. Part 2. Synthesis of lanthanide(III) and yttrium(III) complexes of an 18-membered hexaaza macrocycle, *Inorg. Chim. Acta*, 249, pp. 93-100.

134 Radecka-Paryzek, W. (1985). The template synthesis and characterization of hexaaza 18-membered macrocyclic complexes of cerium(III), praseodymium(III) and neodinium(III) nitrates, *Inorg. Chim. Acta*, 109, pp. L21-L23.

135 Lisowski, J. (1999). ^{1}H and ^{13}C NMR study of paramagnetic chiral macrocyclic lanthanide complexes, *Magn. Reson. Chem.*, 37, pp. 287-294.

136 Lisowski, J. and Mazurek, J. (2002). Chiral macrocyclic La(III), Ce(III), Pr(III) and Eu(III) complexes with chloride anions, *Polyhedron*, 21, pp. 811-816.

137 Bligh, S. W. A., Choi, N., Cummis, W. J., Evagorou, E. G., Kelly, J. D. and McPartlin, M. (1994). Yttrium(III) and lanthanide(III) metal complexes of an 18-membered hexaaza tetraimine macrocycle. Crystal structure of the gadolinium(III) complex, *J. Chem. Soc., Dalton Trans.*, pp. 3369-3376.

138 Drew, M. G. B., Yates, P. C., Esho, F. S., Grimshaw, J. T., Lavery, A., McKillop, K. P., Nelson, S. M. and Nelson, J. (1988). Dicopper(II) complexes of a binucleating N_4 macrocycle containing mono- and di-atomic bridges; magnetic interactions mediated by alkoxo- and diaza-bridging ligands. Crystal structures of $[Cu_2(L^1)(pz)_2][ClO_4]_2$, $[Cu_2(L^1)(OEt)_2(NCS)_2]$, and $[Cu_2(L^1)(OMe)_2(MeCN)_2][BPh_4]_2$, *J. Chem. Soc., Dalton Trans.*, pp. 2995-3003.

139 Drew, M. G. B., Esho, F. S., Lavery, A. and Nelson, S. M. (1984). Dicobalt(II) complexes of a macrocyclic ligand containing hydroxo-, alkoxo-, phenoxo-, thiolato-, halogeno-, and pseudohalogeno-bridges: structures and magnetic exchange interactions, *J. Chem. Soc., Dalton Trans.*, pp. 545-556.

140 Fenton, D. E., Kitchen, S. J., Spencer, C. M., Tamburini, S. and Vigato, P. A. (1988). Complexes of ligands providing endogenous bridges. Part 5. Solution studies on a novel '3+3' hexamine Schiff-base macrocyclic complex of lanthanum, *J. Chem. Soc., Dalton Trans.*, pp. 685-690.

141 Shakir, M., Nasman, O. S. M. and Varkey, S. P. (1996). Binuclear N_6 22-membered macrocyclic transition metal complexes: synthesis and characterization, *Polyhedron*, 15, pp. 309-314.

142 Abid, K. K. and Fenton, D. E. (1984). The synthesis of macrocyclic lanthanide complexes derived from 2,5-furandial-dehyde and a, ? -alkanediamines, *Inorg. Chim. Acta*, 82, pp. 223-226.

143 Abid, K. K. and Fenton, D. E. (1984). Lanthanide complexes of some macrocyclic Schiff bases derived from pyridine-2,6-dicarboxaldehyde and α, ω-primary diamines, *Inorg. Chim. Acta*, 95, pp. 119-125.

144 Sessler, J. L., Mody, T. D. and Lynch, V. (1992). Synthesis and X-ray characterization of a uranyl(VI) Schiff base complex derived from a 2:2 condensation product of 3,4-diethylpyrrole-2,5-dicarbaldehyde and 1,2-diamino-4,5-dimethoxybenzene, *Inorg. Chem.*, 31, pp. 527-531.

145 Aruna, V. A. J. and Alexander, V. (1996). Synthesis of lanthanide (III) complexes of a 20-membered hexaaza macrocycle, *J. Chem. Soc., Dalton Trans.*, pp. 1867-1873.

[146] James, S., Kumar, D. S. and Alexander, V. (1999). Synthesis of lanthanide(III) complexes of 20-membered octaaza and hexaaza Schiff-base macrocycles, *J. Chem. Soc., Dalton Trans.*, pp. 1773-1777.

[147] Kahwa, I. A., Selbin, J., Hsieh, T. C. -Y. and Laine, R. A. (1986). Synthesis of homodinuclear monocyclic complexes of lanthanides and phenolic Schiff bases, *Inorg. Chim. Acta*, 118, pp. 179-185.

[148] Bullita, E., Casellato, U., Guerriero, P. and Vigato, P. A. (1987). Lanthanide complexes with macrocyclic and macrocyclic Schiff bases, *Inorg. Chim. Acta*, 139, pp. 59-60.

[149] Fenton, D. E., and Rossig., (1985). Metal complexes of Schiff base macrocycles having pendant arms bearing ligating groups, *Inorg. Chim. Acta*, 98, pp. L29-L30.

[150] Adams, H., Bailey, N. A., Dwyer, M. J. S., Fenton, D. E., Hellier, P. C., Hempstead, P. D. and Latour, J. M. (1993). Synthesis and crystal structure of a first generation model for the trinuclear copper site in ascorbate oxidase and of a dinuclear silver precursor, *J. Chem. Soc., Dalton Trans.*, pp. 1207-1216.

[151] Benetollo, F., Bombieri, G., Adeyga, A. M., Fonda, K. K., Gootee, W. A, Samaria, K. M. and Vallarino, L. M. (2002). Lanthanide(III) complexes of six-nitrogen-donor macrocyclic ligands with benzyl-type peripheral substituents, and crystal structure of $[La(NO_3)_2(H_2O)(C_{36}H_{38}N_6)](NO_3)(H_2O)$, *Polyhedron*, 21, pp. 425-433.

[152] Dumont, A., Jacques, V. and Desreux, J. F. (2000), New synthons for the synthesis of lanthanide containing macrocyclic Schiff bases featuring substituents available for tethering, *Tetrahedron*, 56, pp. 2043-2052.

[153] Lehn, J. -M. (1980). Dinuclear cryptates: dimetallic macropolycyclic inclusion complexes, *Pure Appl. Chem.*, 52, pp. 2441-2459.

[154] Wilson, L. J. and Rose, N. J. (1968). Geometrically specific multidentate ligands and their complexes. I. A nickel(II) complex of the potentially heptadentate Schiff base derived from 2,2', 2'-triaminotriethylamine and 2-pyridinecarboxaldehyde, *J. Am. Chem. Soc.*, 90, pp. 6041-6045.

[155] Cabral, F., Murphy, B. and Nelson, J. (1984). Complexes of macrocyclic ligand having mono- and binucleating capability: binuclear Mn(II), Fe(II), Co(II), NI(II), ZN(II) complexes, *Inorg. Chim. Acta*, 90, pp. 169-178.

[156] Ngwenya, M. P., Martell, A. E. and Reilbenspies, J. (1990). Template synthesis of a novel macrobicyclic ligand and the crystal structure of its unique dinuclear copper(I) complex, *J. Chem. Soc., Chem. Commun.*, pp. 1207-1208.

[157] Avecilla, F., Bastida, R., de Blas, A., Fenton, D. E., Macias, A., Rodriguez-Blas, A., Garcia-Granda, S. and Corza-Suarez, R. (1997). Metal template synthesis of lanthanide cryptates. Crystal structure of a dysprosium cryptate, *J. Chem. Soc., Dalton Trans.*, pp. 409-413.

[158] Platas, C., Avecilla, F., de Blas, A., Geraldes, C. F. G. C., Rodriguez-Blas, T., Adams, H. and Mahia J. (1999). ^{1}H NMR in solution and solid state structural study of lanthanide(III) cryptates, *Inorg. Chem.*, 38, pp. 3190-3199.

[159] Platas, C., Avecilla, F., de Blas, A., Rodriguez-Blas, T., Geraldes, C. F. G. C., Toth, E., Merbach, A. E. and Bünzli, J. -C. G. (2000). Mono- and bimetallic lanthanide(III) phenolic cryptates obtained by template reaction: solid state

structure, photophysical properties and relaxivity, *J. Chem. Soc., Dalton Trans.,* pp. 611-618.

[160] Avecilla, F., de Blas, A., Bastida, R., Fenton, D. E., Mahia, J., Macias, A., Platas, C., Rodriguez, A. and Rodriguez-Blas, T. (1999). The template synthesis and X-ray crystal structure of the first dinuclear lanthanide(III) iminophenolate cryptate, *J. Chem. Soc., Chem. Commun.,* pp. 125-126.

[161] Lintvedt, R. L., Glick, M. D., Tomlonovic, B. K., Gavel, D. P. and Kuszaj, J. M. (1976). Synthesis, structure, and magnetism of polynuclear chelates. Structural and magnetic comparison of in-plane and out-of-pane exchange in three polynuclear copper complexes, *Inorg. Chem.,* 15, pp. 1633-1645.

[162] Fenton, D. E. and Lintvedt, R. L. (1978). Binuclear copper(II) complexes as mimics for type 3 coppers in metalloenzymes. 1. The importance of cooperative interactions between metals in the reversible multielectron transfer in bis(1,3,5-triketonato)dicopper(II) complexes, *J. Am. Chem. Soc.,* 100, pp. 6367-6375.

[163] Okawa, H., Furutachi, H. and Fenton, D. E. (1998). Heterodinuclear metal complexes of phenol-based compartmental macrocycles, *Coord. Chem. Rev.,* 174, pp. 51-75.

[164] Pilkington, N. H. and Robson, R. (1970). Complexes of binucleating ligands. III. Novel complexes of a macrocyclic binucleating ligand, *Aust. J. Chem.,* 23, pp. 2225-2236.

[165] Hoskins, B. F. and Williams, G. A. (1975). The crystal structure determination of a binuclear copper(II) complex of a tetra-Schiff base macrocycle. *Aust. J. Chem.,* 28, pp. 2607-2614.

[166] Gagné, R. R., Koval, C. A., Smith, T. J. and Cimolino, M. C. (1979). Binuclear complexes of macrocyclic ligands. Electrochemical and spectral properties of homobinuclear $Cu^{II}Cu^{II}$, $Cu^{II}Cu^{I}$, and $Cu^{I}Cu^{I}$ species including an estimated intramolecular electron transfer rate, *J. Am. Chem. Soc.,* 101, pp. 4571-4580.

[167] Spiro, C. L., Lambert, S. L., Smith, T. J., Duesler, E. N., Gagné, R. R. and Hendrickson, D. N. (1981). Binuclear complexes of macrocyclic ligands: variation of magnetic exchange interaction in a series of six-coordinate iron(II), cobalt(II), and nickel(II) complexes and the X-ray structure of a binuclear iron(II) macrocyclic ligand complex, *Inorg. Chem.,* 20, pp. 1229-1237.

[168] Lacroix, P., Kahn, O., Theobald, F., Leroy, J. and Wakselman, C. (1988). Role of the CF_3 attractive group on the electrochemical and magnetic properties of copper(II) dinuclear compounds with Robson-type binucleating ligands, *Inorg. Chim. Acta,* 142, pp. 129-134.

[169] Aono, T., Wada, H., Yonemura, M., Ohba, M., Okawa, H. and Fenton, D. E. (1997). Effect of ring size in macrocyclic dinuclear manganese(II) complexes upon their structure, properties and reactivity towards H_2O_2, *J. Chem. Soc., Dalton Trans.,* pp. 1527-1531.

[170] Okawa, H. and Kida, S. (1972). Binuclear metal complexes. III. Preparation and properties of mononuclear and binuclear copper(II) and nickel(II) complexes of new macrocycles and their related ligands, *Bull. Chem. Soc. Jpn,* 45, pp. 1759-1764.

171   Lambert, S. L., Spiro, C. L., Gagné, R. R. and Hendrickson, D. N. (1982). Binuclear complexes of macrocyclic ligands. Variation of magnetic exchange interaction in a series of heterobinuclear Cu^{II}-M^{II} complexes, *Inorg. Chem.*, 21, pp. 68-72.

172   Lambert, S. L. and Hendrickson, D. N. (1979). Magnetic exchange interactions in binuclear transition-metal complexes. 20. Variation in magnetic exchange interaction for a series of metal(II) complexes of a binuclear ligand, *Inorg. Chem.*, 18, pp. 2683-2686.

173   Tadokoro, M., Okawa, H., Matsumoto, N., Koiakawa, M. and Kida, S. (1991). Template synthesis, structure and characterization of $Ni^{II}_2Pb^{II}$ and $Cu^{II}_2Pb^{II}$ complexes of macrocycles with a N_4O_2 donor set, *J. Chem. Soc., Dalton Trans.*, pp. 1657-1663.

174   Wada, H., Aono, T., Motoda, M., Ohba, M., Matsumoto, N. and Okawa, H. (1996). Macrocyclic heterodinuclear NiMn and CuMn complexes: crystal structure and electrochemical behaviour, *Inorg. Chim. Acta*, 246, pp. 13-21.

175   Hori, A., Yonemura, M., Ohba, M. and Okawa, H. (2001). Template synthesis of macrocyclic dinuclear Cu^{II} complexes and conversion into mononuclear complexes by site-selective copper elimination, *Bull. Chem. Soc. Jpn*, 74, pp. 495-503.

176   Yonemura, M., Matsumara, Y., Furutachi, H., Ohba, M., Okawa, H. and Fenton, D. E. (1997). Migratory transmetalation in diphenoxo-bridged $Cu^{II}M^{II}$ complexes of a dinucleating macrocycle with $N(amine)_2O_2$ and $N(imine)_2O_2$ metal-binding sites, *Inorg. Chem.*, 36, pp. 2711-2717.

177   Mandal, S. K. and Nag, K. (1983). Binuclear metal complexes. 1. Synthesis, characterization, and electrochemical studies of dicopper(II) complexes with 4-methyl-2,6-di(acyl/benzoyl)phenol, *Inorg. Chem*, 22, pp. 2567-2572.

178   Tadokoro, M., Sakiyama, H., Matsumoto, N., Kodera, M., Okawa, H. and Kida, S. (1992). Template synthesis of copper (II) lead (II) complexes of new binucleating macrocycles with dissimilar co-ordination sites, *J. Chem. Soc., Dalton Trans.*, pp. 313-317.

179   Luneau, D., Savariault, J., Cassoux, P. and Tuchagues, J. (1988). Polynuclear manganesse(II) complexes with Robson-type ligands. Synthesis, characterization, molecular structure and magnetic properties, *J. Chem. Soc., Dalton Trans.*, pp. 1225-1235.

180   Casellato, U., Fregona, D., Sitran, S., Tamburini, S. and Vigato, P. A. (1985). New acyclic and cyclic Schiff bases derived from 2,6-diformyl-4-chlorophenol and their interaction with uranyl(VI), copper(II) and nickel(II) ions, *Inorg. Chim. Acta*, 110, pp. 181-190.

181   Ohtsuka, S., Kodera, M., Motoda, K., Ohba, M. and Okawa, H. (1995). Dinuclear $Cu^{II}M^{II}$ (M = Co, Ni, Cu or Zn) and $Cu^{II}Cu^{I}$ complexes of a phenol-based dinucleating macrocycle with dissimilar N_2O_2 and N_2O_2S sites, *J. Chem. Soc., Dalton Trans.*, pp. 2599-2604.

182   Zhou, Y., Ge, Y., Wang, M., Liu, J. and Gou, S. (1998). Preparation and electrochemical behavior of a macrocyclic heterodinuclear $Fe^{III}Co^{III}$ complex, *Supramolecular Sci.*, 5, pp. 515-517.

183 Brooker, S. (2001). Complexes of thiophenolate-containing Schiff-base macrocycles and their amine analogues, *Coord. Chem. Rev.,* 222, pp. 33-56 and references.

184 Brooker, S., Croucher, P. D. and Roxburgh, F. M. (1996). Controlled synthesis and reversible oxidation of a thiolate-bridged macrocyclic dinickel(II) complex, *J. Chem. Soc., Dalton Trans.,* pp. 3031-3037.

185 Brooker, S., Croucher, P. D., Davidson, T. C., Dunbar, G. S., Beck, C. U. and Subramanian, S. (2000). Controlled thiolate coordination and redox chemistry: synthesis, structure, axial-binding and electrochemistry of dinickel(II) dithiolate macrocyclic complexes. *Eur. J. Inorg. Chem.,* pp. 169-179.

186 McKee, V. and Tandon, S. S. (1989). An octacopper(II) complex with 5-oxo and tripod-like perchlorate ligands; formation and X-ray structure of the $[Cu_4(L)O(ClO_4)_2].2H_2O$ dimer, *Inorg. Chem.,* 28, pp. 2901-2902.

187 Bell, M., Edwards, A. J., Hoskins, B. F., Kachab, E. H. and Robson, R. (1989). Synthesis and x-ray crystal structures of tetranickel and tetrazinc complexes of a macrocyclic tetranucleating ligand, *J. Am. Chem. Soc.,* 111, pp. 3603-3610.

188 Edwards, J., Hoskins, B. F., Kachab, E. H., Markiewicz, A., Murray, K. S. and Robson, R. (1992). Synthesis, X-ray crystal structures and magnetic properties of tetranickel complexes of a macrocyclic tetranucleating ligand, *Inorg. Chem.,* 31, pp. 3585-3591.

189 Tandon, S. S., Thompson, L. K. and Bridson, J. N. (1992). A novel antiferromagnetically coupled, dimeric, dodecanuclear copper(II) complex involving two macrocyclic ligands each with a benzene-like arrangement of six copper centers. *J. Chem. Soc., Commun.,* pp. 911-913.

190 McKee, V. and Tandon, S. S. (1991). Synthesis and characterisation of a series of tetra- and octa-copper(II) complexes with macrocyclic ligands, *J. Chem. Soc., Dalton Trans.,* pp. 221-229.

191 Edwards, A. J., Hoskins, B. F., Kachab, E. H., Markiewicz, A., Murray, K.S. and Robson, R. (1992). Synthesis, x-ray crystal structures, and magnetic properties of tetranickel complexes of a macrocyclic tetranucleating ligand, *Inorg. Chem.,* 31, pp. 3585-3591.

192 Edwards, A. J., Hoskins, B. F., Robson, R., Wilson, J. C., Moubaraki, B. and Murray, K. S. (1994). Crystal structures of a tatranucleating macrocycle and its dimanganese(III) derivative and magnetism of the latter, *J. Chem. Soc., Dalton Trans.,* pp. 1837-1842.

193 Motoda, K. -I., Sakiyama, H., Matsumoto, N., Okawa, H. and Fenton, D. E. (1995). Macrocyclic 'dimer-of-dimers' type tetranuclear copper(II) complexes with two bridging hydroxy groups in a face-to-face manner, *J. Chem. Soc., Dalton Trans.,* pp. 3419-3425.

194 Tandon, S. S., Thompson, L. K., Bridson, J. N. and Benelli, C. (1995). Hexanuclear and dodecanuclear macrocyclic copper(II) and nickel(II)complexes with almost planar "benzene-like" metal arrays, *Inorg. Chem.,* 34, pp. 5507-5515.

195 Rybak-Akimova, E. V., Busch, D. H., Kahol, P. K., Pinto, N., Alcock, N. W. and Clase, H. J. (1997). Dicopper complexes with a dissymmetric dicompartmental

Schiff base-oxime ligand: synthesis, structure, and magnetic interaction, *Inorg. Chem.,* 36, pp. 510-520.

[196] Okawa, H., Toku, T., Nonaka, Y., Muto, Y. and Kida, S. (1973). Binuclear metal complexes. VI. Syntheses and properties of binuclear copper(II) complexes of 2,6-bis[N-(β-dialkyl-aminoethyl)iminomethyl]-4-methylphenol, *Bull. Chem. Soc. Jpn,* 46, pp. 1462-1465.

[197] Okawa, H., Kida, S., Muto, Y. and Toku, T. (1972). Binuclear metal complexes. IV. The preparation and properties of binuclear copper(II) complexes of Schiff bases derived from 2,6-diformyl-4-methylphenol and glycine or alanine, *Bull. Chem. Soc. Jpn, 45,* pp. 2480-2484.

[198] Mallah, T., Kahn, O., Gouteron, J., Jeannin, S., Jeannin, Y. and Connor, C. J. O. (1987). Crystal structures and magnetic properties of dinuclear copper(II) complexes 2,6-bis(N-(2-pyridylmethyl)formimidoyl)-4-methylphenolate with azido and cyanato-oxigen exogenous ligands, *Inorg. Chem.,* 26, pp. 1375-1380.

[199] Mallah, T., Boillot, M. L., Kahn, O., Gouteron, J., Jeannin, S. and Jeannin Y. (1986). Crystal structures and magnetic properties of μ-phenolato copper(II) binuclear complexes with hydroxo, azido, and cyanoto-O exogenous bridges, *Inorg. Chem.,* 25, pp. 3058-3065.

[200] Mandal, S. K. and Nag, K. (1984). Dinuclear metal complexes. Part 3. Preparation and properties of hydroxo-bridged dicopper(II) complexes, *J. Chem. Soc. Dalton Trans.,* pp. 2141-2149.

[201] Casellato, U., Guerriero, P., Tamburini, S. and Graziani, R. (1987). Actinide complexes with Schiff bases: the first crystal structure of a binuclear thorium(IV) complex with a pentadentate compartmental ligand, *Inorg. Chim. Acta,* 139, pp. 61-63.

[202] Casellato, U. Fregona, D. Sitran, S. Tamburini, S. Vigato P. A. and Zanello, P.(1984). Preparation and properties of mono, homo and heterobinuclear complexes with a new heptadentate Schiff base ligand, *Inorg. Chim. Acta,* 95, pp. 309-316.

[203] Blicke, F. F. (1954). The Mannich reaction in organic reactions, ed.by John Wiley&Sons, Vol.I., pp. 303.

[204] Hellman, H. and Opitz, G. (1960). α-Amino-alkylierung, *Verlag Chemie,* Gmbh .

[205] Bertoncello, K., Fallon, G. D., Hodgkin, J. H., and Murray, K. S. (1988). Exogenous bridging and nonbridging in copper(II) complexes of a binucleating 2,6-bis((N-methylpiperazino)methyl)-4-chlorophenolate ligand. Crystal structures and magnetic properties of bis(μ-acetato), dinitrito, and bis(azido) complexes. Possible relevance to the type 3 depleted laccase active site, *Inorg. Chem.,* 27, pp.4750-4758.

[206] Luben, M. and Feringa, B. L. (1994).A new method for the synthesis of nonsymmetric dinucleating ligands by aminomethylation of phenols and salicylaldehydes, *J .Org. Chem.,* 59, pp. 2227-2233.

[207] Hodgkin, J. H. (1984). New Mannich base ligands, *Austr. J. Chem.,* 37, pp.2371-2378.

[208] Sargeson, A. M. (1979). Caged metal ions, *Chem. Br.,* 15, pp. 23-27.

209 Curtis, N. F. (1968). Macrocyclic coordination compounds formed by condensation of metal-amine complexes with aliphatic carbonyl compounds, *Coord. Chem. Rev.,* 3, pp. 3-47.

210 Harrowfield, M. and Sargeson, A. M. (1979). Synthesis and reactivity of coordinated imines derived from 2-keto acids, *J. Am.Chem. Soc.,* 101, pp. 1514-1520.

211 Harrowfield, J. M. and Sargeson, A. M. (1974). Reactions of coordinated nucleophiles. Intramolecular imine formation, *J. Am Chem. Soc. ,* 96, pp. 2634-2635.

212 Gainsford, A. R., Pizer, R. D., Sargeson, A. M. and Whimp, P. O. (1981). Intramolecular carbinolamine and imine formation with cobalt(III)-amine complexes. Synthesis, structure, and reactivity, *J. Am. Chem. Soc.,* 103, pp. 792-805.

213 Geue, R. G., Snow, M. R., Springborg, J., Hertl, A. J., Sargeson, A.M. and Taylor, D. (1976). Condensation of formaldehyde with chelated glycine and ethylenediamine: a new macrocycle synthesis; X-ray structures of [α-hydroxymethylserine-bis(ethylenediamine0cobalt(III)]$^{2+}$ and [α-hydroxymethylserine-1,4,8,11-tetra-bis(ethylenediamine)cobalt(III)]$^{2+}$ and [α-hydroxymethylserine-1,4,8,11-tetra-aza-6,13-dioxacyclotetradecanecobalt(III)]$^{2+}$ ions, *J. Chem. Soc., Chem. Commun.,* pp. 285-287.

214 Creaser, I. I., Geue, R. J., Harrowfield, J. M., Herlt, A. J., Sargeson, A. M., Snow, M. R. and Springborg, A. M. (1982). Synthesis and reactivity of aza-capped encapsulated Co(III) ions, *J. Am. Chem. Soc.,* 104, pp. 6016-6025.

215 Watt, G. W. and Alexander, P. W. (1968). Reactions of deprotonated ligands. V.Deprotonated tris (ethylenediamine)rhodium(III) ion, *Inorg. Chem.,* 7, pp. 537-542.

216 Geue, R. J., McCarty, M. G., Sargeson, A. M., Skelton, B. W. and White, A. H. (1985). Stereospecific Template synthesis of a macrotetracyclic hexaazacryptate: X-ray crystal structure of (9,17-dimethyl-13-nitro-1,3,5,7,11,15-hexaazatetracyclo-[11.5.1.1^{3,9}.1^{5,1}] henicosane)cobalt(III) chloride, *Inorg. Chem.,* 24, pp. 1607-1609.

217 Arnold, A. P., Bhula, R., Chen, X., Geue, R. J. and Jackson, W. G. (1999). Precursors to new molecular tube ligands. Double-capped trinuclear cobalt complexes of aminoethanethiol, *Inorg. Chem.,* 38, pp. 1966-1970.

218 Comba, P., Curtis, N. F., Lawrance, G. A., Sargeson, A. M., Skelton, B. W. and White, A. H. (1986). Template synthesis involving carbon acids. Synthesis and characterization of (3,10-Dimethyl-3,10-dinitro-1,4,8,11-tetraazacyclotetradecane) copper(II) and (1,9-diamino-5-methyl-5-nitro-3,7-diazanonane)copper(II) Cations and nitro group reduction products, *Inorg. Chem.,* 25, pp. 4260-4267.

219 Suh, M. P., Shin, W., Kim, D. and Kim, S. (1984). Preparation and molecular structure of the nickel(II) perchlorate complex of a hexadentate macrobicyclic ligand, 1,3,6,8,10,13,16,19-Octaazabicyclo[6.6.6]eicosane, *Inorg. Chem.,* 23, pp. 618-620.

220 Suh, M. P. and Kim, D. (1985). Template synthesis and characterization of a nickel(II)   complex   with

tris(((aminoethyl)amino)methyl)amineosemisepulchrate)nickel(2+), *Inorg. Chem.*, 24, pp. 3712-3714.

[221] Suh, M. P., Shin, W., Kim, H. and Koo, C. H. (1987). Nickel(II) complexes of novel ligands containing a tetraazabicyclononane ring: syntheses and structures of [3,7-bis(2-aminoethyl)-1,3,5,7-tetraazabicyclo[3.3.3]nonane]nickel(II)perchlorate and (8-methyl-1,3,6,8,10,13,15-heptaazatricyclo[13.1.1.1^{13,15}]octadecane)nickel(II)perchlorate, *Inorg. Chem.*, 26, pp. 1846-1852.

[222] Bernhardt, P. V., Curtis, L. S., Curtis, N. F., Lawrance, G. A., Skelton, B. W. and White, A. H. (1989). Condensations of ethane-1,2-diamine, formaldehyde and ammonia(or nitroethane) directed by nickel(II) or copper(II). Crystal structure of (8-methyl-8-nitro-1,3,6,10,13,15-hexaazatricyclo[13.1.1.1^{13,15}]octadecane)nickel(II), *Aust. J. Chem.*, 42, pp. 797-811.

[223] Comba, P., Hambley, T. W.and Lawrance, G. A. (1985). Template synthesis, crystal structure, and spectroscopic characterization of [N,N'-Bis(2-pyridylmethylene)-1,3-diamino-2-methyl-2-nitropropane]copper(II) perchlorate, *Helv. Chim. Acta*, 68, pp. 2332-2341.

[224] Lawrance, G. A., Manning T. M. and O'Leary, M. A. (1988). Metal-directed synthesis of the new potentially pentadentate aminoalcohol ligand 5-amino-5-methyl-3,7-diazanonan-1,9-diol based on ethanolamine, *Polyhedron*, 7, pp. 1263-1266.

[225] Brush, J. R., Magee, R. J., O'Connor, M. J., Teo, S. B., Geue, R. J. and Snow, M. R. (1973). Nature of the copper(II) complex formed in the reaction of formaldehyde with bis((S)-serinato)copper(II), *J. Am.Chem.. Soc.*, 95, pp. 2034-2035.

[226] Dempsey, A. and Phipps, D. A. (1979). Coordination catalysis: tautomeric vs. carbanion mechanisms in the racemisation of L-alanine induced by pyruvate and Zn^{2+} ions, *Inorg. Chim. Acta.*, 36, pp. L425-L427.

[227] Casella, L., Pasini, A., Ugo, R. and Visca, M. (1980). Reactions of amino-acids co-ordinated to metal ions. Part1. Investigation of the condensation of formaldehyde and metal-co-ordinated glycine, *J. Chem. Soc., Dalton Trans*, pp. 1655-1663.

[228] Teo, S. B. and Teoh, S. G. (1980). Nature of the copper(II) complex formed in the reaction of formaldehyde with bis(glycinato)copper(II), *Inorg. Chim. Acta.*, 44, pp. L269-L270.

[229] Teo, S. B., Teoh, S. G. and Snow M. R. (1984). Condensation of bis(glycinato)nickel(II) with formaldehyde and ammonia: x-ray structure of bis[3N, 7N-(1,3,5,7-tetraazabicyclo[3.3.1]nonyl)diacetato]nickel(II), *Inorg. Chim. Acta.*, 85, pp. L1-L2.

[230] Balla, J., Bernhardt, P. V., Buglyo, P., Comba, P., Hambley, T. W., Schmidlin, R., Stebler, S. and Varnagy, K. (1993). Chiral quadridentate ligands based on amino acids: template syntheses and properties of the free ligands and their transition-metal complexes, *J. Chem. Soc., Dalton Trans*, pp. 1143-1149.

[231] Thewalt, U. and Bugg, C. E. (1972). Röntgenographische Charakterisierung des Reaktionsproduktes von bis(S-amino-dithionitrrito)-nickel(II) mit Ammoniak, Formaldehyd und Methanol, *Chem. Ber.* 105, pp. 1614-1620.

[232] Izatt, R. M. and Christensen, J. J. (1978). (Eds.), Synthetic multidentate macrocyclic ligands, Academic Press, New York.pp. 273-290

[233] Christensen, J. J. and Izatt, R. M. (1987). (Eds.), Synthesis of macrocycles, the design of selective complexing agents, Wiley, New York.

[234] Bhula, R., Osvath, P. and Weatherburn, D. C. (1988). Complexes of tridentate and pentadentate macrocyclic ligands, *Coord. Chem. Rev.*, 91, pp. 89-213.

[235] Dietrich, B., Viout, P. and Lehn, J.-M.(1993). Macrocyclic Chemistry, Weinheim.pp.45-345.

[236] De Sousa Healy, M. and Rest, A. J. (1978). Template reactions. *Adv. Inorg. Chem. Radiochem.*, 1, pp. 21-40.

[237] Rossignoli, M., Allen, C. C., Hambley, T. W., Lawrance, G. A. and Maeder, M. (1996). "Swollen" macrocycles: palladium(II)-directed template syntheses of pendant-arm 14-, 16-, and 18-membered macrocycles, *Inorg.Chem.*, 35, pp. 4961-4966.

[238] Rossignoli, M., Bernhardt , P. V., Lawrance, G. A. and Maeder, M. (1997). Gold(III) template synthesis of a pendant-arm macrocycle, *J. Chem. Soc. Dalton Trans.*, pp. 323-327.

[239] Xiao, H., Lawrance, G. A. and Hambley, T. W. (1988). Platinum(II) template synthesis of a pendant-arm macrocycle, *Austr. J. Chem.*, 41, pp. 871-874.

[240] Hancock, R. D. (1986). Macrocycles and their selectivity for metal ions on the basis of size, *Pure & Appl.Chem.*, 58, pp. 1445-1452.

[241] Bernhardt, P. V. and Lawrance, G. A. (1990). Complexes of polyaza macrocycles bearing pendent coordination groups, *Coord. Chem. Rev.*, 100, pp. 297-343.

[242] Curtis, N. F., Gainsford, G. J., Hambley, T. W., Lawrance, G. A., Morgan K. R. and Siriwardena, A. (1987). Hexaco-ordination of a diamino-substituted tetra-azamacrocycle to cobalt(III) and nickel(II): compressed Co-N bonds, *J. Chem. Soc. Chem. Commun*, pp. 295-297.

[243] Bernhardt, V., Lawrance, G. A. and Hambley , T. W. (1989). 6,13-Diamino-6,13-dimethyl-1,4,8,11-tetra-azacyclotetradecane, L^7, a new potentially sexidentate polyamine ligand. Variable co-ordination to cobalt(III) and crystal structure of the complex$[Co(L^7)]Cl_2[ClO_4]$, *J. Chem. Soc. Dalton Trans.*, pp. 1059-1065.

[244] Börzel, H., Comba, P., Pritzkow, H. and Sickmuler, A. F. (1998) Preparation, structure, and electronic properties of a low-spin iron(II) hexaamine compound, *Inorg.Chem.*, 37, pp. 3853-3857.

[245] Bernhardt, V., Hambley, T. W. and Lawrance, G. A. (1989). Co-ordination of 6,13-diamino-6,13-dimethyl-1,4,8,11-tetra-azacyclotetradecane to iron(III). The first fully characterised hexa-amin-iron(III) complex, *J. Chem. Soc. Dalton Trans.*, pp. 553-554.

[246] Bernhardt, V., Comba, P., Hambley T. W. and Lawrance, G. A. (1991). Coordination of the sexidentate macrocycle 6,13-dimethyl-1,4,8,11-tetraazacyclotetradecane-6,13-diamine to iron(III), *Inorg. Chem.*, 30, pp. 942-946.

[247] Bernhardt, V., Hambley, T. W. and Lawrance, G. A. (1990). Rhodium(III) complexes of the ambidentate macrocycle 6,13-diamino-6,13-dimethyl-1,4,8,11-tetra-azacyclotetradecane. Crystal structure of the sexidentate complex, *J. Chem. Soc. Dalton Trans.*, pp. 983-987.

248   Bernhardt, P. V., Lawrance, G. A., Patalinghug, W. C., Skelton, B. W., White , A. H., Curtis, N. F., Siriwardena, A. (1990). Co-ordination of 6,13-dimethyl-1,4,8,11-tetra-azacyclotetradecane-6,13-diamine to platinum(II) and palladium(II). Syntheses, characterisation, and X-ray crystal structures of the perchlorate salts of both complexes, *J. Chem. Soc. Dalton Trans.*, pp. 2853-2858.

249   Bernhardt, P. V., Lawrance, G. A., Comba, P., Martin, L. L. and Hambley, T. W. (1990). Synthesis, physical properties and X-ray crystal structure of an oxovanadium(IV) complex of the pendant-arm macrocycle 6,134-dimethyl-1,4,8,11-tetra-azacyclotetradecane-6,13-diamine, *J. Chem. Soc. Dalton Trans.*, pp. 2859-2862.

250   Bernhardt, V., Comba, P., Curtis, N. F., Hambley, T. W. , Lawrance, G. A., Maeder, M. and Siriwaredena, A. (1990), Coordination of the "pendant-arm"macrocycle 6,13-diamino-6,13-dimethyl-1,4,8,11-tetraazacyclotetradecane to chromium(III). Crystal structure and physical properties of the hexacoordinated complex ion, *Inorg. Chem.*, 29, pp. 3208-3213.

251   Bernhardt, V., Lawrance, G. A., Maeder, M., Rosignoli, M. and Hambley , T. W. (1991). Sexidentate co-coordination of the pendant-arm macrocycle 6,13-dimethyl-1,4,8,11-tetraazacyclotetradecane-6,13-diamine(L^1) to zinc(II). Crystal structure of [ZnL^1][ClO$_4$]$_2$·H$_2$O, *J. Chem. Soc. Dalton Trans.*, pp. 1167-1170.

252   Bernhardt, V., Comba, P., Hambley, T. W., Lawrance, G. A., Varnagy, K. (1992). Isolation and Complexation of the cis isomer of the pendant arm macrocycle 6,13-dimethyl-1,4,8,11-tetraazacyclo-tetradecane-6,13-diamine, *J. Chem. Soc. Dalton Trans.*, pp. 355-359.

253   Bernhardt, P. V., Comba, P. and Hambley, T. W. (1993). Complexation of cis-6,13-dimethyl-1,4,8,11-tetraazacyclotetradecane-6,13-diamine with the first row transition metal ions cobalt(III), chromium(III), and nickel(II), *Inorg. Chem.*, 32, pp. 2804-2809.

254   Curtis, N. F. and Sirivardena, A. (1991). Compounds of cobalt(III) with trans-6,13-diamino-6,13-dimethyl-1,4,8,11-tetraazacyclotetradecane, *Austr. J. Chem.*, 44, pp. 1041-48.

255   Curtis, N. F., Robinson, W. T. and Wetherburn, D. C. (1992). Structural studies of seven compounds of cobalt(III) with trans-6,13-diamino-6,13-dimethyl-1,4,8,11-tetraazacyclotetradecane, *Aust. J. Chem.*, 45, pp. 1663-1680.

256   Bernhardt, P. V., Hetherington, J. C. and Jones, L. A. (1996). N-methylation of diamino-substituted macrocyclic complexes: intramolecular cyclisation, *J. Chem. Soc. Dalton Trans.*, pp. 4325-4330.

257   Comba, P., Curtis, N. F., Lawrance, G. A., O'Leary, M. A., Skelton, B. W. and White, A. H. (1988). Comparisons of thirteen- to sixteen- membered tetra-azacycloalkane copper(II) complexes derived from template syntheses involving nitroethane and formaldehyde. Crystal structures of (10-methyl-10-nitro-1,4,8,12-tetra-azacyclopentadecane)copper(II)   and   (3-methyl-3-nitro-1,5,9,13-tetra-azacyclohexadecane)copper(II) perchlorates, *J. Chem. Soc. Dalton Trans.*, pp. 2145-2151.

258   Lawrance, G. A., Rossignoli, M., Skelton, B. W. and White, A. H. (1987). Metal-directed synthesis involving formaldehyde and nitroethane, and crystal structure

of the copper(II) complex of the 14-membered macrocycle 6-methyl-6-nitro-1,4,8,11-tetraazacyclotetradecane, *Aust. J. Chem.*, 40, pp. 1441-1449.

259 Bayada, A., Lawrance, G. A., Maeder, M. and O'Leary, M. A. (1994). Metal-directed synthesis of aminobenzyl polyaza macrocycles: candidates for attachment to polymers and biomolecules, *J. Chem. Soc., Dalton Trans,* pp. 3107-3111.

260 Curtis, N. F., Li, X. and Watherburn, D. C. (1993). Compounds of iron(III) with trans-1,4,8,11-tetraazacyclotetradecane-6,13-dicarboxylic acid: structures of trans-(trans-1,4,8,11-tetraazacyclotetradecane-6,13-dicarboxylato)iron(III)perchlorate and trans- (trans-1,4,8,11-tetraazacyclotetradecane-6,13-dicarboxylato)iron(III) trans-(trans-1,4,8,11-tetraazacyclotetradecane-6,13-dicarboxylic acid)dichloroiron(III) trans-1,4,8,11-tetraazoniacyclotetradecane-6,13-dicarboxylic acid dichloride tetraperchlorate tetrahydrate, *Inorg. Chem.*, 32, pp. 5838-5843.

261 Lawrance, G. A. and O'Leary, M. A. (1987). Macrocyclic tetraamines from reaction of the (1,10-diamino-4,7-diazadecane)copper(II) cation with formaldehyde and the carbon acids nitroethane and diethylmalonate: variability in reactivity, *Polyhedron,* 6, pp. 1291-1294.

262 Lawrance, G. A., Skelton, B. W., White, A. H. and Wilkes, E. N. (1991). Tetraazacycloalkanes with pendant carboxylates. Copper(II)-directed syntheses and crystal structure of(ethyl 1,5,9,13-tetraazabicyclo[11.2.2]heptadecane-7-carboxylate)copper(II) perchlorate, *Aust. J. Chem.*, 44, pp. 1511-1522.

263 Bernhardt, P. V., Lawrance, G. A., Skelton, B. W. and White, A. H. (1990). Facile metal-directed synthesis and crystal structure of a new amino acid ester, methyl 3-[(2-aminoethyl)amino]-2-[(2-aminoethyl)aminomethyl]propionate, as the copper(II) complex *Aust. J .Chem.*, 43, pp. 399-404.

264 Li, X.and Curtis, N. F. (1992).Compounds of (1,4,8,11-tetraazacyclotetradecane)copper(II) with 6- and 13-carbamoyl, -methylcarbamoyl, -ethylcarbamoyl and –hydrazinocarbonyl substituents, *Aust. J. Chem.*, 45, pp. 1087-1094.

265 Hambley, T. W., Lawrance, G. A., Maeder, M. and Wilkes, E. N. (1992). Synthesis of a new potentially sexidentate pendant-arm macrocyclic polyamino acid and co-ordinate to cobalt(III), *J. Chem. Soc., Dalton Trans,* pp.1283-1289.

266 De Blas, A., De Santis, G., Fabbrizzi, L., Licchelli, M., Manotti Lanfredi, A. M., Morosini, P., Pallavicini, P. and Ugozzoli, F. (1993). Amides and sulfoamides: efficient molecular padlocks for the template synthesis of azacyclam(1,3,5,8,12-pentaazacyclotetradecane) macrocycles, *J. Chem. Soc., Dalton Trans.*, pp. 1411-1416.

267 De Blas, A., De Santis, G., Fabbrizzi, L., Licchelli, M., Manotti Lanfredi, A. M., Pallavicini, P., Poggi, A. and Ugozzolli, F. (1993). Pyridines with an appended metallocyclam subunit. Versatile building blocks to supramolecular multielectron redox systems, *Inorg. Chem.*, 32, pp. 106-113.

268 De Santis, G., Fabbrizzi, L., Licchelli, M., Mangano, C. and Pallavicini, P. (1993). The copper(I) complex of a metallocyclam-functionalized phenantroline: a poorly stable species that is very resistant to oxidation, *Inorg. Chem.*, **32**, pp. 3385-3387.

[269] De Blas, A., De Santis, G., Fabbrizzi, L., Licchelli, M. and Pallavicini, P. (1993). Novel routes to functionalized cyclam-like macrocycles, *Pure & Appl. Chem.*, 65, pp. 455-459.

[270] Lampeka, Y. D., Gavrish, S. P., Maloshtan, I. M., Dalley, N. K., Lamb, J. D. and Nazarenko, A. Y. (1998). New cyclam-type copper(II) complexes with amide molecular padlock: synthesis, properties and crystal structure, *Inorg. Chim. Acta*, 282, pp. 142-148.

[271] Bernhardt, P. B. and Hayes, E. J. (1998). Aminotriazines as locking fragments in macrocyclic synthesis, *Inorg. Chem.*, 37, pp. 4214-4219.

[272] Comba, P. Lampeka, Y. D., Nazarenko, A.Y., Prikhod'ko, A. I., Pritzkow, H. And Taraszewska, J. (2002). Cooperative effects in the binding of substrates to nickel(II) and nickel(III) complexes with bis(macrocyclic) ligand, *Eur. J. Inorg. Chem.*, pp. 1871-1882.

[273] Goedken, V. L. and Peng, S.-M. (1973). Template condensation: metal ion directed syntheses of macrocyclic, tricyclic, and quadricyclic metal complexes from butane-2,3-dione dihydrazone and formaldehyde, *J. Chem. Soc., Chem. Commun.*, pp. 62-63.

[274] Suh, M. P.and Kang, S. -G. (1998). Synthesis and properties of nickel(II) and copper(II) complexes of 14-membered hexaaza macrocycles, 1,8-dimethyl- and 1,8-diethyl-1,3,6,8,10,13-hexaazacyclotetradecane, *Inorg. Chem.*, 27, pp. 2544-2546.

[275] a) Hay, R. W., Armstrong, J. M. and Hassan, M. M. (1992). Facile synthesis of the nickel(II) complex of a new pendant arm macrocycle, *Transition Met. Chem.*, 17, pp. 270-271. b) Hay, R. W., Crayston, J. A., Cromie, T. J., Lightfoot, P.and de Alwis, D. C. L. (1997). The preparation, chemistry and crystal structure of the nickel(II) complex of N-hydroxyethylazacyclam[3-(2'-hydroxyethyl)-1,3,5,8,12-penta-azacyclotetradecane nickel(II) perchlorate]. A new electrocatalyst for CO_2 reduction, *Polyhedro*, 16, pp. 3557-3563.

[276] Suh, M. P., Shim, B. Y. and Yoon, T. S. (1994). Template syntheses and crystal structures of nickel(II) complexes of hexaaza macrocyclic ligands with pendant functional groups: formation of a coordination polymer, *Inorg. Chem.*, 33, pp. 5509-5514.

[277] Suh, M. P., Lee, E. Y. and Shim, B. Y. (1998). Synthesis and properties of nickel(III) complexes of hexaazamacrocyclic ligands, *Inorg. Chim. Acta*, 269, pp. 337-341.

[278] Kang, S. –G., Ryu, K., Jung, S. –K. And Kim, J. (1999). Template synthesis, crystal structure, and solution behavior of a haxaaza macrocyclyc nickel(II) complex containing two N-aminoethyl pendant arms, *Inorg. Chim. Acta*, 293, pp. 140-146.

[279] Anastasi, D., Curtis, N. F., Gladkikh, O. P., Goode, T. J. C. and Weatherburn, D. C. (1998). Copper(II) compounds of 5-alkyl-3,5,7-triazanonane-1,9-diamines and 3,10-bisalkyl-1,3,5,8,10,12-hexaazacyclotetradecanes; the Structure of 1R,5S,8R,12S-{3,10-bis(2-hydroxypropyl)-1,3,5,8,10,12-hexaazacyclotetradecane-N^1, N^5, N^8, N^{12}}copper(II) perchlorate, *Aust. J. Chem.*, 51, pp. 673-679.

280 Peng, S. -M., Gordon, G. C. and Goedken, V. L. (1978). Template condensations: metal-ion-directed syntheses of macrocyclic complexes from 1,3 butanedione dihydrazone and aldehydes or ketones, *Inorg. Chem.*, 17, pp. 119-126.

281 Peng, S. -M. and Goedken, V. L. (1978). Tricyclic complexes,$[M(C_{14}H_{24}N_8O_2)L_2]X_2$, obtained from the condensation of formaldehyde with octaazamacrocyclic complexes, $[M(C_{10}H_{20}N_8]^{n+}$: synthesis, studies, and X-ray structural characterization, *Inorg.Chem.*, 17, pp. 820-828.

282 Lawrance, G. A., Maeder, M., O'Leary, M. A., Skelton, B. W. and White A. H., (1991). Macrocycle formation from copper(II) directed condensation of linear tetraamines with formaldehyde alone and with added 2-aminoethanol, *Aust. J. Chem.*, 44, pp. 1227-1236.

283 Comba, P., Lawrance, G. A., Rossignoli, M., Skelton, B., W. and White, A. H. (1988). Synthesis of a cis-N_2S_2 donor macrocycle by metal-directed condensation of formaldehyde, nitroethane and a diaminodithiaalkane. Crystal structure of the (6-methyl-6-nitro-1,11-dithia-4,8-diazacyclotetradecane)-copper(II) perchlorate product, *Aust. J. Chem.*, 41, pp. 773-781.

284 Wei, G., Allen, C. C., Hambley, T. W., Lawrance, G. A. and Maeder, M.(1997). Hydrogenolysis of a pendant alcohol dithiadiazamacrocycle. Crystal structure of the 6-amino-6-methyl-1,11-dithia-4,8-diazacyclotetradecane hydrogenolysis product as a chlorocobalt(III) complex, *Inorg. Chim. Acta*, 261, pp. 197-200.

285 Lawrance, G. A., O'Leary, M. A., Skelton, B. W., Woon, F-H. and White, A. H. (1988). Reactions of (10-methyl-10-nitro-1,4,8,12-tetraaza-cyclopentadecane)copper(II) ion in aqueous hydrochloric acid, including zinc reduction. Crystal structure of (2-methyl-1,4,8,11-tetraazacyclotetradec-1-ene)copper(II)perchlorate, *Aust. J. Chem.*, 41, pp. 1533-1544.

286 Lawrance, G. A., Martinez, M., Skelton, B. W. and White, A. H. (1991). Quinquedentate coordination of 6-methyl-1,11-dithia-4,8-diazacyclotetradecan-6-amine to cobalt(III). X-ray crystal structure of the $[Co(N_3S_2)(OOCCH_3)](ClO_4)_2 \cdot H_2O$ complex. *Aust. J. Chem.*, 44, pp. 113-121.

287 Wainwright, K. P. (1980). Bridging alkylation of saturated polyaza macrocycles: a mean for imparting structural rigidity, *Inorg. Chem.*, 19 , pp. 1396-1398.

288 Wainwright, K. P., Ramasubbu, A. (1982). Template syntheses of chiral tetradentate ligands derived from L-amino acids. Structural and spectroscopic characterization of the free ligands and of their copper(II) complexes, *J. Chem. Soc. Chem. Commun.*, pp. 277-283.

289 Hancock, R. D., Patrick, G., Wade, P. W. and Hosken, G. D. (1993). Structurally reinforced macrocyclic ligands, *Pure & Appl.Chem.*, 65, pp. 473-476.

290 Hancock, R. D., Ngwenya, M. P., Wade, P. W., Boeyens, J. C. A. and Dobson, S. M. (1989). The synthesis of complexes of novel structurally reinforced tetraaza-macrocyclic ligands of high ligand field strenght. A structural and molecular mechanics study, *Inorg. Chim. Acta*, 164, pp. 73-84.

291 Cotton, F. A. and Wilkinson, G. Advanced Inorganic Chemistry, Wiley-Interscience, New York, 5-th edn., 1988, pp. 344.

292 Suh, M. P., Kang, S. -G., Goedken, V. L. and Park, S. -H. (1991). Template condensation reactions of formaldehyde with tetraamines and ethylenediamine:

preparation, properties, and structures of nickel(II) and copper(II) complexes of hexaaza macrotricyclic ligands 1,3,6,8,11,14-hexaazatricyclo[12.2.1.1$^{8.11}$]octadecane and 1,3,6,8,12,15-hexaazatricyclo[13.3.1.1$^{8.12}$]eicosane, *Inorg. Chem.*, 30, pp. 365-370.

293   Suh, M. P., Shin, W., Kang, S.-G., Lah, M.S. and Chung, T.-M. (1989). Template condensation of formaldehyde with triamines. Synthesis and characterization of nickel(II) complexes with the novel hexaaza macrotricyclic ligands 1,3,6,9,11,14-hexaazatricyclo[12.2.1.1$^{6.9}$]octadecane and 1,3,6,10,12,15-hexaazatricyclo[13.3.1.1$^{6.10}$] eicosane, *Inorg. Chem.*, 28, pp. 1602-1605.

294   Suh, M. P., Kim, I. S., Kim, M. J. and Oh, K. Y. (1992), Synthesis, properties and X-ray structures of nickel(I) complexes of polyaza macrotricyclic ligands, *Inorg. Chem.*, 31, pp. 3620-3625.

295   Suh, M. P., Lee, Y. J. and Jeong, T. (1995). Properties and crystal structure of a four-co-ordinate nickel(I) complex with the macrotricycle 1,3,6,8,12,15-hexaazatricyclo[13.3.1.1$^{8.12}$]icosane, *J. Chem. Soc., Dalton Trans.*, pp. 1577-1581.

296   Suh, M. P., Choi, J., Kang, S. -G. and Shin, W. (1989). Synthesis, characterization and X-ray structure of the octahedral nickel(II) complex of a pentadentate hexaaza macrobicyclic ligand: chloro(9-(methoxymethyl)-1,4,6,9,11,14-hexaazabicyclo-[12.2.1]heptadecane)nickel(II) perchlorate, *Inorg. Chem.*, 28, pp. 1763-1765.

297   Suh, M. P., Kim, I. S., Cho, S.-J. and Shin, W. (1994). Syntheses of square-planar Nickel-(II) and –(I) complexes of an octaaza macrohexacyclic ligand and crystal structure of the nickel(II) complex, *J.Chem.Soc., Dalton Trans.*, pp. 2765-2769.

298   Lawrance, G. A., Manning, T. M., Skelton, B. W. and White, A. H. (1988). Synthesis and characterization of dinickel(II) and dipalladium(II) complexes of the macrocyclic binucleating ligand 3,13-dimethyl-3,13-dinitro-1,5,11,15-tetra-azacycloeicosane-8,18-dithiol(L^5). Crystal structure of the complex [Ni$_2$(L^5-2H)]{NO$_2$}$_2$·3.5 H$_2$O, *J. Chem. Soc., Chem. Commun.*, pp. 2491-2495.

299   Comba, P., Lawrance, G. A., Manning, T. M., Markiewicz, A., Murray, K. S., O'Leary, M. A., Skelton, B. W and White, A. H. (1990). Synthesis, crystal structure and magnetism of a macrocyclic binuclear dicopper(II) amino alcohol complex from a metal-directed reaction involving formaldehyde and nitroethane, *Aust. J. Chem.*, 43, pp. 69-78.

300   Lawrance, G. A., Maeder, M., Manning, T. M., O'Leary, M. A., Skelton, B. W. and White, A. H. (1990). Synthesis and characterization of dinickel(II) and dipaladium(II) complexes of the macrocyclic binucleating ligand 3,13-dimethyl-3,13-dinitro-1,5,11,15-tetra-azacycloeicosane-8,18-dithiol(L^5). Crystalstructure of the complex [Ni$_2$ (L^5-2H)][NO$_2$]$_3$.3H$_2$O, *J. Chem. Soc., Dalton Trans.*, pp. 2491-2495

301   Rosokha, S. V. and Lampeka, Y. D. (1991). A new one-pot synthesis of bis(macrocyclic) complexes; preparation and characterization of nickel complexes with bis(pentaazamacrocyclic) ligands, *J. Chem Soc., Chem. Commun.*, pp. 1077-1079.

302   Rosokha, S. V., Lampeka, Y. D. and Maloshtan, I. M. (1993). Synthesis and properties of a new series of bis(macrocyclic) dicopper(II,II) and dinickel(III,III)

complexes based on the 14-membered pentaaza unit, *J. Chem. Soc., Dalton Trans*, pp. 631-636.

[303] Kang, S.-G., Jung, S.-K., Kwean, J. K. and Kim, M.-S. (1993). Template synthesis of a new dinuclear nickel(II) complex of a bis(macrobicyclic) ligand, *Polyhedron*, 12, pp. 353-356.

[304] Kang, S. -G., Ryu, K., Suh, M. P. and Jeong, J. H. (1997). Template synthesis and crystal structure of a novel mononuclear nickel(II) complex with a face-to-face bis(macrocyclic) ligand, *Inorg. Chem.*, 36, pp. 2478-2481.

[305] Suh, M. P. and Kim, S. K. (1993). Synthesis of dinickel(I) complexes of bismacrocyclic ligands, *Inorg. Chem.*, 32, pp. 3562-3564.

[306] Barefield, E. K., Freeman, G. M. and Van Derveer, D. G. (1986). Electrochemical and structural studies of nickel(II) complexes of N-alkylated cyclam ligands: X-ray structuresof trans-$[Ni(C_{14}H_{32}N_4)(OH_2)_2]Cl_2 \cdot 2H_2O$ and $[Ni(C_{14}H_{32}N_4)](O_3SCF_3)_2$, *Inorg. Chem.*, 25, pp. 552-558.

[307] Comba, P. and Hilfenhaus, P. (1995). One-step template synthesis and solution structures of bis(macrocyclic) octaamine dicopper(II) complexes, *J. Chem. Soc., Dalton Trans*, pp. 3269-3274.

[308] Gobi, K.V. and Ohsaka, T. (1998). Electrochemical and spectral properties of novel dinickel(II) and dicopper(II) complexes with N,N-linked bis(pentaazacyclotetradecane. *Electrochimica Acta*, 44, pp. 269-278.

[309] Bernhardt, P. V., Comba, P., Gahan, L. R. and Lawrance, G. A. (1990). Dicopper(II) complexes of spiro macrobicycles with aza and thia-aza donor sets, *Aust. J. Chem.*, 43, pp. 2035-2044.

[310] Harrowfield, J. M., Herlt, A. J., Sargeson, A. M. (1980). Complexes with complicated chelate ligands, *Inorg. Synth.*, 20, pp. 85-100.

[311] Creaser, I. I., Harrowfield, J. M., Herlt, A. J., Sargeson, A. M., Springborg, R. J.Geue, and Snow, M. R. (1977). Sepulcrate: a macrobicyclic nitrogen cage for metal ions, *J. Am. Chem. Soc.*, 99, pp. 3181-3182.

[312] Harrowfield, J. M., Herlt, A. J., Lay, P. A. and Sargeson, A. M. (1983). Synthesis and properties of macrobicyclic amine complexes of rhodium(III) and iridium(III), *J. Am. Chem. Soc.*, 105, pp. 5503-5505.

[313] Boucher, H. A., Lawrance, G. A., Lay, P. A., Sargeson, A. M., Bond, A. M., Sangster, D. F. and Sullivan, J. C. (1983). Macrobicyclic (hexaamine)platinum(IV) complexes: synthesis, characterization, and electrochemistry, *J. Am. Chem. Soc.*, 105, pp. 4652-4661.

[314] Ramasami, T., Endicott, J.F. and Brubaker, G. R. (1983). Photophysics of a macrobicyclic hexaminechromium(III) complex and the mechanism for excited-state relaxation in chromium(III)-amine complex, *J. Phys. Chem.*, 87, pp. 5057-5059.

[315] Sargeson, A. M. (1984). Encapsulated metal ions, *Pure & Appl. Chem.*, 56, pp. 1603-1619.

[316] Barttolomey, G. A., Clark, I. J., Creaser, I. I., Engelhardt, L. M., Geue, R. J., Hagen, K. S., Harrowfield, J. M., Lawrance, G. A., Lay, P. A., Sargeson, A. M., See, A. G., Skelton, B. W., White, A. H. and Wilner, F. R. (1994). The Synthesis and

structure of encapsulating ligands: properties of bicyclic hexamines, *Aust. J. Chem.*, 47, pp. 143-179.

[317] Balahura, R.J., Ferguson, G., Ruhl, B.L. and Wilkins, R. G. (1983). Kinetics of the reduction of nitro to hydroxylamine groups by dithionite in a cobalt(II) cryptand complex. X-ray analysis of [1,8-bis(hydroxyamino0-3,6,10,13,16,19-hexaazabicyclo[6.6.6] eicosane]cobalt(III)chloride Tetrahydrate, *Inorg. Chem.*, 22, pp. 3990-3992.

[318] Bernhardt, P. V., Bygot, A. M. T., Geue, R. J., Hendry, A. J, Korybut-Daszkiewicz, B. R., Lay, P. A., Pladziewicz, J. R., Sargeson, A. M. and Willis, A. C. (1994). Stabilization of cobalt cage conformers in the solid state and solution, *Inorg. Chem.*, 33, pp. 4553-4561.

[319] Creaser, I. I., Kamorita, T., Sargeson, A. M., Willis, A. C. and Yamanari, K. (1994). New macrocyclic complexes derived from cobalt(III) cage complexes, *Aust. J. Chem.*, 47, pp. 529-544.

[320] Geue, R. J., Osvath, P., Sargeson, A. M., Acharya, K. R., Noor, S. B., Row, T. N. G. and Venkatesan, K. (1994). The Reaction of a nitro-capped cobalt(III) cage complex with base: the crystal structure of a contracted cage complex, and the mechanism of its formation, *Aust. J. Chem.*, 47, 511-527.

[321] Creaser, I. I., Lydon, J. D., Sargeson, A. M., Horn, E. and Snow, M. R. (1984). A zinc-alkyl caged cobalt(III) derivative, *J. Am. Chem. Soc.*, 106, pp. 5729- 5730.

[322] Geue, R. J., Hendry, A. J. and Sargeson, A. M. (1989). Configurational and conformational effects on electron transfer rates, *J. Chem. Soc. Chem. Commun.*, pp. 1646-1647.

[323] Yamaoka, H. (1977). A. C. Polarographic study of some pentaamminecobalt(III) complexes on a dropping mercury electrode in aqueous acid perchlorate solution, *Collect. Czech. Chem. Commun.*, 42, pp. 2845-2849.

[324] Bond, A. M., Lawrance, G. A., Lay, P. A. and Sargeson, A. M. (1983). Electrochemistry of macrobicyclic (hexaamine)cobalt(III) complexes. Metal-centered and substituent reductions, *Inorg. Chem.*, 22, pp. 2010-2021.

[325] Hamershoi, A., Geselowitz D. and Taube, H. (1984). Redetermination of the hexaaminecobalt(III/II) electron-self-exchange rate, *Inorg. Chem.*, 23, pp. 979-982.

[326] Dwyer, F. P. and Sargeson, A. M. (1961). The rate of electron transfer between the tris-(ethylenediamine)cobalt(II) and cobalt(III) ions, *J. Phys. Chem.*, 65, pp. 1892-1894.

[327] Creaser, I. I., Sargeson, A. M. and Zanella, A. W. (1983). Outer-sphere electron-transfer reactions involving caged cobalt ions, *Inorg. Chem.*, 22, pp. 4022-4029.

[328] Hammershøi, A. and Sargeson, A. M. (1983). Macrotricyclic hexaamine cage complexes of cobalt(III): synthesis, characterization, and properties, *Inorg. Chem.*, 22, pp. 3554-3561.

[329] Geue, R. J., McCarthy, M. G., Sargeson, A. M., Horn, E. and Snow, M. R. (1986). Stabilization of a carbanion by co-ordination to a metal centre. The structure of 11-methylamino-6,16-dinitro-4,8,14,18,21-penta-azatetracyclo-[14.4.2.1$^{3.19}$. 1^{9,13}]tetracos-4-enato(6)cobalt(III)tetrachlorozincate hydrate, *J. Chem. Soc. Chem. Commun.*, pp. 848-850.

330 Belinski, J. A., Squires, M. E., Kuchna, J. M., Bennett, B. A. and Grzybowski, J.,J. (1988). The Synthesis and characterization of mononuclear and binuclear iron(II) clathrochelate complexes derived from 2,3-butanedione oxime hydrazone, *J. Coord. Chem.*, 19, pp. 159-169.

331 Höhn, A., Geue, R. J., Sargeson, A. M. and Willis, A. C. (1989). Phospha-capped Cobalt(III) Cage Molecules: Synthesis, Properties, and Structure, *J. Chem. Soc. Chem. Commun.* pp. 1644-1645.

332 Höhn, A., Geue, R. J., Sargeson, A. M. and Willis, A. C. (1989). Stabilization of an unusual cvonformation of an encapsulated metal ion cobalt(methylarsasarcophagine): synthesis and structure, *J. Chem. Soc. Chem. Commun.*, pp. 1648-1649.

333 Gahan, L. R., Hambley, T. W., Sargeson, A. M. and Snow, M. R. (1982). Encapsulated metal ions: synthesis, structure, and spectra of nitrogen-sulfur ligand atom cages containing cobalt(III) and cobalt(II), *Inorg. Chem,* 21, pp. 2699-2706.

334 Gahan, L. R., Lawrance, G. A. and Sargeson, A. M. (1984). Electrochemistry of macrobicyclic mixed nitrogen-sulfur donor complexes of cobalt(III), *Inorg. Chem.*, 23, pp. 4369-4376.

335 Dubs, R. V., Gahan, L. R. and Sargeson, A. M. (1983). Rapid electron self-exchange involving low-spin cobalt(II) and cobalt(III) in an encapsulated cage complex, *Inorg. Chem.*, 22, pp. 2523-2527.

336 Comba, P., Engelhardt, L. M., Harrowfield, J. M., Lawrance, G. A., Martin, L. L., Sargeson, A. M. and White, A. H. (1985). Synthesis and characterization of a stable hexa-amine vanadium(IV) cage complex, *J. Chem. Soc., Chem. Commun.*, pp. 174-176.

337 Comba, P., Creaser, I. I., Gahan, L. R., Harrowfield, J. M., Lawrance, G. A., Martin, L. L., Mau, A. W. H., Sargeson, A. M., Sasse, W. H. F. and Snow, M. R. (1986). Macrobicyclic chromium(III) hexaamine complexes, *Inorg. Chem.*, 25, pp. 384-389.

338 Bernhardt, V., Bramley, R., Engelhardt, L. M., Harrowfield, J. M., Hockless, D. C. R., Korybut-Daszkiewicz, B. R., Krausz, E. R., Morgan, T., Sargeson, A. M., Skelton, B. W. and White, A. H. (1995). Copper(II) complexes of substituted macrobicyclic hexaamines: combined trigonal and tetragonal distortions, *Inorg. Chem.*, 34, pp. 3589-3599.

339 Comba, P., Sargeson, A. M., Engelhardt, L. M., Harrowfield, J. M., White, A. H., Horn, E., Snow, M. R. (1985). Analysis of trigonal-prismatic and octahedral preferences in hexaamine cage complexes, *Inorg. Chem.*, 24, pp. 2325-2327.

340 Martin, L. L., Martin, R. L., Murray, K. S. and Sargeson, A. M. (1990). Magnetism and electronic structure of a series of encapsulated first-row transition metals, *Inorg. Chem.*, 29, pp. 1387-1394.

341 Martin, L. L., Martin, R. L. and Sargeson, A. M. (1994). The ligand field $^1A_1$-$^5T_2$ spin crossover with iron(II) encapsulated in hexa-amine cages, *Polyhedron*, 13, pp. 1969-1980.

342 Stratemeier, H., Hitchman, M. A., Comba, P., Bernhardt, P. V. and Riley, M. J. (1991). EPR spectrum and metal-ligand bonding parameters of a low-spin(hexaamine)iron(III) complex, *Inorg. Chem.*, 30, pp. 4088-4093.

[343] Sargeson, A. M. (1996). The potential for the cage complexes in biology, *Coord. Chem. Rev.,* 151, pp. 89-114.

[344] Lever, A.B.P. (1965). The Phtalocyanines, *Adv. Inorg. Radiochem.,* 7, 28-114.

[345] Sessler, J.L., Vivian, A.E., Seidel, D. Burrell, A.K., Hoehner, M., Mody, T.D., Gebauer, A., Weghorn, S.J. and Lynch, V. (2001). Actinide expanded porphyrin complexes, *Coord. Chem. Rev.,* 216-217, pp. 411-434.

[346] Lindoy, L. F. and Busch, D. H. (1971). Complexes of macrocyclic ligands, *Prep. Inorg. React.,* 6, pp.1- 62.

[347] Lenznoff, C. C., Hu, M. and Nolan, K. J. M. (1996). The synthesis of phtalocyanines at room temperature, *Chem. Commun.,* pp. 1245-1246.

[348] Tomoda, H., Saito, S. and Shiraishi, S. (1983). Synthesis of metallophtalocyanines from phtalonitrile with strong organic bases, *Chem. Lett.,* pp. 313-318.

[349] Piechocki, C., Simon, J., Skoulios, A, Guillon, D. and Weber, P. (1982). Discotic mesophases obtained from substituted metallophtalocyanines. Towards liquid crystalline one-dimensional conductors. *J. Am. Chem. Soc.,* 104, 5245-5247.

[350] Ishikawa, N. and Kaizu, Y. (2002). Synthetic, spectroscopic and theoretical study of novel supramolecular structures composed of lanthanide phtalocyanine double-decker complexes, *Coord. Chem. Rev.,* 226, pp. 93-101.

[351] Metz, J., Schneider, O. and Hanack, M. (1984). Synthesis and properties of substituted (phtalocyaninato)iron and –cobalt compounds and their pyridine adducts, *Inorg. Chem.,* 23, pp. 1065-1071.

[352] Yang, C. H., Lin, S. F., Chen, H. L. and Chang, C. T., (1980). Electrosynthesis of the metal phtalocyanine complexes, *Inorg. Chem.,* 19, pp. 3514-3543.

[353] Christie, R. M. and Deans D. D.. (1989). An investigation into the mechanism of the phtalonitrile route to copper phtalocyanines using differential scanning calorimetry, J. Chem Soc. Perkin, 2, pp. 193-1998.

[354] Linstead, R. P. and Whalley, M. (1952). Conjugated macrocycles/. Part XXII. Tetraazaporphyn and its metallic derivatives, *J. Chem. Soc.,* 4839-4845.

[355] Rothemund, P., (1939). Porphyn studies. III. The structure of the porphine ring system, *J. Am. Chem. Soc.,* 61, pp. 2912-2915.

[356] Badger, G. M. and Ward, A. D. (1964). Porphyrins. II. The synthesis of porphyrins from 2-aminomethylpyrroles, *Aust. J. Chem.,* 17, 649-660.

[357] Johnson, A. W., Kay, I. T., Markham, E., Price, R. and Shaw, K. B., (1959). Colouring matters derived from pyrroles. Part II. Improved syntheses of some dipyrromethenes and porphyrins, *J. Chem. Soc.,* pp. 3416-3424.

[358] Johnson, A. W. (1975). Porphyrins and related ring systems, *Chem. Soc. Rev.* 4, pp. 1-26.

[359] Badger, G. M. Harris, R. L. N. and Jones, R. A. (1964). Porphyrins. V. Studies on the cyclization of linear tetrapyrroles, *Aust. J. Chem.,* 17, 1013-1021.

[360] Ulman, A., Fisher, D.and Ibers, J. A. (1982). Synthesis of some 5,10,15, 20-tetraalkylchlorin and tetraalkylporphyrin complexes of transition metals. *J. Heterocycl.Chem.,* 19, pp. 409-423.

[361] Ulman, A., Gallucci, J., Fisher, D. and Ibers, J. A. (1980). Facile Syntheses of tetraalkylchlorin and tetraalkylporphyrin complexes and comparison of the

structures of the tetramethylchlorin and tetramethylporphyrin complexes of nickel(II), *J. Am. Chem. Soc.*, 102, pp. 6853-6854.

362 Wojaczynski, J. and Latos-Grazynski, L. (2000). Poly- and oligometalloporphyrins associated through coordination, *Coord. Chem. Rev.*, 204,pp. 113-171.

363 Broadhurst, M. J., Grigg, R. and Johnson, A. W. (1972). Sulfur extrusion reactions applied to the synthesis of corroles and related systems, *J. Chem. Soc., Perkin I*, pp. 1124- 1130.

364 Johnson A.W., Kay, I. T. and Rodrigo, R. (1963). 2,2'-Bipyrrolic macrocyclic ring systems., *J. Chem. Soc.*, pp. 2336-2342.

365 Kreuter, B., Pfaltz, A., Nordmann, R., Hodgson, K. O., Dunitz, J. D. and Eschenmoser, A. (1976). Versuche zur Redox-Simulation der photochemischen A/D-secocorrin →Corrin-Cycloisomerisierung. Elektrochemische Oxydation von Nickel(II) 1-methyliden-2,2,7,7,12,12,hexamethyl-15-cyan-1.19-secocorrin-perchlorat. *Helv. Chim. Acta*, 59, pp. 924-937.

366 Conlon, M., Johnson, A. W., Overend, W. R., Rajapaksa, D. and Elson, C. M. (1973). Structure and reactions of cobalt corroles, *J. Chem. Soc., Perkin I*, pp. 2281-2285.

367 Bröring M.and Hell, C. (2001). Manganese as a template: a new synthesis of corrole, *Chem. Commun.*, pp. 2336-2337.

368 Lehn, J. -M. (1995). Supramolecular chemistry - concepts and perspectives, VCH, Weinheim.
Lehn, J. -M. (1990). Perspectives in supramolecular chemistry-from molecular recognition towards molecular information processing and self-organization, *Angew. Chem. Int. Ed. Engl.*, 29, pp. 1304-1319.

369 Dietrich-Buchecker, C. O. and Sauvage, J. -P. (1987). Interlocking of molecular threads: from the statistical approach to the templated synthesis of catenands, *Chem.Rev.*, 87, pp. 795-810.

370 Constable, E. C. (1994). Higher oligopyridines as a structural motiv of metallosupramolecular chemistry, *Prog. Inorg. Chem.*, 42, pp. 67-138.

371 Fujita, M. (1999). Self-assembly of [2]catenanes containing metals in their backbones, *Acc. Chem. Res.*, 32, pp. 53-61.

372 Frisch, H. L. and Wasserman, E. J. (1961). Chemical topology, *J. Am. Chem. Soc.*, 83, pp. 3789-3795.

373 Schill, G. (1971). Catenanes, rotaxanes and knots, Academic Press, New York.

374 Wasserman, E. J. (1960). The preparation of interlocking rings: a catenane, *J. Am. Chem. Soc.*, 82, pp. 4433-4434

375 Sauvage, J. -P. (1990). Interlacing molecular threads on transition metals: catenands, catenates and knots, *Acc. Chem. Res.*, 23, pp. 321-327.

376 Hubin, T. J. and Busch, D. H. (2000). Template routes to interlocked molecular structures and orderly molecular entanglements, *Coord. Chem. Rev.*, 200-202, 5, pp. 5-52.

377 Dietrich-Buchecker, C. O., Sauvage, J. -P. and Kintzinger, J. P. (1983). Une nouvelle famille de molecules: les metallo-catenanes, *Tetrahedron Lett.*, 24, pp. 5095-5098.

378 Dietrich-Buchecker, C. O., Sauvage, J. -P. and Kern, J. -M. (1984). Templated synthesis of interlocked macrocyclic ligands: the catenands, *J. Am. Chem. Soc.,* 106, pp. 3043-3045.

379 Cesario, M., Dietrich-Buchecker, C. O., Guilhem, J., Pascard, C. and Sauvage, J. -P. (1985). Molecular structure of a catenand and its copper(I) catenate: complete rearrangement of the interlocked macrocyclic ligands by complexation, *J. Chem. Soc. Chem. Commun.*, pp. 244-247.

380 Livoreil, A., Dietrich-Buchecker, C. O. and Sauvage, J. -P. (1994). Electrochemically triggered swinging of a [2]-catenate, *J. Am. Chem . Soc.,* 116, pp. 9399-9400.

381 Mitchell, D. K. and Sauvage, J. -P. (1988). A topologically chiral [2]catenand, *Angew. Chem. Int. Ed. Engl.,* 27, pp. 930-931.

382 Dietrich-Buchecker, C. O., Sauvage, J. -P. and Weiss, J. (1986). Interlocked macrocyclic ligands: a catenand whose rotation of one ring into the other is precluded by bulky substituents, *Tetrahedron Lett.,* 27, pp. 2257-2260.

383 Mohr, B., Weck, M., Sauvage, J. -P. and Grubbs, R. H. (1997). High-yield synthesis of [2]catenanes by intramolecular ring-closing metathesis, *Angew. Chem. Int. Ed. Engl.,* 36, pp. 1308-1310.

384 Cardenas, D. J., Collin, J. -P., Gaviña, P., Sauvage, J. -P., De Cian, A., Fischer, J., Armaroli, N., Flamigni, L., Vicinelli, V. and Balzani, V. (1999). Synthesis, X-ray structure, and electrochemical and excited-state properties of multicomponent complexes made of a $[Ru(Tpy)_2]^{2+}$ unit covalently linked to a [2]-catenate moiety. Controlling the energy-transfer direction by changing the catenate metal ion, *J. Am. Chem. Soc.,* 121, pp. 5481-5488.

385 Sauvage, J. -P. and Ward, M. (1991). A Bis(terpyridine)ruthenium(II) catenate, *Inorg. Chem.,* 30, pp. 3869-3874.

386 Fujita, M., Ibukuro, F., Hagihara, H. and Ogura, K. (1994). Quantitative self-assembly of a [2]catenane from two preformed molecular rings, *Nature,* 367, 720-723.

387 Fujita, M., Ibukuro, F., Yamaguchi, K. and Ogura, K. (1995). A molecular lock, *J. Am. Chem. Soc,* 117, pp. 4175-4176.

388 Goodgame, D. M. L., Hill, S. P. W. and Williams, D. J. (1993). Ligand-induced formation of a triple helical bridge involving O-donor ligands in dimeric lanthanide complexes, *J. Chem. Soc. Chem. Commun.,* pp. 1019-1021.

389 Raehm, L., Kern, J. -M., Sauvage, J. -P., Hamann, C., Palacin, S. and Bourgoin, J. Ph. (2002). Disulfide- and thiol-incorporating copper catenenes: synthesis, deposition onto gold and surface studies, *Chem. Eur. J.,* 8, pp. 2153-2162.

390 Sauvage, J. -P. and Weiss, J. (1985). Synthesis of dicopper(I) [3]catenates: multiring interlocked coordinating systems, *J. Am. Chem. Soc.,* 107, pp. 6108-6110.

391 Guilhem, J., Pascard, C., Sauvage, J. -P. and Weiss, J. (1988). Solution study and molecular structure of a [3]-catenand: intramolecular interaction between the two peripheral rings, *J. Am. Chem. Soc.,* 110, pp. 8711-8713.

392 Dietrich-Buchecker, C. O., Khemiss, A. and Sauvage, J. -P. (1986). High-yield synthesis of multiring copper(I) catenates by acetylenic oxidative coupling, *J. Chem. Soc., Chem Commun.,* pp. 1376-1378.

393 Seel, C. and Vögtle, F. (2000). Templates, "wheeled reagents" and a new route to rotaxanes by anion complexation: the trapping method, *Chem. Eur. J.,* 6, pp. 21-24.

394 Chambron, J. -C., Heitz, V., Sauvage, J. -P. (1992). A rotaxane with two rigidly held porphyrins as stoppers, *J. Chem. Soc. Chem. Commun.*, pp. 1131-1133.

395 Shery Zhu, S., Carroll, P. J. and Swager, T. M. (1996). Conducting polymetallorotaxanes: a supramolecular approach to transition metal ion sensors, *J. Am. Chem. Soc.,* 118, pp. 8713-8714.

396 Ogino, H. and Ohata, K. (1984). Synthesis and properties of rotaxane complexes. 2. Rotaxanes consisting of α- or β-cyclodextrin threaded by (μ-α,ω-diaminoalkane)bis[chlorobis(ethylenediamine)cobalt(III)] complexes, *Inorg. Chem.*, 23, pp. 3312-3316.

397 Wylie, S. and Macartney, D. H. (1992). Self-assembling metal rotaxane complexes of α-cyclodextrin, *J. Am. Chem. Soc.*, 114, pp. 3136-3138.

398 Collin, J. P., Gaviñá, P. and Sauvage, J. -P. (1996). Electrochemically induced molecular motions in a copper(I) complex pseudorotaxane, *Chem. Commun.*, pp. 2005-2006.

399 Jimenez-Molero, M. C., Dietrich-Buchecker, C. and Sauvage, J. -P. (2002). Chemically induced contraction and stretching of a linear rotaxane dimer, *Chem. Eur.J.*, 8, pp. 1456-1466.

400 Cárdenas, D. J., Gaviñá, P. and Sauvage, J. -P. (1997). Construction of interlocking and threaded rings using two different transition metals as templating and connecting centers: catenanes and rotaxanes incorporating Ru(terpy)$_2$-units in their framework, *J. Am Chem. Soc.*, 119, pp. 2656-2664.

401 Armaroli, N., Balzani, V., Collin, J. -P., Gaviña, P., Sauvage, J. -P. and Ventura, B. (1999). Rotaxanes incorporating two different coordinating units in their thread: synthesis and electrochemically and photochemically induced molecular motions, *J. Am Chem. Soc.*, 121, pp. 4397-4408.

402 Chambron, J. -C., Dietrich-Buchecker, C. O., Nierengarten, J. -F. and Sauvage, J. -P. (1993). Transition metal directed threading of molecular strings into coordinating rings, *J. Chem. Soc. Chem. Commun.*, pp. 801-804.

403 Solladié, N., Chambron, J. -C. and Sauvage, J. -P. (1999). Porphyrin-stoppered [3]- and [5]rotaxanes, *J. Am Chem. Soc.*, 121, pp. 3684-3692.

404 Solladié, N., Chambron, J. -C., Dietrich-Buchecker, C. O. and Sauvage, J. -P. (1996). Multicomponent molecular systems incorporating porphyrins and copper(I) complexes: simultaneous synthesis of [3]- and [5]rotaxanes, *Angew. Chem. Int. Ed. Engl.*, 35, pp. 906-909.

405 Potts, K. T., Keshavarz, K. M., Tham, F. S., Abruna, H. D. and Arana, C. R. (1993). Metal ion-induced self-assembly of functionalized 2,6-oligopiridines. 2. Copper-containing double-stranded helicates derived from functionalized quaterpyridine and quinquepyridine: redox state-induced transformations and electron communication in mixed-valence systems, *Inorg. Chem.*, **32,** pp. 4422-4435.

406 Potts, K. T., Keshavarz, K. M., Tham, F. S., Abruna, H. D. and Arana, C. R. (1993). Metal ion-induced self-assembly of functionalized 2,6-oligopiridines. 3. Metal-metal interaction and redox state-induced transformations in double-stranded

helicates derived from functionalized quinquepyridine and sexipyridine, *Inorg. Chem.*, **32**, pp. 4436-4449.

[407]   Williams, A. (1997). Helical complexes and beyond, *Chem. Eur. J.*, 3, pp. 15-19.

[408]   Rowan, A. E. and Nolte, R. J. M. (1998). Helical molecular programming, *Angew. Chem., Int. Ed.*, 37, pp. 63-68.

[409]   Shieh, S. -J., Chou, C. -C., Lee, G. -H., Wang, C. -C. and Peng, S. -M. (1997). Linear pentanuclear complexes containing a chain of metal atoms: $[Co^{II}_5(\mu_5\text{-}tpda)_4(NCS)_2]$ and $[Ni^{II}_5(\mu_5\text{-}tpda)_4Cl_2]$, *Angew. Chem. Int. Ed.*, 36, pp. 56-59.

[410]   Constable, E. C., Hannon, M. J., Tocher, D. A. and Ward, M. D. (1990). A single stranded diruthenium(II) helical complex, *J. Chem. Soc., Chem. Commun.*, pp. 621-622.

[411]   Carlucci, L., Ciani, G., Proserpino, D. M. and Sironi, A. (1994). 2D Polymeric silver(I) complexes consisting of markedly undulated sheets of squares. X-Ray crystal structures of $[Ag(ppz)_2](BF_4)$ and $[Ag(ppz)_2](PF_6)$ (pyz = piperazine, pyz = pyrazine), *Inorg. Chem.*, 34, pp. 5698-5700.

[412]   Constable, E. C., Chotalia, R. and Tocher, D. A. (1992). The first example of a mono-helical complex of 2,2':6',2'':6'',2''':6''',2''''-sexipyridine; preparation, crystal and molecular structure of bis(nitrato-*O,O'*)(2,2':6',2'':6'',2''':6''',2''''-sexipyridine)europium(III) nitrate, *J. Chem. Soc., Chem. Commun.*, pp. 771-773.

[413]   Piguet, C., Bernardinelli, C., Boyquet, B., Quattropani, A. and Williams, A.F. (1992). Self-assembly of double and triple helices controlled by metal ion stereochemical preference, *J. Am. Chem. Soc.*, 114, pp. 7440-7451.

[414]   Munakata, M., Wu, L. P., Kuroda-Sowa, T., Maekawa, M., Moriwaki, K. and Kitagawa, S. (1997). Two types of new polymeric copper(I) complexes of pyrazinecarboxamide having channel and helical structures, *Inorg. Chem.*, 36, pp. 5416-5418.

[415]   Krämer, R., Lehn, J. -M. and Marquis-Rigault, A. (1993). Self –recognition in helicate self-assembly: spontaneous formation of helical metal complexes from mixtures of ligands and metal ions, *Proc. Natl. Acad. Sci. USA.*, 90, pp. 5394-5398.

[416]   Stiller, R. and Lehn, J. -M. (1998). Synthesis and properties of silver(I) and copper(I) helicates with imine-bridged oligopyridine ligands, *Eur. J. Inorg. Chem.*, pp. 977-982.

[417]   Struckmeier, G., Theuralt, U. and Fuhrhop, J. -H. (1976). Structures of zinc octaethyl formylbiliverdinate hydrate and its dehydrated bis-helical dimer, *J. Am. Chem. Soc.*, 98, pp. 278-279.

[418]   Potts, K. T., Gheysen Raiford, K. A. and Keshavarz, K. M. (1993). Metal-ion-induced self-assembly of functionalized 2,6-oligopyridines. 1. Ligand design, synthesis, and characterization, *J. Am. Chem. Soc.*, 115, pp. 2793-2807.

[419]   Ruttimann, S., Piguet, C., Bernardinelli, G., Bocquet, B. and Williams, A. F. (1992). Self-assembly of dinuclear helical and nonhelical complexes with copper(I), *J. Am. Chem. Soc.*, 114, pp. 4230-4237.

[420]   Bernardinelli, G., Piguet, C. and Williams, A. F. (1992). The first self-assembled dinuclear triple-helical lanthanide complex: synthesis and structure, *Angew. Chem. Int. Ed. Engl.*, 31, pp. 1622-1624.

421 Pfiel, A. and Lehn, J. -M., (1992). Helicate self-organisation: positive cooperativity in the self-assembly of double-helical metal complexes, *J. Chem. Soc. Chem. Commun.*, pp. 838-840.

422 Harding, M. M., Koert, U., Lehn, J. -M., Marquis-Rigault, A., Piguet, C. and Siegel, J. (1991). Synthesis of unsubstituted and 4,4'-substituted oligobipyridines as ligand strands for helicate self-assembly, *Helv. Chim. Acta*, 74, pp. 594-610.

423 Constable, E. C., Edwards, A. J., Hannon, M. J. and Raithby, P. R. (1994). Double-helical complexes from simple 2,2':6',2''-terpyridines; the crystal and molecular structure [$Cu_2(Ph_2tpy)_2$][PF_6]$_2$ (Ph_2tpy = 6,6''-diphenyl-2,2':6',2''-terpyridine), *J. Chem. Soc., Chem. Commun.*, pp. 1991-1992.

424 Constable, E. C., Hannon, M. J., Martin, A. and Raithby, P. R. (1992). Self-assembly of double-helical complexes of 2,2':6',2'':6'',2''':6''',2''''-quaterpyridine (qtpy); the X-ray crystal structures of [$Cu_2(qtpy)_2$][PF_6]$_2$ and [$Ag_2(qtpy)_2$][BF_4]$_2$, *Polyhedron*, 11, pp. 2967-2971.

425 Constable, E. C., Drew, M. G. B. and Ward, M. D. (1987). Molecular helicity in inorganic complexes; the preparation, crystal and molecular structure of bis(2,2':6',2'':6'',2''':6''',2''''-quinquepyridine)acetatodicopper(II) hexafluorophosphate monohydrate, *J. Chem. Soc., Chem. Commun.*, pp. 1600-1601.

426 Barley, M., Constable, E. C., Corr, S. A., McQueen, R. C. S., Nutkins, J. C., Ward, M. D. and Drew, M. G. B. (1988). Molecular helicity in inorganic complexes; double helical binuclear complexes of 2,2':6',2'':6'',2''':6''',2''''-quinquepyridine (L): crystal structures of [$Cu_2L_2(O_2CMe)$][PF_6]·H_2O and [Cu_2L_2][PF_6]$_3$·2MeCN, *J. Chem. Soc., Dalton Trans.*, pp. 2655-2662.

427 Constable, E. C., Ward, M. D. and Tocher, D. A. (1991). Molecular helicity in inorganic complexes; bi- and tri-nuclear complexes of 2,2':6',2'':6'',2''':6''',2'''':6'''',2'''''-sexipyridine and the crystal and molecular structure of bis(μ-2,2':6',2'':6'',2''':6''',2'''':6'''',2'''''-sexipyridine-?$^3N,N',N''$: ?$^3N''',N'''',N'''''$)dicadmium hexafluorophosphate-acetonitrile (1/4), *J. Chem. Soc., Dalton Trans.*, pp. 1675-1683.

428 Constable, E. C., Ward, M. D. and Tocher, D. A. (1990). Spontaneous assembly of a double-helical binuclear complex of 2,2':6',2'':6'',2''':6''',2'''':6'''',2'''''-sexipyridine, *J. Am. Chem. Soc.*, 112, pp. 1256-1258.

429 Hasenknopf, B., Lehn, J. -M., Kneisel, B. O., Baum, G. and Fenske, D. (1996). Self-assembly of a circular double helicate, *Angew. Chem. Int. Ed. Engl.*, 35, pp. 1838-1840.

430 For example, Stratton, W. J. and Busch, D. H. (1960). The complexes of pyridinaldazine. III. Infrared spectra and continued synthetic studies, *J. Am. Chem. Soc.*, 82, pp. 4834-4839.

431 Hamblin, J., Jackson, A., Alcock, N. W. and Hannon, M. J. (2002). Triple helicates and planar dimers arising from silver(I) coordination to directly linked bis-pyridylimine ligands, *J. Chem. Soc. Dalton Trans.*, pp. 1635-1641.

432 Caradoc-Davies, P. L. and Hanton, L. R. (2001). Formation of a single-stranded silver(I) helical-coordination polymer containing p-stacked planar chiral N_4S_2 ligands, *Chem. Commun.*, pp. 1098-1099.

[433] Constable, E. C., Hanon M. J. and Tocher, D. A. (1993). Dinuclear double helicates incorporating a 1,3-phenylene spacer; the crystal and molecular structure of diacetato-1 ?²*O*-, 2 ?²*O*-bis[μ-1,3-bis(4-methylthio-2,2'-bipyridin-6-yl)benzene-1 ?²*N,N*': 2 ?²*N'',N''*']dinickel bis(hexafluorophosphate), *J. Chem. Soc. Dalton Trans.*, pp. 1883-1890.

[434] Hamblin, J., Childs, L. J., Alcock, N. W. and Hannon, M. J. (2002). Directed one-pot syntheses of enantiopure dinuclear silver(I) and copper(I) metallo-supramolecular double helicates, *J. Chem. Soc. Dalton Trans.*, pp. 164-169.

[435] Garrett, T. M., Koert, U., Lehn, J. -M., Rigault, A., Meyer, D. and Fischer, J. (1990). Self-assembly of silver(I) helicates, *J. Chem. Soc. Chem. Commun.* pp. 557-558.

[436] Krämer, R., Lehn, J. -M., De Cian, A. and Fischer, J. (1993). Self-assembly, structure and spontaneous resolution of a trinuclear triple helix from an oligobipyridine ligand and Ni^{II} ions, *Angew. Chem. Int. Ed. Engl.*, 32, pp. 703-706.

[437] Masood, M. A., Enemark, E. J. and Stack, T. D. (1998). Ligand self-recognition in the self assembly of a [{Cu(L)}₂]²⁺ complex: the role of chirality, *Angew. Chem. Int. Ed. Engl.*, 37, pp. 928-931.

[438] Fletcher, N. C., Keene, F. R., Viebrock, H. and Von Zelewsky, A., (1997). Molecular architecture of polynuclear ruthenium bipyridil complexes with controlled metal helicity, *Inorg. Chem.*, 36, pp. 1113-1121.

[439] Woods, C. R., Benaglia, M., Cozzi, F. and Siegel, J. S. (1996). Enantioselective synthesis of copper(I) bipyridine based helicates by chiral templating of secondary structure: transmission of stereochemistry on the nanometer scale, *Angew. Chem. Int. Ed. Engl.*, 35. pp. 1830-1833.

[440] McMorran, D. A. and Steel, P. J. (1998). The first coordinatively saturated, quadruply stranded helicate and its encapsulation of a hexafluorophosphate anion, *Angew. Chem. Int. Ed. Engl.*, 37, pp. 3295-3297.

[441] Youinou, M. -T., Ziessel, R. and Lehn, J. -M. (1991). Formation of dihelicate and mononuclear complexes from ethane-bridged dimeric bipyridine or phenantroline ligands with copper(I), cobalt(II), and iron(II) cations, *Inorg. Chem.*, 30, pp. 2144-2148.

[442] Lawrence, D. S., Jiang, T. and Levett, M. (1995). Self-assembling supramolecular complexes, *Chem. Rev.*, 95, pp. 2229-2260.

[443] Lam, M. H. W., Lee, D. Y. K., Chiu, S. S. M., Man, K. W. and Wong, W. T. (2000). Synthesis and crystal structures of mono-helical complexes of zinc(II) and europium(III) with 1,2-bis{[3(2-pyridyl)pyrazol-1-yl]etoxy}ethane, *Eur, J. Inorg. Chem.*, pp. 1483-1488.

[444] Potts, K. T., Horowitz, C. P., Fessak, A., Keshavarz, M. K., Nash, K. E. and Toscano, P. J. (1993). Coordination of ethynylpyridine ligands with Cu(I): X-ray structure of a novel, triple-helical, tricuprous complex, *J. Am. Chem. Soc.*, 115, pp. 10444-10445.

[445] Piguet, C., Bunzli, J. C., Bernardinelli, C., Hopfgartner, G. and Williams, A. F. (1993). Self-assembly and photophysical properties of lanthanide dinuclear triple-helical complexes, *J. Am. Chem. Soc.*, 115, pp. 8197-8206.

446 Champman, R. D., Loda, R. T., Riehl, J. P. and Schwartz, R. W. (1984). Spectroscopic investigation of the multidentate coordination equilibrium among conformational isomers of tris(2,2',2''-terpyridyl)europium(III) perchlorate in acetonitrile, *Inorg. Chem.*, 23, pp. 1652-1657.

447 Harrowfield, J. M., Kim. Y., Skelton, B. W. and White, A. H. (1995). Mixed transition metal/lanthanide complexes: structural characterization on solids containing cage amine chromium(III) cations and tris(dipicolinato)lanthanide anions, *Aust. J. Chem.*, 48, 807-823 and references therein.

448 Grenthe, I. (1961). Stability relationships among the rare earth dipicolinates, *J. Am. Chem. Soc.*, 83, pp. 360-364.

449 Renaud, F., Piguet, C., Bernardinelli, G., Bunzli, J. -C. G. and Hopfgartner, G. (1997). In search for mononuclear helical lanthanide building blocks with predetermined properties: triple-stranded helical complexes with N,N,N',N'-tetraethylpyridine-2,6-dicarboxamide, *Chem. Eur. J.*, 3, pp. 1646-1659.

450 Rigault, S., Piguet, C., Bernardinelli, G. and Hopfgartner, G. (1998). Lanthanide-assisted self-assembly of an inert, metal-containing nonadentate tripodal receptor, *Angew. Chem. Int. Ed.*, 37, pp. 169-172.

451 Piguet, C. and Bunzli, J. -C.G. (1999). Mono- and polymetallic lanthanide-containing functional assemblies: a field between tradition and novelty, *Chem. Soc. Rev.*, 28, pp. 347-358.

452 Piguet, C., Bünzli, J. -C. G., Bernardinelli, C. G., Bochet, G. and Froidevaux, P. (1995). Design of luminescent building blocks for supramolecular triple-helical lanthanide complexes, *J. Chem. Soc., Dalton Trans.*, pp. 83-97.

453 Rigault, S., Piguet, C. and Bunzli, J. -C. G. (2000). The solution structure of supramoleculare lanthanide triple helices revisited: application of crystal-field independent paramagnetic NMR techniques to mono- and di-metallic complexes, *J. Cherm. Soc., Dalton Trans.*, pp. 2045-2053.

454 Telfer, S. G., Bocquet, B. and Williams, A. F. (2001). Thermal spin crossover in binuclear iron(II) helicates: negative coorperativity and a mixed spin state in solution, *Inorg. Chem.*, 40, pp. 4818-4820.

455 Piguet, C., Hopfgartner, G., Boquet, B., Schaad, O., Wiliams, A. F. (1994). Self-assembly of heteronuclear supramolecular helical complexes with segmental ligands, *J. Am. Chem. Soc.*, 116, pp. 9092-9102.

456 Martin, N., Bunzli, J. -C. G., McKee, V., Piguet, C. and Hopfgartner, G. (1998). Self-assembled dinuclear lanthanide helicates: substantial luminescence enhancement upon replacing terminal benzimidazole groups by carboxamide binding units, *Inorg. Chem.*, 37, pp. 577-589.

457 Elhabiri, M., Scopelliti, R., Bunzli, J. -C. G. and Piguet, C. (1999). Lanthanide helicates self-assembled in water: a new class of higly stable and luminescent dimetallic carboxylates, *J. Am. Chem. Soc.*, 121, pp. 10747-10762.

458 Iglesias, C. P., Elhabiri, M., Hollenstein, M., Bunzli, J. -C. G. and Piguet, C. (2000). Effect of a halogenide substituent on the stability and photophysical properties of lanthanide triple-stranded helicate with ditopic ligands derived from bis(benzimidazolyl)pyridine, *J. Cherm. Soc., Dalton Trans.*, pp. 2031-2043.

[459] Kersting, B., Meyer, M., Owers, R. E. and Raymond, P. K. N. (1996). Dinuclear cathecolate helicates: their inversion mechanism, *J. Am. Chem. Soc.,* 118, pp. 7221-7222.

[460] Albrecht, M. and Kotila, S. (1995). Formation of a „*meso*-helicate" by self-assembly of three bis(catecholate) ligands and two titanium(IV) ions, *Angew. Chem. Int. Ed. Engl.,* 34, pp. 2134-2137.

[461] Huynh, H. V., Schulze-Isfort, C., Seidel, W.W., Lüger, T., Fröhlich, R., Kataeva, O., Hahn, F.E. (2002). Dinuclear complexes with bis(benzenedithiolate) ligands, *Chem. Eur. J.,* 8, pp. 1327-1335.

[462] Duhme, A. -K., Dauter, Z., Hider, R. C. and. Pohl, S. (1996). Complexation of molybdenum by siderophores: synthesis and structure of the double-helical *cis*-dioxomolibdenum(VI) complex of a bis(catecholamide) siderophore analogue, *Inorg. Chem.,* 35, pp. 3059-3061.

[463] Enemark, E. J. and Stack, T. D. P. (1996). Spectral and structural characterization of two ferric coordination modes of a simple bis(catecholamide) ligand: metal-assisted self-assembly in a siderophore analog, *Inorg. Chem.,* 35, pp. 2719-2720.

[464] Caulder, D. L., Powers, R. E., Parac, T. N. and Raymond, K. N. (1998). The self-assembly of a predesigned tetrahedral M_4L_6 supramolecular cluster, *Angew. Chem. Int. Ed.,* 37, pp. 1840-1843.

[465] Scherer, M., Caulder, D. L., Johnson, D. W. and Raymond, K. N. (1999). Triple helicate-tetrahedral cluster interconversion controlled by host-guest interactions, *Angew. Chem.Int. Ed.,* 38, pp. 1588-1591.

[466] Albrecht, M., Napp, M., Schneider, M., Weis, P. and Fröhlich, R. (2001). Kinetic *versus* thermodynamic control of the self-assembly of isomeric double stranded dinuclear titanium(IV) complexes from a phenylalanine-bridged dicatechol ligand, *Chem. Commun.,* pp. 409-410.

[467] Caulder, D. L. and Raymond, K. N. (1997). Supramolecular self-recognition and self-assembly in gallium(III) catecholamide triple helices, *Angew. Chem. Int. Ed. Engl.,* 36, pp. 1440-1441.

[468] Meyer, M., Kersting, M. B., Powers, R. E. and Raymond, K. N. (1997). Rearrangement reactions in dinuclear triple helicates, *Inorg. Chem.,* 36, pp. 5179-5191.

[469] Libman, J., Tor, Y., Shanzer, A. (1987). Helical ferric ion binders, *J. Am. Chem. Soc.,* 109, 5880-5881.

[470] Hasenknopf, B., Lehn, J. -M., Boumediene, N., Dupont-Gervais, A., Van Dorsselaer, A., Kneisel, B. and Fenske, D. (1997). Self-assembly of tetra- and hexanuclear circular helicates, *J. Am. Chem. Soc.,* 119, pp. 10956-10962.

[471] Mamula, O., Von Zelewsky, A. and Beradinelli, G. (1998). Completaly stereospecific self-assembly of a circular helicate, *Angew. Chem. Int. Ed. Engl.,* 37, pp. 289-293.

[472] Mamula, O., Monlien, F. J., Porquet, A., Hopfgartner, G., Merbach, A. E. and Von Zelewsky, A. (2001). Self-assembly of multinuclear coordination species with chiral bipyridine ligands: silver complexes of 5,6-CHIRAGEN (*o,m,p*-xylidene) ligands and equilibrium behaviour in solution, *Chem. Eur. J.,* 7, pp. 533-538.

473 Dietrich-Buchecker, C. and Sauvage, J. -P. (1989). A synthetic molecular trefoil knot, *Angew. Chem. Int. Ed. Engl.*, 28, pp. 189-192 ; Du, S. M., Stollar, B. D. and Seeman, N. C. (1995). A synthetic DNA molecule in three knotted topologies. *J. Am. Chem. Soc.*, 117, pp. 1194-1200.

474 Dietrich-Buchecker, C., Guilhem, J., Pascard, C. and Sauvage, J. -P. (1990). Structure of a synthetic trefoil knot coordinated to two copper(I) centers, *Angew. Chem. Int. Ed. Engl.*, 29, pp. 1154-1156.

475 Dietrich-Buchecker, C., Sauvage, J. -P., De Cian, A. and Fischer, J. (1994). High-yield synthesis of a dicopper(I) trefoil knot containing 1,3-phenylene groups as bridges between the chelate units, *J. Chem. Soc., Chem. Commun.*, pp. 2231-2232.

476 Rapenne, G., Dietrich-Buchecker, C. and Sauvage, J. -P. (1996). Resolution of a molecular trefoil knot, *J. Am. Chem. Soc.*, 118, pp. 10932-10933.

477 Rapenne, G., Dietrich-Buchecker, C. and Sauvage, J. -P. (1999). Copper(I)- or iron(II)-templated synthesis of molecular knots containing two tetrahedral or octahedral coordination sites, *J. Am. Chem. Soc.*, 121, pp. 994-1001.

478 Carina, R. F., Dietrich-Buchecker, C. and Sauvage, J. -P. (1996). Molecular composite knots, *J. Am. Chem. Soc.*, 118, pp. 9110-9116.

479 Woods, C. R., Benaglia, M., Toyota, S., Hardcastle, K. and Siegel, J. S. (2001). Trinuclear copper(I)-bipyridine triskelion: template/bascule control of coordination complex stereochemistry in a trefoil knot precursor, *Angew. Chem. Int. Ed. Engl.*, 40, pp. 749-751.

480 Leininger, S., Olenyuk, B. and Stang, P. J. (2000). Self-assembly of discrete cyclic nanostructures mediated by transition metals. *Chem. Rev.*, 100, pp. 853-908.

481 Beer, P. D., Berry, N., Drew, M. G. B., Fox, O. D., Padilla-Tosta, M. E. and Patell, S. (2001). Self-assembled dithiocarbamate-copper(II) macrocycles for electrochemical anion recognition, *Chem. Commun.*, pp. 199-200.

482 Escarty, F., Miranda, C., Lamarque, L., Latorre, J., Garcia-Espana, E., Kumar, M., Arán, V. J. and Navarro, P. (2002). Cu^{2+} -Induced formation of cage-like compounds containing pyrazole macrocycles, *Chem. Commun.*, pp. 936-937.

483 Youinou, M. -T., Rahmouni, N., Fischer, J. and Osborn, J. A. (1992). Self-assembly of a Cu_4 complex with coplanar copper(I) ions: synthesis, structure, and electrochemical properties, *Angew. Chem. Int. Ed. Engl.*, 31, pp. 733-735.

484 Fox, O. D., Drew, M. G. B., Wilkinson, E. J. S. and Beer, P. D. (2000). Cadmium- and zinc-directed assembly of nano-sized, resorcarene-based host architectures which strongly bind C_{60}, *Chem. Commun.*, pp. 391-392.

485 Stang, P. J. and Cao, D. H. (1994). Transition metal based cationic molecular boxes. Self-assembly of macrocyclic platinum(II) and palladium(II) tetranuclear complexes, *J. Am. Chem. Soc.*, 116, pp. 4981-4982.

486 Fujita, M., Umemoto, K., Yoshizawa, M., Fujita, N., Kusukawa, T. and Biradha, K. (2001). Molecular panelling *via* coordination, *Chem. Commun.*, pp. 509-518 and references therein.

487 Fujita, M., Yazaki, J. and Ogura, K. (1990). Preparation of a macrocyclic polynuclear complex, [(en)Pd(4,4'-bpy)]$_4$(NO$_3$)$_8$, which recognizes an organic molecule in aqueous media, *J. Am. Chem. Soc.*, 112, pp. 5645-5647.

[488] Fujita, M., Oguro, D., Miyazawa, M., Oka, H., Yamaguchi, K. and Ogura, K. (1995). Self-assembly of ten molecules into nanometre-sized organic host frameworks, *Nature*, 378, pp. 469-471.

[489] Park, S. J. and Hong, J. -J. (2001). Self-assembled nanoscale capsules between resorcin[4]arene derivatives and Pd(II) or Pt(II) complexes, *Chem. Commun.*, pp. 1554-1555.

[490] Baxter, P.,N. W., Lehn, J. -M., Kneisel, B. O. and Fenske, D. (1997). Multicomponent self-assembly: preferential generation of a rectangular [2x3]G grid by mixed-ligand recognition, *Angew. Chem. Int. Ed. Engl.*, 36, pp. 1978-1981.

[491] Cotton, A. A., Lin, C. and Murillo, C. A. (2001). Supramolecules structures based on dimetal units: simultaneous utilization of equatorial and axial connections, *Chem. Commun.*, pp. 11-12.

[492] Cotton, A. A., Daniels, L. M., Lin, C. and Murillo, C. A. (1999). Square and triangular arrays based on Mo_2^{4+} and Rh_2^{4+} units, *J. Am. Chem. Soc.*, 121, pp. 4538-4539.

[493] Cotton, A. A., Daniels, L. M., Lin, C. and Murillo, C. A. (1999). *Chem. Commun.*, pp. 841-

[494] Baxter, P. N. W., Hanan, G. S. and Lehn, J. -M. (1996). Inorganic arrays *via* multicomponent self-assembly: the spontaneous generation of ladder architectures, *Chem. Commun.*, pp. 2019-2020.

[495] Baxter, P. N. W., Lehn, J. -M., Fischer, J. and Youinou, M. -T. (1994). Self-assembly and structure of a [3x3] inorganic grid from nine silver ions and six ligand components, *Angew. Chem. Int. Ed. Engl.*, 33, pp. 2284-2287.

[496] Baxter, P. N. W., Lehn, J. -M., Baum, G. and Fenske, D. (2000). Self-assembly and structure of interconverting multinuclear inorganic arrays: a [4x5]-Ag_{20}^I grid and an Ag_{10}^I quadruple helicate, *Chem. Eur. J.*, 6, pp. 4510-4517.

[497] Hanan, G. S., Arana, C. R., Lehn, J. -M. and Fenske, D. (1995). Synthesis, structure, and properties of dinuclear and trinuclear rack-type Ru^{II} complexes, *Angew. Chem. Int. Ed. Engl.*, 34, pp. 1122-1124.

[498] Hanan, G. S., Volkmer, D., Schubert, U. S., Lehn, J. -M., Baum, G. and Fenske, D. (1997). Coordination arrays: tetranuclear cobalt(II) complexes with [2x2]-grid structure, *Angew. Chem. Int. Ed. Engl.*, 36, pp. 1842-1844.

[499] Garcia, A. M., Romero-Salguero, F. J., Bassani, D. M., Lehn, J. -M., Baum, G. and Fenske, D. (1999). Self-assembly and characterization of multimetallic grid-type lead(II) complexes, *Chem. Eur. J.*, 5, pp. 1803-1808.

[500] Sun, W. -Y., Fei, B.–L., Okamura, T., Tang, W. -X. and Ueyama, N. (2001). Construction and characterization of organic-inorganic hybridized molecules with infinite 2D grid network and 1D zigzag chain structures, *Eur. J. Inorg. Chem.*, pp. 1855-1861.

Index

"soft" cations 112
1,5-diazacyclooctane 154
18-membered hexaaza macrocycles ..91
18-membered macrocycle 139
acid padlock 129, 160
alkylation 161, 279
amelide 148
amino-acid complexes 134
aminotriazines 148
ammeline 148
antiferomagnetic interaction 121
aromatic diamines 47, 78, 85
asymmetric cyclic ligands 111
bridging sulfur 38
cadmium(II) 140, 141
cadmium(II) complexes 21, 263
cage complexes170, 172, 173, 174, 175,
    178, 181, 184, 185, 186, 282, 283,
    284
capping reaction127, 131, 135, 180,
    183
carbinolamine intermediates 25
carbon acid 126
catenand 209, 217, 286
catenane209, 210, 212, 215, 218, 219,
    224, 285, 286
chelate effect 5, 8, 9, 257
chiro-optical phenomena 172
chlorin 199
chrom(III) 140
chromium(III) 36, 276, 281, 283, 291
chromium(VI) 187
clatrochelates 180
cobalt(0) 177
cobalt(I) 174
cobalt(II)25, 26, 29, 31, 32, 36, 39, 48,
    50, 64, 65, 82, 98, 172, 174, 180,

184, 196, 197, 199, 200, 205, 207,
    255, 261, 262, 265, 266, 269, 282,
    283, 290, 294
cobalt(II) directed condensation 25
cobalt(III)128, 129, 136, 140, 141, 153,
    167, 168, 172, 174, 175, 178, 179,
    183, 184, 205, 207, 273, 275, 276,
    277, 279, 282, 283
compartmental ligands40, 46, 96, 97,
    99, 116
contracted cage 171, 173, 282
coordinated carbinolamine 130
coordinated imine127, 128, 129, 130,
    136
coordinated nucleophiles. 128, 129, 273
copper complex 25, 36, 146, 200
copper(I)193, 195, 196, 199, 209, 210,
    212, 215, 216, 217, 218, 222, 224,
    226, 227, 232, 243, 245, 246, 247,
    254, 286, 287, 288, 290, 293
copper(II)26, 30, 31, 32, 36, 38, 40, 41,
    43, 48, 49, 50, 64, 65, 73, 74, 78, 82,
    83, 91, 98, 99, 101, 104, 105, 106,
    108, 109, 111, 113, 114, 116, 118,
    119, 120, 121, 123, 124, 125, 131,
    132, 133, 134, 135, 136, 138, 139,
    140, 141, 142, 145, 146, 149, 150,
    152, 153, 154, 155, 156, 158, 159,
    160, 163, 164, 166, 185, 187, 189,
    193, 195, 199, 200, 202, 204, 259,
    260, 261, 262, 264, 265, 269, 270,
    271, 272, 273, 274, 276, 277, 278,
    279, 280
copper(II) complexes26, 32, 41, 43, 50,
    104, 120, 121, 123, 124, 125, 142,
    154, 166, 259, 260, 262, 265, 269,
    271, 272, 276, 278, 279, 280

copper(III) 205
cryptand................................90, 95, 96
cryptate 94, 268, 269
cryptate tethering effect.....................5
cyclam138, 140, 141, 147, 156, 165, 278, 281
cysteinamine....................................183
diimine complexes...........................48
dinitrociclam...................................138
discotic mesophases........................189
double-helical structures.................229
dysprosium 59, 60, 79, 96, 268
encapsulated complexes 167, 170
equilibrium template effect.................4
erbium...59, 60
europium(III) 187
fused ring................. 152, 157, 158, 180
fused small-rings 157
gadolinium...........................59, 60, 267
gadolinium(III) 187
glyoxal...............................35, 42, 48
gold(I)..150
gold(III) 139
hafnium...188
Hexadentate ligands.........................32
hexadentate tripodal ligand.............133
holmium 59, 60
homopiperazine 154, 155
interlocking rings.....209, 216, 223, 285
intramolecular condensation............127
iridium (III)......................................174
iron(II)29, 33, 34, 35, 37, 54, 55, 57, 60, 65, 140, 180, 186, 241, 246, 260, 263, 265, 269, 275, 283, 290, 291, 293
Iron(II) complex 35
iron(III)..196
iron(IV)..205
Jahn-Teller...............................141, 185
kinetic1, 2, 3, 8, 13, 15, 22, 25, 93, 127, 128, 138, 139, 141, 220, 257
lanthan(III)..92
lanthanide61, 64, 75, 76, 77, 78, 79, 80, 81, 82, 85, 86, 87, 88, 91, 92, 95, 264, 265, 266, 267, 268, 269

lanthanide ions 61, 82, 88, 264
lanthanum...........59, 60, 76, 82, 85, 267
large fused-ring 167
lead(II) 103, 108, 111, 265
ligands containing sulfur donor atoms 112
lutetium 96, 190, 191
macrobicyclic cages 93
macrocyclic effect.............. 5, 8, 13, 74
macropolycyclic molecules.............. 93
manganese(II)39, 54, 55, 58, 101, 114, 123, 185, 187, 206, 216, 263, 264, 269
melamine................................ 148, 149
Molecular organisation....................... 4
negative template 14
neodymium 59, 60
nickel(II)3, 4, 7, 10, 11, 14, 18, 19, 21, 22, 23, 26, 27, 29, 30, 31, 32, 39, 40, 41, 42, 48, 49, 50, 53, 64, 65, 67, 69, 71, 73, 74, 93, 105, 111, 113, 114, 131, 132, 137, 138, 139, 140, 141, 145, 146, 149, 150, 156, 157, 158, 159, 160, 161, 162, 165, 180, 195, 196, 199, 200, 204, 205, 257, 258, 259, 261, 262, 265, 266, 268, 269, 270, 273, 274, 275, 276, 278, 280, 281, 285
o-aminobenzaldehyde 257, 266
oxovanadium(IV).............. 37, 140, 276
oxovanadium(V) 36
palladium(II)20, 131, 139, 140, 159, 187, 275, 276
palladium(II) complexes 18, 19, 250
pendant arm90, 91, 140, 143, 147, 268, 276, 278
phtalonitrile187, 190, 191, 193, 195, 284
piperazine............................... 153, 154
platinum(II) 139, 140, 187, 276
platinum(III).................................... 175
Porphirins........................... 216, 221
positive template 14
praseodymium............... 53, 59, 60, 267
preorganising ability...................... 126

pyrroles 196, 284
reinforced macrocycles 156
relationship between the size of the metal ion 138
rhodium(II) 175
rhodium(III)140, 141, 175, 178, 187, 273, 281
Ruthenium(II) phtalocyanine 187
samarium 59, 60, 187
sarcophagine 169, 184
self-condensation 73, 266
semispulchrate 133
sepulchrate 130, 168, 169, 184
seven-coordination 54
side-off binucleating ligands 124
silver(I) complex 90, 94
size of the macrocyclic ring 50
solid-state template synthesis 30
spin-spin-pairing 120
Spiro aza- 166
spiro thiaaza- macrobicycles 166
stepwise template reaction 102, 108
stereoselectivity 141
steric strain 10, 11, 49, 154

structural reinforcement 153
sulfur donor 22, 112
Supramolecular structures 190
terapyrroles 198
terbium complex 89
tetradentate macrocycles 52
tetrapyrrenes 198
thermodynamic1, 4, 9, 13, 25, 34, 47, 74, 86, 138, 139, 185, 258, 292
thorium(IV) 187
titanium 188
transmetallation 67, 91, 94, 125
tridentate macrocyclic ligands 52
triple helix conformation 96
unsymmetrical macrocycles 103
uranium 38
uranyl 191
ytterbium 59, 60, 187
yttrium 52, 59, 60, 75, 262, 264, 267
zinc(II)21, 23, 36, 39, 113, 140, 188, 202, 205, 249, 259, 260, 262, 265, 276, 290
zirconium 188